Partially Ordered Systems

V.A. Belyakov

Diffraction Optics of Complex-Structured Periodic Media

With 86 Illustrations

Springer-Verlag
New York Berlin Heidelberg London Paris
Tokyo Hong Kong Barcelona Budapest

o 4872162

PHYSICS

V.A. Belyakov
All-Union Surface and Vacuum Research Center
Andreevskaya nab. 2
117334 Moscow
Russian Federated Republic

Library of Congress Cataloging-in-Publication Data
Belyakov, V. A. (Vladimir Alekseevich)
    Diffraction optics of complex-structured periodic media / V.A.
Belyakov.
      p. cm. — (Partially ordered systems)
    Revised and augmented translation of 1988 Russian edition.
    Includes bibliographical references and index.
    ISBN 0-387-97654-X (U.S. : alk. paper). — ISBN 3-540-97654-X
(Germany : alk. paper)
      1. Diffraction. 2. Liquid crystals — Optical properties.
    3. Matter — Effect of radiation on. 4. X-rays — Diffraction.
    I. Title. II. Series.
    QC415.B37   1992
    535.4 — dc20   91-22809

Printed on acid-free paper.

Original Russian edition: *Difraktsionnaya optika periodycheskykh sred slozhnoy struktury*. Moskva: Nauka, 1988.

Photocomposed from a L^{A}T$_E$X file.
Printed and bound by Edwards Brothers, Inc., Ann Arbor, MI.
Printed in the United States of America.

9 8 7 6 5 4 3 2 1

ISBN 0-387-97654-X Springer-Verlag New York Berlin Heidelberg
ISBN 3-540-97654-X Springer-Verlag Berlin Heidelberg New York

Preface to the English Edition

Little time has elapsed since 1988 when a Russian edition of this book was published, but most of the problems touched on here were under intensive investigation then. This indicates that the subject of the book was and is quite urgent, and as a consequence, a number of new publications on the subject have appeared since the first edition. It is why in the course of preparing a new edition, I decided that many new results ought to be reflected in it.

Because I wanted to retain the general style of the original book and its overall features, the problem of integrating new material was solved in the following way. Three chapters, devoted to the fields of the most significant achievements, were added: on polarization phenomena in X-ray optics, on X-ray magnetic scattering, and on Mössbauer filtration of synchrotron radiation. Only minor alterations were made to the other chapters; their end-of-chapter reference sections were also updated.

I hope that the English edition of this book will give easier access to my Western colleagues to the original Soviet results. Soviet authors contributed quite significantly to the subject of this book and look forward to developing joint investigations of the coherent interaction of different kind of radiations with matter. I would also like to express my gratitude to the Springer-Verlag editing staff for collaboration in this project.

Moscow   V.A. Belyakov

Preface to the Russian Edition

Rapid progress in experimental techniques in the past 15–20 years, coupled with the ever-increasing variety of the objects of investigation, have advanced the study of a series of phenomena which were beyond the scope of researchers and which are very important for various applications.

Among such phenomena are the numerous unusual effects observed in coherent interaction of various electromagnetic and corpuscular radiations with regular (spatially periodic) media having complicated structures. These are polarization characteristics of light scattering, the unusually high gyrotropy of chiral liquid crystals, which changes its sign depending on frequency, nuclear scattering of gamma and synchrotron radiation sensitive to magnetic order in crystals, various anomalies in angular and frequency distributions of the coherent radiation of fast charged particles, and similar effects in nonlinear generation of high-order optical harmonics in periodic media, etc.

It should be emphasized that since specific objects, types of radiation, and conditions under which the above effects manifest themselves are very diverse and, at first glance, have quite a different nature, they are often considered unrelated phenomena. In many studies, such separate considerations of these phenomena are detrimental, and instead of reinventing the wheel and ignoring the experience and results of other distant fields of research, investigators have only to examine, in fact, closely related ones. It is worthy to remind the reader here of the words written of L. Brillouin and M. Parodi some 30 years ago in the foreword to their now famous *Wave Propagation in Periodic Structures*. "No matter how diverse the problems are, their mathematical treatment results in equations having a series of common properties and similar solutions. This deep relation between quite different problems was very well known already to Kelvin and Rayleigh. But then it has been forgotten and only in recent works this very important statement has attracted attention anew." Unfortunately, the above affirmation sounds as urgent today, in application to modern problems, as it did then.

This book presents from one standpoint and, in fact, within the context of one approach a theoretical description of various phenomena—diffraction of Mössbauer gamma radiation by magnetically ordered crystals, optics of chiral liquid crystals, coherent radiation of fast charged particles in periodic media with a complicated structure, some problems of nonlinear optics for periodic media, and dynamical diffraction of thermal neutrons by perfect magnetically ordered crystals. The main experimental results, if any, are explored to the extent necessary for the illustration of general qualitative effects and their comparison with theoretical calculations.

This book also includes introductory material on wave propagation in one-dimensional periodic media and the approximations describing the interaction of radiation with periodic media. The main focus of the book is the application of the kinematical and dynamical approximations of the theory of scattering to the above-mentioned phenomena, which permits one to obtain an analytical solution of the problems under consideration. The numerical methods for solving the corresponding problems are beyond the scope of this book, although the results obtained by such methods (with corresponding references) are given in some instances to illustrate the experiment or theory under consideration. This is explained not only by the fact that for most problems, the above-mentioned methods yield results with sufficient accuracy, but also by the fact that these methods most clearly illustrate the common physical nature of the phenomena under consideration. We do not concentrate on the scattering of low-energy electrons in crystals, which, in fact, is closely to the problems considered in our book. Space limitations decided this, but more important, the above approximation fails to describe with sufficient accuracy the interaction between low-energy electrons and crystals. Also, extensive literature, reviews and monographs already exist on this problem.

I believe that the consideration of all the above problems in one book will be useful for both specialists in the field and beginners, since at present corresponding data are scattered in numerous original publications or highly specialized monographs. The bulk of this book presents a first attempt to generalize the problems of diffraction optics for various media with periodicities of a nonscalar nature, but at the same time, some problems considered here may also be found in recent monographs. Thus, some problems of the dynamical theory of diffraction of Mössbauer gamma radiation are stated by V.G. Baryshevskii in his monograph, *Nuclear Optics of Polarized Media* (Minsk, Izd-vo BGU, 1976) and by M.A. Andreeva and R.N. Kuz'min in *Mössbauer Gamma Optics* (Moscow, Izd-vo MFU, 1982). Problems concerning the optics of chiral cholesteric liquid crystals in an experimental and theoretical context may be found in *Optics of Cholesteric Liquid Crystals*, written by the author together with A.S. Sonin (Moscow, Nauka, 1982). The radiation of uniformly moving charged particles in media with the periodicity of scalar characteristics not associated with channeling is considered by G.M. Garibyan and C. Yang in *X-Ray Transition Radiation* (Erevan, Izd-vo AN Arm. SSR, 1983) and V.L. Ginzburg and V.N. Tsytovich in *Transition Radiation and Transition Scattering* (Moscow, Nauka, 1984).

Quite a large part of this book is based on the results obtained at the Theoretical Laboratory of the All-Union Research Institute for Physical-

Technical and Radiotechnical Measurements and in the Laboratory of Solid State Theory of the All-Union Surface and Vacuum Research Centre and published at different times in coauthorship with Yu. M. Aivazian, R.Ch. Bokun, A.G. Grigorian, V.E. Dmitrienko, A.A. Grigor'ev, I.V. Maslov, S.M. Osadchii, V.P. Orlov, E.V. Smirnov, and N.V. Shipov. The author is very grateful to all of these individuals.

The author also wishes to thank V.E. Dmitrienko and M.V. Fedorov, who have read the manuscript, for their valuable remarks.

Contents

Symbol Index

η, ξ, γ, μ, ζ, δ, k, ϵ, ω: variables in Mathieu equation, 1–4

ϵ_i, δ: principal value of dielectric tensor and dielectric anisotropy, 4

$\hat{\epsilon}$: dielectric tensor, 6

p: cholesteric pitch, 6

τ: reciprocal lattice vector, 6

K, k: wave numbers, 7

q: wave vector in a homogeneous medium, 7

$\mathbf{n}_{\pm}$: circular polarization unit vector, 7

r: ratio of refractive indices, 12

R: reflection coefficient, 14

$\mathbf{k}_0$, $\mathbf{k}_1$; $\mathbf{e}_0$, $\mathbf{e}_1$: wave and polarization vectors, 26

f: scattering amplitude, 26

$F(\ldots)$: structure amplitude, 26

$\hat{T}$: coherent scattering tensor, 27–28

ρ: polarization matrix, 28

$\hat{\epsilon}_s$: Fourier transform of dielectric tensor, 29–30

2θ: scattering angle, 30

$\Delta\alpha$: deviation from the Bragg condition, 38

$f(\mathbf{k})$: Lamb–Mössbauer factor, 49

ξ_i: Stokes parameters, 57

P: polarization factor, degree of polarization, 60

CLC: cholesteric liquid crystal, 80–81

S: order parameter, 81

T: transmission coefficient, 96

D: circular dichroism, 97

C^*: chiral smectic, 112

BPI, BPII, BPIII: different types of blue phases of liquid crystals, 113

$d(k_p)$: thickness of unit layer, 146

e^{-W}: Debye–Waller factor, 146

Introduction

The interaction of radiation with matter is the main source of information on the structure of matter on microscopic and large-scale levels. A special place in the structural studies of matter is given to the processes of the coherent interaction of radiation with matter. This relates, first and foremost, to the study of the condensed state. Experimental techniques based on the coherent interaction of photons, electrons, neutrons, etc. with various objects provide the information forming the basis of modern concepts on the structure and dynamics of molecules, crystals, liquids, and amorphous solids. The achievements in this field are well known.

X-ray scattering yields information on the structure of crystals and molecules, including very important biological ones such as proteins and DNA. Scattering of thermal neutrons also gives, in addition to structural information, some data on the dynamics of the object under study (its excitation spectrum and magnetic structure). Coherent scattering of electrons and ions has been successfully used for the investigation of surface phenomena and various properties of thin layers. Coherent radiation of fast charged particles in condensed media, in particular, Cherenkov radiation, is directly related to the physical characteristics of the media. As is well known, this phenomenon is successfully used in high-energy physics for the detection of energetic particles. The great importance of the above-mentioned phenomena in physics and their numerous applications explain the ever-increasing interest in these phenomena and the need for their thorough investigation on a large scale.

In particular, the interaction of various types of radiation with periodic media has been studied at all the stages of development of physics. This problem is considered in numerous articles and monographs. At the same time, the further progress of physics—a widening of the range of investigations, on the one hand, and the trend toward a more and more detailed study of the condensed state of matter based on a microscopic approach, on the other—establishes new problems on the theory of the interaction of radiation with matter at each new stage of its development.

The current state of the research and application of radiation interaction with matter, in particular, with periodic media, is characterized by an increase in the number of methods and objects and the greater importance of such studies in different fields. If earlier it had been X-ray diffraction which was considered the main method for studying periodic media (crystals), at present, along with widespread electron and neutron diffraction techniques, other diffraction methods have gradually evolved: those associated with the application of synchrotron radiation, Mössbauer gamma radiation, beams of charged particles, laser radiation in the optical and infrared wavelength ranges. In the optical range one naturally keeps in mind the objects for which the scale of spatial periodicity is of the order of the light wavelength. These are, for example, chiral liquid crystals, crystals with a periodic domain structure, artificial heterogeneous structures, and finally, periodic inhomogeneities of properties induced in a substance by a strong laser wave.

Along with an increased number of objects to which the conventional theory of the interaction of radiation with periodic systems is applied, new aspects of the theory itself have been developed. If earlier on the main focus was media with the simplest periodic properties (periodic scalar characteristics), recently important progress has been achieved in the study of objects with complicated (nonscalar) spatially periodic properties (magnetic structures, liquid crystals, etc.). We also address the emergence of nonlinear optics of periodic media.

The coherent interaction of radiation with the periodic media of a complicated structure (with nonscalar scattering from individual scatterers), being analogous in its general features to that with simple media (e.g., X-ray diffraction [1], radiation of uniformly moving fast charged particles in crystals [2,3]), is characterized by some specific features. These features are associated mainly with more complicated polarization characteristics of the radiation interaction with the object and more specific details about the angular (frequency) distribution of coherent scattering directly related to the structural characteristics of the corresponding media. Thus, in the optics of chiral liquid crystals [4], complicated polarization characteristics of selective scattering are observed, whereas in the diffraction of Mössbauer gamma radiation [5,6], the existence of magnetic (nuclear) diffraction maxima has been experimentally confirmed and they are widely used in investigations.

All this suggests new possibilities for studying the structure of condensed media. We are witnessing today the emergence of magnetic Mössbauer diffraction and magnetic X-ray diffraction, useful techniques which complement magnetic neutron diffraction, the only direct method for the determination of the magnetic structure of crystals. Mössbauer diffraction also permits one to study collective effects in the interaction of gamma quanta with atomic nuclei. The further development of X-ray diffraction holds promise for studying the effects associated with the presence of a small nonscalar component in X-ray scattering amplitudes due to the magnetic moments of atoms, anisotropy of atomic bonding, or thermal vibrations in the crystal lattice. Future achievements in this field depend on the application of synchrotron radiation in the corresponding investigations.

Analogous properties should be expected for the coherent radiation of fast charged particles moving with a constant velocity in periodic media.

The angular distribution of radiation emitted by a particle in a periodic medium may drastically differ from the corresponding distribution of radiation from a "dielectric boundary" (transition radiation). Diffraction of the emitted radiation in a periodic medium may result in the fact that along with the direction permitted for radiation in a homogeneous medium, radiation of the same frequency may also be emitted in the direction related to the initial one by the Bragg condition. In a perfect crystal, the enhancement of radiation with a certain frequency emitted by a particle may take place and also frequency (angular) bands forbidden for emission appear.

These problems have been thoroughly studied for simple periodic media (with the periodicity of scalar characteristics) both in this theoretical and experimental aspects [2,3,7]. Moreover, X-ray transition radiation in layered media is used for detecting ultrarelativistic particles [3]. However, the corresponding studies, especially experimental ones, for media with complicated structures (nonscalar characteristics of individual scatterers) are only at their very beginning.

All the above illustrates the importance of the presentation and generalization from a unified standpoint of recent theoretical and, partly, experimental achievements (Chapt. 1–5, 8–10) on the interaction of various kinds of electromagnetic radiation and fast charged particles with periodic media. The primary focus is achievements in new directions of research related to the interaction of radiation with media having a complicated structure and nonscalar scattering characteristics. The material is presented in such a way as to emphasize the common physical grounds of different, at first glance, phenomena. The general concepts and methods are illustrated by concrete phenomena—diffraction of Mössbauer gamma quanta from magnetically ordered crystals, optics of chiral liquid crystals, and Cherenkov radiation of charged particles in periodic media.

Some problems of nonlinear optics of periodic media [4,8] and of dynamical scattering of neutrons from magnetically ordered crystals (presented in Chapt. 6 and 7) are also closely related to the above phenomena, so their physics may be described by similar mathematical methods. Thus, in the case of nonlinear optical generation of higher-order harmonics in periodic media (similar to the case of fast particle radiation), redistribution of the generated wave propagation directions may take place and it may be enhanced in comparison with the case of the homogeneous medium.

The dynamical diffraction of thermal neutrons from magnetically ordered crystals and its mathematical description are very similar to the dynamical diffraction of Mössbauer radiation. This is true, first and foremost, of the polarization characteristics of scattering.

At the same time, the experimental study of the above-mentioned effects of nonlinear optics of periodic media and dynamical scattering of thermal neutrons from perfect magnetically ordered crystals is at its very beginning. The numerous effects predicted by the theory are still waiting for experimental confirmation. However, quite recently important progress in the neutron investigations of perfect magnetic crystals was achieved (see Chapt. 7).

References

1. Z.G. Pinsker: *Dynamical Scattering of X-Rays in Crystals* (Berlin, Springer Verlag, 1978).

2. V.L. Ginzburg, V.N. Tsytovich: *Transition Radiation and Transition Scattering* (Moscow, Nauka, 1984) (in Russian).

3. G.M. Garibyan, Yan Shi: *X-Ray Transition Radiation* (Yerevan, AN Arm. SSR, 1983) (in Russian).

4. V.A. Belyakov, A.S. Sonin: *Optics of Cholesteric Liquid Crystals* (Moscow, Nauka, 1982); V.A. Belyakov, V.E. Dmitrienko: *Optics of Chiral Liquid Crystals* Soviet Scientific Reviews, Section A (Ed. I.M. Khalatnikov, Harwood Academic Publishers, GmbH), v. 13, part I (1989).

5. V.A. Belyakov: UFN **115**, 553 (1975) [Sov. Phys. - Usp. **18**, 267 (1975)].

6. M.A. Andreeva, R.I. Kuz'min: *Mössbauer Gamma Optics* (Moscow, MGU, 1982) (in Russian).

7. V.G. Baryshevskii: *Channeling, Radiation and Reactions in Crystals at High Energies* (Minsk, BGU, 1982) (in Russian).

8. S.M. Arakelyan, Yu.S. Chilingaryan: *Nonlinear Optics of Liquid Crystals* (Moscow, Nauka, 1984) (in Russian).

1

Waves in Media with One-Dimensional Periodicity (Exact Solution)

1.1   Layered Medium

Our focus is the interaction between radiation and periodic media of complicated structures. Nevertheless, we begin our consideration with simple periodic systems, since their properties are typical of wave propagation in all media; their results are general and may become the zeroth approximation for more complicated systems.

1.1.1   Harmonic Modulation of Media

Consider the propagation of an electromagnetic wave in an infinite medium with one-dimensional periodicity. Let the dielectric permittivity of the medium be modulated along the periodicity direction (z axis). Then the wave equation reduces to the differential equation of the form [1,2]

$$d^2\psi/dz^2 + f(z)\psi = 0 \qquad (1.1)$$

where $f(z)$ is a periodic function.

In the general case of an arbitrary periodic function $f(z)$, infinite number of harmonics in the expansion of $f(z)$ into a Fourier series, Eq. (1.1) is Hill's equation. The solutions of Hill's equation are analyzed in the general form [3] but for simplicity we analyze its solutions for a specific form of $f(z)$.

In the specific case where $f(z)$ is a harmonic function, one harmonic in the Fourier expansion of $f(z)$, equation (1.1) reduces to the Mathieu equation:

$$d^d u/d\xi^2 + (\eta + \gamma \cos^2 \xi)u = 0. \qquad (1.2)$$

According to Floquet's theorem, the general solution of Eq. (1.2) has the form

$$u(\xi) = D_1 A(\xi)e^{\mu\xi} + D_2 B(\xi)e^{-\mu\xi} \qquad (1.3)$$

where D_1 and D_2 are arbitrary constants and $A(\xi)$ and $B(\xi)$ are periodic functions, with period equal to the period of property modulation in the

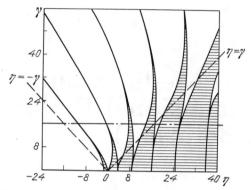

FIGURE 1.1. Stability diagram for the Mathieu equation.

medium. Expressions of form (1.3) describe the spatial variation of an electromagnetic wave—in the derivation of Eq. (1.1) it is assumed that time dependence is described by $e^{i\omega t}$. The solution is regarded as the superposition of two waves propagating in opposite directions.

Omitting a detailed analysis [3], we concentrate on the parametric dependence of solutions to Eq. (1.2) on η and γ. We distinguish two cases: (i) μ entering equation (1.3) as an imaginary quantity which corresponds to nondecaying monochromatic waves and (ii) μ as a complex or real quantity which corresponds to decaying waves. The analysis of equation (1.2) is illustrated in Fig. 1.1 where η and γ are plotted on coordinate axes. Open regions correspond to values of η and γ for which μ is either a complex or real quantity where the wave is decaying, whereas the hatched regions correspond to values of η and γ for which μ is an imaginary quantity. Thus, the hatched regions correspond to transmission bands and open regions stop bands. If γ infinitely increases and $\eta > -\gamma$, the hatched regions become narrower tending, in the limit, to straight lines parallel to the line $\eta = -\gamma$. For $\eta < -\gamma$, i.e., below the line $\eta = -\gamma$, there are no propagating modes. At the boundaries between open and hatched regions the real part of μ goes to zero. At the η axis the value of γ is zero. This corresponds to the case of an unmodulated homogeneous medium for which there are no forbidden bands. In the vicinity of axis $\eta, \gamma << \eta$, which corresponds to a small modulation of medium properties (dielectric constant in our case), wave propagation is often described rather accurately by the approximate solutions of the Maxwell equation for a periodic medium. Two neighboring curves limiting adjacent hatched regions have a common datum point and point of tangency of $(n-1)$th order on the η axis, with abscissa $\eta = n^2$ (n is an integer). Therefore, for $n = 1$ the neighboring curves originate from one point; for $n = 2$ they have a common vertical tangent; for $n = 3$ there is a common vertical tangent and curvature, etc. Note that parameter η, determined by the mean value of dielectric permittivity, ϵ_0 and parameter γ, determined by the amplitude of the spatial modulation of dielectric per-

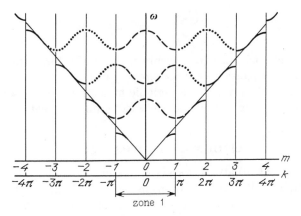

FIGURE 1.2. Brillouin diagram.

mittivity ϵ_1, are proportional to wave frequency squared ω^2, but their ratio (for fixed modulation amplitude ϵ_1 and ϵ_0) is independent of frequency. Therefore, for fixed values of ϵ_0 and γ_1, the values of μ for different frequencies ω are determined by the points of the stability diagram on the line $\eta/\gamma = \epsilon_0/\epsilon_1 = \text{const}$; the particular case of such a line, $\eta = \gamma$, is shown in Fig. 1.1. Some portions of the corresponding lines are within stop bands. This means that not all the waves (with arbitrary frequencies) propagate in the periodic medium - there are some frequency ranges, the so-called "forbidden bands", which have no propagating modes. In addition, the stability diagram shows that the number of "forbidden bands" is infinite and that their frequency width increases with modulation amplitude.

The dispersion curve (relation between ω and μ) which corresponds to the above band structure has the shape shown in the Brillouin diagram (Fig. 1.2). Some portions of the dispersion curve, in the first Brillouin zone ($|\mu| \leq \pi/d$ where d is the modulation period), demonstrate another important result known in optics of periodic media: the position of the break in the dispersion curve, i.e., the position of the forbidden band in momentum space, does not change with amplitude ϵ_1; however, the value of such a break, i.e., the forbidden band frequency width, depends on ϵ_1. We note that the Brillouin diagram shows only the real part of k. As follows from the stability diagram, k is real in the transmission band, whereas it is a complex quantity, ($k = i\mu\pi$), in the forbidden band. Therefore, in the range of the break on the Brillouin diagram, the imaginary component of k, the real component being fixed, depends on frequency and goes to zero at the forbidden band boundary. From the general form of the solution of the Mathieu equation (1.3), it follows that since functions A and B are periodic in an electromagnetic wave in a periodic medium, each component of Eq. (1.3) may be written as

$$\psi = \sum_{s=-\infty}^{s=+\infty} A_s \exp[i(k + s\tau)z], \tag{1.4}$$

where A_s are coefficients, equal, within a constant factor, to the coefficients for the Fourier expansion of function $A(\xi)$ in (1.3). That is, generally, a monochromatic plane wave in a periodic medium is not a plane wave but an infinite superposition of plane waves determined by equation (1.3).

1.1.2   Properties of Eigenwaves

In many practical situations, for small modulation amplitude of dielectric permittivity, a small number of terms in expansion (1.3) are important, i.e., only some A_s coefficients are large whereas other coefficients are smaller, at least by a factor of δ, where $\delta = \epsilon_1/\epsilon_0$ is the relative modulation amplitude of dielectric permittivity or dielectric anisotropy. In the simplest case, where k is far from the forbidden band, expansion (1.4) is reduced to one term. This means that it is a good approximation to represent an electromagnetic field as a plane wave with a certain wave vector k in those regions. If k is inside the forbidden band, or in close proximity to it, an electromagnetic field cannot be represented by a plane wave, irrespective of small modulation amplitude. Thus, expansion (1.4) must be extended to two terms, with $s = 0$ and $s = n$ where n is the number of the forbidden band to which k belongs. Such a situation corresponds to electromagnetic wave diffraction by a periodic structure and approximation (1.4) with two terms is the two-wave approximation. The equation relating wave vectors of two plane waves is the Bragg condition,

$$k_n = k + n\tau. \tag{1.5}$$

For small modulation amplitude ($\delta \leq 0.1 - 0.01$) the two-wave approximation is often satisfactory from the practical standpoint and is used to quantitatively describe the interaction of radiation with periodic media. And, in the case of harmonic modulation of dielectric permittivity (and small amplitude δ) the frequency width of forbidden bands rapidly decreases with band number, n, as δ^n.

For the general form of spatial modulation of the dielectric properties of a medium i.e., if $f(z)$ in Eq. (1.1) is an arbitrary periodic function, the problem reduces to the solution of Hill's equation and yields the same qualitative results as the previously analyzed solution of Mathieu's equation [2,3]. Hence, without a detailed analysis for the general form of the periodic function $f(z)$ in equation (1.1), we conclude that forbidden band frequency width may not depend on band number n in a regular way. It is essentially determined by the value of the nth harmonic of the Fourier expansion of $f(z)$ and generally increases with harmonic amplitude. In particular cases, for certain ratios between different Fourier harmonics, the frequency width of individual bands may be equal to zero [2].

Summarizing the results of the brief analysis of electromagnetic wave propagation in media with the simplest form of dielectric, property periodicity, modulation of scalar dielectric permittivity, we emphasize that, even in this simplest situation, there is no exact analytical solution to the problem. The basic principles are:

1. A monochromatic wave in a periodic medium is a superposition of an infinite number of plane waves.

2. There is an infinite number of "forbidden" frequency bands in wave propagation.

3. If a wave propagates along the direction of a medium periodicity, its polarization characteristics coincide with those of an isotropic homogeneous medium. That is, wave characteristics are independent of polarization.

The important but specific practical case of small modulation amplitude of dielectric properties permits simplification of the general statement above: the one plane wave approximation may be adequate in regions outside the forbidden band, whereas the two-wave approximation applies inside and in close proximity to it. If the modulation of dielectric properties is harmonic, the forbidden band frequency width decreases with zone number n as δ^n.

The general features of electromagnetic wave propagation in arbitrary media are the same as in periodic modulations of scalar dielectric permittivity of media. However, for arbitrary periodic media and for waves propagating at an angle to the periodicity direction, some specific properties apply which concern the polarization characteristics of the waves and fine details of the forbidden-band structure. The mathematical construct used to describe wave propagation in periodic media with complicated structure is more sophisticated than that in the simplest case considered above. Nonetheless, the simple qualitative results are useful to understand the physics of more complicated situations.

1.2 Optics of Cholesterics (Exact Solution)

An example of unusual optical properties caused by structure of unidimensional periodicity is shown by cholesteric liquid crystals, the so-called cholesterics [4]. Neglecting the details in our consideration, we should like to underline here another unique aspect of optics of cholesterics and similar media: The Maxwell equations have been solved analytically and exactly for light propagation along the axes of cholesterics and chiral smectics [5–6]. The obtained exact solution is simple and is the only example of a simple exact analytical solution of the Maxwell equations for periodic media. No other simple exact solution is known—not even for periodic structures more simple than cholesterics (see, e.g., Sect. 1.1).

1.2.1 Dielectric Properties of Cholesterics

A cholesteric is a locally uniaxial medium for electromagnetic radiation in the optical range. The anisotropy direction of the medium varies (rotates) in space along some direction normal to the local anisotropy axis in accordance with harmonic law (see Fig. 1.3). This direction, the optical axis of a cholesteric, is also called the cholesteric axis. The dielectric permittivity tensor $\hat{\epsilon}(r)$ of a cholesteric depends on the coordinates as follows [4]:

$$\hat{\epsilon}(r) = \begin{pmatrix} \bar{\epsilon} + \bar{\epsilon}\delta \cos\varphi(z) & \pm\bar{\epsilon}\delta \sin\varphi(z) & 0 \\ \pm\bar{\epsilon}\sin\varphi(z) & \bar{\epsilon} - \bar{\epsilon}\delta \cos(z) & 0 \\ 0 & 0 & \epsilon_3 \end{pmatrix} \quad (1.6)$$

where $\bar{\epsilon} = (\epsilon_1+\epsilon_2)/2$, $\delta = (\epsilon_1-\epsilon_2)/(\epsilon_1+\epsilon_2)$, and $\epsilon_1, \epsilon_2 = \epsilon_3$ are the principal values of the dielectric permittivity tensor. The Z axis is directed along the cholesteric axis, the rotation angle $\varphi(Z)$ of the local anisotropy direction around the cholesteric axis is linearly related to coordinate z and $\varphi(z) = \tau z/2$ where $\tau = 4\pi/p$. Quantity p is the period or pitch of the cholesteric helix and is equal to the distance along the cholesteric axis within which the direction of the local anisotropy makes a complete revolution about that axis. The period of dielectric properties of a cholesteric coincides with a half-pitch $p/2$ of the helix since the anisotropy directions rotated by a half-turn and the former ones are physically equivalent. Two signs in tensor (1.6) correspond to two geometric possibilities—left- and right-handed helices. Thus, spatial changes in the dielectric permittivity tensor of a cholesteric reduce to a harmonic rotation ($\varphi \sim z$) of two principal axes of the tensor $\hat{\epsilon}(r)$ with the variation of the coordinate along the cholesteric axis. (See Fig. 1.3.)

1.2.2 Eigenwaves

In the absence of external currents and charges, the Maxwell equation for a cholesteric has the form

$$\text{rot } \mathbf{E} = c^{-1}\partial\mathbf{B}/\partial t, \quad \text{div } \mathbf{D} = 0, \quad \mathbf{D} = \hat{\epsilon}\mathbf{E}$$
$$\text{rot } \mathbf{H} = c^{-1}\partial\mathbf{D}/\partial t, \quad \text{div } \mathbf{B} = 0, \quad \mathbf{B} = \hat{\mu}\mathbf{H}. \quad (1.7)$$

Assuming the magnetic permittivity tensor, $\mu = 1$, for a plane wave propagating along the optical axis of a cholesteric, axis Z, the field is independent of coordinates x and y and equations (1.7) for the electric field of the wave yields

$$\partial^2\mathbf{E}/\partial z^2 = \hat{\epsilon}(z)c^{-2} \cdot \partial^2\mathbf{E}/\partial t^{-2}. \quad (1.8)$$

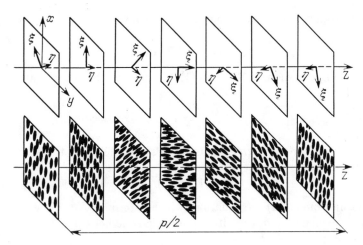

FIGURE 1.3. Schematic representation of a cholesteric structure (below) and the corresponding change of spatial orientations of the principal axes of the dielectric tensor (above).

Notice that for propagation direction parallel to the helical axis, an electromagnetic wave is strictly transverse i.e., vector $\mathbf{E}$ is normal to axis Z.

We seek the solution of equation (1.8) for fixed wave frequency ω. As shown in Sect. 1.1., in a spatially periodic medium (as in a cholesteric), the simplest solution of the Maxwell equation is not a plane wave, e^{ikz}, as in a homogeneous medium, but a Bloch wave of the form

$$\mathbf{E}(z,t) = \sum_s \mathbf{E}_s e^{i(\mathbf{K}+s\tau)z} e^{-i\omega t}. \tag{1.9}$$

Thus, the problem reduces to a search for amplitude E_s and wave vectors K in the superposition (1.9). As shown in Sec. 1.1, in the general case of a medium with one-dimensional periodicity, an infinite number of amplitudes E are nonzero.

In a cholesteric, specific polarization properties of scattering, determined by the dielectric permittivity tensor (1.6) result in the fact that if a wave propagates along the optical axis, only two amplitudes E_s in the exact solution (1.9) are nonzero while all the other amplitudes vanish. Then the solution of equation (1.8) is sought in the form

$$\mathbf{E}(z,t) = e^{-i\omega t}(E^+ \mathbf{n}_+ e^{i\mathbf{K}^+ z} + E^- \mathbf{n}_- e^{i\mathbf{K}^- z}), \tag{1.10}$$

where $\mathbf{n}_\pm = (\hat{\mathbf{x}} \pm i\hat{\mathbf{y}})/\sqrt{2}$ are the unit vectors of circular polarizations (hereinafter $\hat{\mathbf{a}} = \mathbf{a}/|\mathbf{a}|$) and K^+ and K^- are related by the Bragg condition

$$K^+ - K^- = \tau. \tag{1.11}$$

We introduce the following notation for convenience

$$K^+ = \eta^+ q, \quad K^- = \eta^- q, \quad q = \omega\sqrt{\bar{\epsilon}}/c. \tag{1.12}$$

Substituting equation (1.10) into (1.8), produces the set of equations for E^+ and E^-

$$[1 - (\eta^+)^2]E^+ + \delta E^- = 0, \delta E^+ + [1 - (\eta^-)^2]E^- = 0. \tag{1.13}$$

The condition for a nontrivial solution of system (1.13) is the zero value of its determinant

$$[1 - (\eta^+)^2][1 - (\eta^-)^2] - \delta^2 = 0. \tag{1.14}$$

We obtain from (1.14) the equations which determine K^+ and K^- as functions of frequency or q, the pitch of a cholesteric helix p and the dielectric anisotropy parameter δ:

$$K_j^+/q = \eta_j^+ = q^{-1}(\tau/2 \pm k^\pm), \quad \eta_j^- = \eta_j^+ - 2\tilde{\lambda},$$
$$k^\pm = q[1 + \tilde{\lambda}^2 \pm (4\tilde{\lambda}^2 + \delta^2)^{1/2}]^{1/2}, \quad \tilde{\lambda} = \tau/2q. \tag{1.15}$$

Expression (1.15) for the fixed value of frequency ω determines four values of $\eta_j^\pm$, i.e., four values of wave vectors $K_j^\pm$. Four corresponding solutions of system (1.13) and expression (1.10) determine eigenwaves in a cholesteric for the propagation direction coinciding with its optical axis.

Let us number the eigensolutions in the following way: $j = 1, 4$ for signs "+" and "−" before K^+ in the first equation in (1.15) and $j = 2, 3$ for signs "+" and "−" before K^- in the same expression.

The amplitude ratio, E^-/E^+, in the jth eigensolution is determined by the following expression (Fig. 1.4)

$$\xi_j = (E^-/E^+) = \delta/[(\eta_j^-)^2 - 1]. \tag{1.16}$$

1.2.3 Properties of Eigensolutions

It follows from equation (1.10) that, eigensolutions are superpositions of two circularly polarized plane waves. Depending on the sign of K^+ and K^- in equation (1.10), these are either two waves with opposite circular polarizations propagating in the same direction (same signs of K^+ and K^-) or two waves with the same circular polarization propagating in opposite directions (different signs of K^+ and K^-). We should like to remind here that the changes in the sign of the wave vector in the expression describing the circular wave is equivalent to the simultaneous change of propagation direction and sense of the circular polarization of the wave. The existence of eigensolutions in which the constituent plane waves propagate in opposite directions reflects the possibility that light diffracts from the spatial structure of cholesterics.

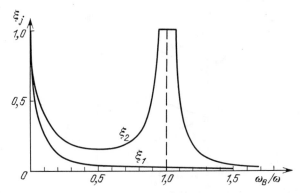

FIGURE 1.4. Amplitude ratio $\xi_j = |E^-/E^+|$ in eigensolutions versus frequency.

From equations (1.10), (1.12) and (1.15) it follows that solutions 1 and 4 containing k^+ in (1.15) correspond to two nondiffracting modes in the periodic structure of a cholesteric: one propagating along the Z axis, the other in the opposite direction. Solutions 2 and 3 relate to modes which experience diffraction in a cholesteric.

We focus on the physical characteristics of eigensolutions and their frequency dependence in the case when the dielectric anisotropy parameter of a cholesteric $\delta \ll 1$. In solution 1, K^+ and K^- have the same signs for all frequencies whereas the ratio of one amplitude (E^+ or E^-) to the other is of the order of δ (Fig. 1.4) everywhere except the high-frequency limit $\omega/c \gg \tau/\delta$ where both amplitudes are equal. Therefore, solution 1 corresponds to a wave circularly polarized in the direction opposite the screw sense of the cholesteric helix, within the accuracy of δ, for all the frequencies except $\omega/c > \tau/\delta$. For $\omega/\tau \gg \tau/\delta$, i.e., in the limit of very short waves, solution 1 corresponds to a plane-polarized wave, the polarization plane of which is rotated, with period p, matching the screw sense of the helix and the variation in the z coordinate. The polarization plane is parallel to the local orientation of the principal axes of the dielectric permittivity tensor $\hat{\epsilon}$ which corresponds to the least principal value, ϵ_2. The wavevector of the eigensolution in this limit is $q(1 - \delta/2)$. As mentioned previously, solution 4 is analogous to solution 1 but describes wave propagation in the opposite direction. Depending on frequency, wave vectors K^+ and K^- in solution 2 may be real with the same signs, real with opposite signs, or complex quantities. The region of complex K^+ and K^+ is close to the Bragg frequency $\omega_B = \tau c/2\sqrt{\epsilon}$ in the range $\omega_B/\sqrt{1+\delta} < \omega < \omega_B\sqrt{1-\delta}$; it corresponds to a "forbidden band" and pure imaginary values of K^-. That is, in a cholesteric, no waves propagage in this frequency range if their circular polarization matches the screw sense of the cholesteric helix. Independently of δ, in the forbidden band $|E^-/E^+| = 1$, and selective reflection of light associated with diffraction occurs in cholesterics. Outside

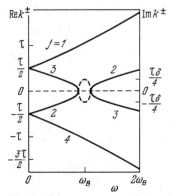

FIGURE 1.5. Roots of the dispersion equation (1.14) versus frequency (solid lines denote $Re\,K$, dashed lines indicate $Im\,K$ [see (1.15)].

the forbidden band, except in the high-frequency limit, the ratio of E^- or E^+ to the other amplitude is of the order of δ and solution 2 corresponds, with sufficient accuracy, to the wave circularly polarized in the sense of the cholesteric helix. In the limit of very short waves $\omega c \gg \tau/\delta$, this solution becomes a plane-polarized wave with the polarization plane normal to that of solution 1 which rotates with period p and varies with z. In this limit, the wave vector of solution 2 is equal to $q(1+\delta/2)$. Solution 3 has the same properties as solution 2 with one exception: outside the forbidden band it describes wave propagation in the opposite direction. The frequency dependence of k^+ and k^- in eigensolutions is illustrated in Fig. 1.5 which shows that the dispersion law $K(\omega)$ for solutions 1 and 4 is close to that for a wave propagating in a homogeneous medium. The dispersion law for solutions 2 and 3 is diffractive in nature and has a forbidden band (cf. Fig. 1.2). In summary, the features of light progation in a cholesteric: selective diffraction reflection is expressed by light propagating along the optical axis whose circular polarization matches the screw sense of the cholesteric helix; it is essential that only first-order diffraction reflection occurs and no higher-order reflection is possible which corresponds to $|s| = 1$ in relationships of type (1.5). The absence of higher-order reflections for light propagation along the optical axis of a cholesteric is also illustrated by dispersion curves in Fig. 1.5 with only one forbidden band for each of solutions 2 and 3.

1.3   Solution of Boundary Problem

The description of experimentally observed optical characteristics in the interaction of light with periodic media requires that we know the properties of the eigenwaves in the corresponding media, and find the solution of the Maxwell equations that satisfy certain boundary conditions, i.e., to solve the boundary problem. The solutions of the boundary problem for periodic media are illustrated with the following example of a cholesteric.

1.3.1   Planar Cholesteric Texture

Solution of the Boundary Problem. Using the eigensolutions obtained previously, we solve the problem of light reflection and transmission in a cholesteric sample of finite thickness. Consider the case of normal incidence onto a plane parallel plate of thickness L with optical axis normal to the sample surface. That is, we consider the optical properties of an ideal planar texture for normal incidence of light.

We must determine which and to what degree eigensolutions of system (1.13) are "excited" by the incident wave to find the wave amplitudes of light reflected from and transmitted through the cholesteric. For normal incidence of light, the boundary conditions (continuous tangential components of electric and magnetic fields at the boundary) reduce to continuous electric and magnetic fields on sample surfaces since the wave in a cholesteric is strictly tansverse. Therefore, the boundary conditions are written in the form

$$\mathbf{E}^e(0) + \mathbf{E}^r(0) = \mathbf{E}(0), \quad \mathbf{E}^t(L) = \mathbf{E}(L)$$
$$\mathrm{rot}[\mathbf{E}^e(z) + \mathbf{E}^r(z)]_{z=0} = [\mathrm{rot}\,\mathbf{E}(z)]_{z=0}$$
$$\mathrm{rot}\,\mathbf{E}(z)_{z=L} = \mathrm{rot}\,\mathbf{E}(z)_{z=L}. \tag{1.17}$$

The electric fields of the incident $\mathbf{E}^e$, reflected $\mathbf{E}^r$ and transmitted $\mathbf{E}^t$ waves and the wave $\mathbf{E}$ propagating in the sample are expressed in the form

$$\mathbf{E}^e = e^i(q_e z - \omega t)(E_e^+ \mathbf{n}_+ + E_e^- \mathbf{n}_-)$$
$$\mathbf{E}^r = e^{-i(q_e z + \omega t)}(E_r^+ \mathbf{n}_- + E_r^- \mathbf{n}_-)$$
$$\mathbf{E}^t = e^{i(q_e z - \omega t)}(E_t^+ \mathbf{n}_+ + E_t^- \mathbf{n}_-)$$
$$\mathbf{E} = e^{-i\omega t} \sum_{j=1}^{4} E_j^+ (e^{iK_j^+ z} \mathbf{n}_+ + \xi_j e^{iK_j^- z} \mathbf{n}_-) \tag{1.18}$$

where $\mathbf{n}_+, \mathbf{n}_-$ are the unit vectors of the circular polarizations and q_e is the wave vector of light outside the sample. When waves propagate in the positive direction of the axis, (particularly incident waves), $\mathbf{n}_+$ and $\mathbf{n}_-$ describe the left and right-handed circular polarizations, respectively whereas $\mathbf{n}_+$ and $\mathbf{n}_-$ describe the right and left-handed circular polarizations respectively when reflecting waves propagate in the opposite direction. This fact is reflected in expression for $\mathbf{E}^r$ in (1.18).

Substituting (1.18) into (1.17), equating each factor before $\mathbf{n}_+$ and $\mathbf{n}_-$ to zero, and transforming, we find that reflected and transmitted amplitudes are expressed in terms of the amplitudes of eigenwaves E_j excited in the sample

$$E_r^+ = (1/2) \sum_j \xi_j (1 - K_j^- / q_e) E_j^+,$$

$$E_t^+ = (1/2) \sum_j e^{i(K_j^+ - q_e)L} (1 + K_j^+ / q_e) E_j^+,$$

$$E_r^- = (1/2) \sum_J (1 - K_j^+ / q_e) E_j^+,$$

$$E_t^- = (1/2) \sum_j \xi_j e^{i(K_j^- - q_e)L} (1 + K_j^- / q_e) E_j^+. \qquad (1.19)$$

The amplitudes of eigenwaves E_j^+ are given by the following set of equations

$$\sum_j (1 + K_j^+ / q_e) E_j^+ = 2E_e^+, \quad \sum_j e^{iK_j^+ L} (1 - K_j^+ / q_e) E_j^+ = 0,$$

$$\sum_J \xi_j (1 + K_j^- / q_e) E_j^+ = 2E_e^-, \quad \sum_j \xi_j e^{iK_j^- L} (1 - K_j^- / q_e) = 0. \quad (1.20)$$

Thus, the case of light transmission and reflection for normal incidence of a beam onto a planar cholesteric texture reduces to the solution of a nonuniform set of four linear equations. Its solution and Eqs. (1.15) and (1.16) for parameters ξ_j and $K_j^\pm$ provide the exact description of optical characteristics in the general case of a planar layer. In particular, it describes their dependence on frequency, light polarization and the dielectric properties of the medium outside the cholesteric.

Although the exact expressions for reflected and transmitted wave amplitudes are complicated, their structures are simple. Thus, the expression for the amplitude E_r^+ of the diffracted, circularly polarized wave is written as the following determinant ratio

$$E_r^+ = [\det D]^{-1} \begin{vmatrix} & & & & E_e^+ \\ & D & & & E_e^- \\ & & & & 0 \\ & & & & 0 \\ -- & -- & -- & -- & -- \\ {}^r\alpha_{51}^+ & {}^r\alpha_{52}^+ & {}^r\alpha_{53}^+ & {}^r\alpha_{54}^+ & 0 \end{vmatrix} \qquad (1.21)$$

where D is the matrix of system (1.20) and the elements of the fifth row are given by the following equation

$$^r\alpha_{5j}^+ = \xi_j (1 - r\eta_j^-) \qquad (1.22)$$

where $r = \sqrt{\bar{\epsilon}/\epsilon}$, ϵ is the dielectric permittivity outside the cholesteric. Expressions for E_r^-, E_t^+, and E_t^- are also derived from Eq. (1.21) if the elements of the fifth row are substituted by

$$^r\alpha_{5j}^- = (1 - r\eta_j^+), \quad {}^t\alpha_{5j}^+ = e^{i(K_j^+ - q_e)L}(1 + r\eta_j^+)$$
$$^t\alpha_{5j}^- = \xi_j e^{i(K_j^- - q_e)L}(1 + r\eta_j^-) \qquad (1.23)$$

respectively. In the general case, the exact solutions of the boundary problem (1.19)–(1.23) are cumbersome and it is more convenient to analyze them by numerical methods. The form of solution (1.21) is most convenient since it is used directly in computer calculations.

Analytical expressions are obtained directly from the general solution for some particular cases or by using simplifying assumptions. We present such cases to clarify the analysis and physical aspects of the problem.

1.3.2 Reflection from Thick Layers

Assume that thickness L of a planar cholesteric layer is very large, so that in solving the boundary problem, we may say that the liquid crystal occupies a half-space. The physical condition for such a simplification is the small distance over which light is absorbed in the cholesteric relative to layer thickness. The obtained results are applicable, however, over a wider range. Thus, for light frequencies in the selective reflection band, the corresponding condition becomes a small value of extinction length relative to layer thickness L. Extinction length is the distance over which the diffracted wave is attenuated by a factor of e due to diffraction reflection.

If light is incident from the outside half-space onto the half-space filled with the cholesteric, only two eigensolutions of those described by (1.13)–(1.15) are "excited" in the sample, which correspond to wave propagation into the cholesteric depth, i.e., to solutions 1 and 2. Solutions 3 and 4 which describe light propagation in the opposite direction cannot be "excited". This is proved by assuming that there is an arbitrary weak absorption in a cholesteric—a small imaginary addition to $\bar{\epsilon}$. It can be shown, using (1.20), that the amplitudes of the corresponding solutions decrease exponentially with thickness L. The physical explanation for the absence of excitation in the half-space of solutions 3 and 4 is that if the excitation did occur, the amplitudes of the corresponding fields would have increased infinitely towards the cholesteric depth; this is inconsistent with the finite values of the incident wave amplitude.

Thus taking into account that amplitudes E_j^+ and E_j^- for a half-space have nonzero values in (1.20) and (1.19) for solutions 1 and 2 only, i.e., $j = 1, 2$ and the cited four-order systems reduce to second-order systems. These are equations (1.19) and (1.20) without exponential factors

$$\sum_{j=1,2}(1 + r\eta_j^+)E_j^+ = 2E_e^+, \quad \sum_{j=1,2}\xi_j(1 + r\eta_j^-)E_j^+ = 2E_e^-$$
$$E_r^+ = 1/2 \sum_{j=1,2}\xi_j(1 - r\eta_j^-)E_j^+, \quad E_r^- = 1/2 \sum_{j=1,2}(1 - r\eta_j^+)E_j^+. \qquad (1.24)$$

The solution of this system is represented in another form (1.21). Simplification occurs in the reduction of determinant rank by two orders (1.21). The corresponding expressions are obtained by cancelling the third and fourth columns and rows of the determinant (1.21). For example, the amplitude of reflected wave is described as

$$E_r^+ = \begin{vmatrix} 1+\tilde{\eta}_1^+ & 1+\tilde{\eta}_2^+ & E_e^+ \\ \xi_1(1+\tilde{\eta}_1^-) & \xi_2(1+\tilde{\eta}_2^-) & E_e^- \\ \xi_1(1-\tilde{\eta}_1^-) & \xi_2(1-\tilde{\eta}_2) & 0 \end{vmatrix} [\det D]^{-1} \qquad (1.25)$$

where

$$\tilde{\eta}_j^+ = r\eta_j^+,$$

$$\det D = \begin{vmatrix} 1+\tilde{\eta}_1^+ & 1+\tilde{\eta}_2^+ \\ \xi_1(1+\tilde{\eta}_1^-) & \xi_2(1+\tilde{\eta}_2^-) \end{vmatrix}.$$

Similarly, using (1.21), we obtain the expression for amplitude E_r^-:

$$E_r^- = [\det D]^{-1} \begin{vmatrix} 1+\tilde{\eta}_1^+ & 1+\tilde{\eta}_2^+ & E_e^+ \\ \xi_1(1+\tilde{\eta}_1^-) & \xi_2(1+\tilde{\eta}_2^-) & E_e^- \\ 1-\tilde{\eta}_1^+ & 1-\tilde{\eta}_2^+ & 0 \end{vmatrix}. \qquad (1.26)$$

Equations (1.25) and (1.26) determine the reflection coefficient and polarization of the reflected wave for an arbitrary polarization of incident light. Since the eigenwaves in a cholesteric are superpositions of circularly polarized waves, we consider first the reflection of circularly polarized waves from a cholesteric, i.e., assume that either E_e^+ or E_e^- is zero. It is possible to find four amplitudes from equations (1.25) and (1.26), two E_r^{++} and E_r^{+-} which describe light reflected from diffracting and nondiffracting polarization into a wave with diffracting polarization in (1.25) when E_e^- and E_e^+ are set equal to zero in (1.25). Two other amplitudes, E_r^{-+} and E_r^{--}, describe the reflection of waves with the same polarization into waves with nondiffracting polarization given by equation (1.26) when E_e^- and E_e^+ are set to zero.

The reflection coefficients R^+ for circularly polarized diffracting waves and R^- for circularly polarized nondiffracting waves are expressed in terms of the introduced amplitudes

$$R^+ = |E_e^+|^{-2}(|E_r^{++}|^2 + |E_r^{-+}|^2), \quad R^- = |E_e^-|^{-2}(|E_r^{+-}|^2 + |E_r^{--}|^2). \quad (1.27)$$

Expressions (1.25)–(1.27) produce the exact solution to the problem of light reflected from a half-space filled with a cholesteric for an arbitrary relationship of cholesteric dielectric properties to surrounding medium.

Generally, reflection from the sample occurs according to two mechanisms—diffraction and dielectric (associated with the jump in the dielectric constant at the dielectric-medium interface). But in order to reveal the diffraction properties of reflection typical of periodic media, particularly in cholesterics, the analysis of obtained results should begin with the simulation where the averaged dielectric constant $\bar{\epsilon}$ of the cholesteric coincides with that of the surrounding medium, ϵ.

1.3.3 The Case Where Medium and Cholesteric Have Equal Dielectric Constants

All formulae are simplified by substitution of $\tilde{\eta}$ by η.

Assuming that δ is small, equations (1.25) and (1.26), the reflected wave amplitudes, in the zeroth approximation with respect to δ are:

$$E_r^{++} = -\xi_2 E_e^+, \, E_r^{-+} = E_r^{+-} = E_r^{--} = 0. \qquad (1.28)$$

The results show that, in the zeroth approximation with respect to δ, reflection from a cholesteric is experienced only by light with left or right-handed circular polarization. The corresponding reflection coefficient R^+ is

$$R^+ = |\xi_2|^2 = |E_2^- / E_2^+|^2 \qquad (1.29)$$

and is governed by the amplitude ratio of waves E^- and E^+ in the second eigensolution. In the forbidden band $|\xi_2| = 1$ we find that, within the considered approximation with respect to δ, only the wave with diffracting polarization experiences reflection from a cholesteric in the frequency range $\omega_B \sqrt{1 + \delta} < \omega < \omega_B \sqrt{1 - \delta}$, with reflection coefficient $R^+ = 1$. If a cholesteric reflects a wave with diffracting polarization (i.e., left or right-handed circular polarization), the reflected and incident wave share polarization of the same sense. However, reflection from the dielectric boundary changes the sense of the circular polarization, i.e., the left-handed circular polarization changes to right-handed and vice versa.

In summary, we conclude that, in the zeroth approximation with respect to δ, reflection from a cholesteric occurs inside the selective reflection band; independent of incident-wave polarization, reflected light is of circular polarization and matches the sense of cholesteric helix. The value of the reflection coefficient in the selective reflection band is independent of frequency and determined by the fraction of diffraction polarization present in the incident beam, i.e.,

$$R = R^+[1/2(1 - P) + P|\mathbf{en}_+^*|^2] \qquad (1.30)$$

where e is the incident wave polarization vector and P is the degree of polarization.

The obtained formulae describe light reflection from thick cholesteric layers, when $\bar{\epsilon} = \epsilon$ and the anisotropy parameter is small (Fig. 1.6, curve for $r = 1$). The corresponding expressions do not account for light reflection from the cholesteric boundary. We emphasize that light reflection from the boundary occurs also in the case where the averaged dielectric constant $\bar{\epsilon}$ of the cholesteric coincides with that of the external medium, but is insignificant for small values of δ; such reflections from the boundary are described by equations (1.25) and (1.26). If we consider first-order terms with respect to δ, the amplitudes of circularly polarized reflected waves are

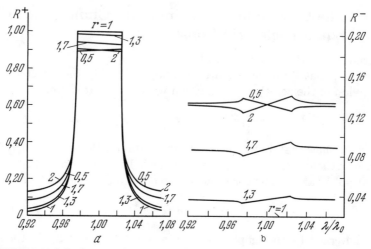

FIGURE 1.6. Calculated reflection spectra of a cholesteric for normal incidence of light with the jump in dielectric permittivity (different r) at the sample boundary for $\delta = 0.05$. (a) Right-handed circular polarization of the incident light; (b) left-handed circular polarization of the incident light [8].

given by

$$E_r^{++}/E_e^+ = -\xi_2(1 - \Delta\omega/\omega_B), \quad E_r^{+-}/E_e^- = \delta\xi_2/4$$
$$E_r^{-+}/E_e^+ = \delta\xi_2/4, \quad E_r^{--}/E_e^- = \delta/8(1 + (\Delta\omega)/2\omega_B)) \qquad (1.31)$$

where $\Delta\omega = \omega - \omega_B$.

It follows from (1.31) that in the selective reflection band $R^- \approx 5\delta^2/64$, i.e., the reflection coefficient of nondiffracting circular polarization is independent practically of frequency and is of the order of δ^2. For diffracting polarization, the reflection coefficient differs from unity (by a value of the order of δ) and its proportional to δ^2 outside the selective reflection band when the frequency of light deviates from the boundary of the selective reflection band by $\Delta\omega > \delta\omega_B$. (Fig. 1.6, curve for $r = 1$.) Thus, the diffractive nature of reflection in a cholesteric appears in the frequency range of the selective reflection band. Outside it, the reflection coefficient is determined by the difference in dielectric properties on both sides of the interface. It is independent of the polarization and is of the order of δ^2, i.e., it is proportional to the difference between dielectric constants of the cholesteric and the medium squared.

We emphasize that the properties of diffraction reflection studied on a cholesteric are common to all periodic media. In particular, they are also characteristic of the previously described periodic medium with scalar dielectric permittivity. Thus, the curve for $r = 1$ in Fig. 1.6 qualitatively describes reflection from a half-space with the sole difference that now reflection coefficient is independent of wave polarization and coincides with

R^+, the reflection coefficient of light with diffracting polarization in a cholesteric.

1.3.4 Dependence of Reflection on Polarization

In previous sections, we considered light reflection with circular polarization which is independent of cholestic surface anisotropy axis orientation, i.e., of the director **n** orientation. The reflection characteristics of noncircularly polarized reflected light, however, depend on cholesteric surface director **n**. This principle is illustrated best by plane-polarized incident light.

Expressions (1.25) and (1.26) give plane-polarized light reflection coefficient dependence on director orientation at cholesteric surface:

$$R(\xi) = R_{un} - |E_e|^{-2}[\sin 2\xi Re(E^{++}E^{+-} + E^{-+}E^{--}) \\ + \cos 2\xi Im(E^{++}E^{+-} + E^{-+}E^{--})] \qquad (1.32)$$

where ξ is the angle between the director orientation on the cholesteric surface and the polarization plane of light and $R_{un} = |E_e|^{-2}(|E^{++}|^2 + |E^{--}|^2 + |E^{+-}|^2 + |E^{-+}|^2)/2$ is the reflection coefficient of an unpolarized beam.

It follows from Eq. (1.32) [see, e.g., (1.31)] that reflection coefficient modulation amplitude depends on the orientation of the polarization plane of light and may reach a value of the order of dielectric anisotropy δ whereas the reflection coefficient maximum and the minimum are determined by the director orientation on the sample surface. Hence, the possibility exists to determine both molecule (director) orientation on the sample surface and the value for the dielectric anisotropy δ of the cholesteric using polarization dependence on reflection coefficient. Thus, within the linear approximation with respect to δ, orientations of the polarization plane corresponding to maximum and minimum reflection are independent of frequency in the selective reflection band for $\delta > 0$ and are given by the expressions $\xi = -\pi/4$ and $\xi = \pi/4$ for the right-handed cholesteric helix and $\xi = \pi/4$ and $\xi = -\pi/4$ for the left-handed helix. Expressions (1.31) and (1.32) produce the formula for the plane-polarized light reflection coefficient in the selective-reflection band

$$R(\xi) = (1/2)(1 - 2\Delta\omega/\omega_B) \mp (\delta/4)\sin 2\xi \qquad (1.33)$$

where the upper sign corresponds to the right-handed and the lower to the left-handed sense of the cholesteric helix.

1.3.5 Effect of Dielectric Boundaries

We considered the situation where cholesteric boundary effects on reflected light are minimal and reflection from a cholesteric is determined by diffraction from the periodic structure alone (i.e., boundary effect vanishes if

$\delta \to 0$). This analysis facilitates studying diffraction properties of reflection from cholesterics which manifest themselves most clearly in the polarization properties and frequency characteristics of reflection.

Experimental measurements must meet the condition $\bar{\epsilon} = \epsilon$. The common and clearest case is one in which the mean dielectric constant $\bar{\epsilon}$ of a cholesteric differs from the dielectric constant of the external medium. Large differences between medium ϵ and mean cholesteric dielectric constant $\bar{\epsilon}$, ($|\bar{\epsilon} - \epsilon| > \delta$), affect reflection characteristics due to the strong frequency dependence of diffraction reflection phase in the selective reflection band and the interference between it and dielectric reflection. Diffraction reflection phase changes by π over the frequency range of the selective reflection band (1.16) and this explains strong interference effects.

A qualitative illustration of the variation in the reflection characteristics from a half-space filled with a cholesteric when $\bar{\epsilon} \neq \epsilon$ follows. (See Ref. [4] for more details.)

Analysis of general formulae (1.25–1.27) indicates that light reflection from a cholesteric with nondiffracting circular polarization (R^-) is due to the difference between $\bar{\epsilon}$ and ϵ alone (within the zeroth approximation with respect to δ). In this approximation, reflection coefficients R^+ and R^- are independent of frequency in the selective-reflection band and their dependence on frequency is described by terms of nonzero-order in δ. If anisotropy δ is small, the approximation of reflection coefficients of circularly-polarized waves as independent of frequency in the selective-reflection band is satisfactory [8–10]. In this case the simplified equations for R^+ and R^- apply:

$$R^+ = \frac{1 + 14r^2 + r^4}{(1+r)^4}, \quad R^- = \left(\frac{1-r}{1+r}\right)^2$$
$$r = \sqrt{\bar{\epsilon}/\epsilon}. \tag{1.34}$$

The reflection coefficient of unpolarized light,

$$R_{un} = R^+/2 + R^-/2 = (1 + 6r^2 + r^4)(1+r)^{-4} \tag{1.35}$$

is independent of frequency as well.

Within the same approximation, however, reflection of plane-polarized light shows strong reflection coefficient frequency dependence in the selective reflection band as r deviates from unity (Fig. 1.7)

$$R_L = 2|(2r\xi_2)(1+r)^{-2} \pm 1/2(1-r)(1+r)^{-1}|^2$$
$$+ 1/2(1-r)^2(1+r)^{-2} \tag{1.36}$$

where ξ_2 is determined by (1.16), the upper sign relates to light polarized along the director on the cholesteric surface and lower sign to the polarization normal to the director (assuming anisotrophy in δ is positive).

Since the previously mentioned frequency dependence is due to the phase frequency dependence of the wave experiencing diffraction reflection [see

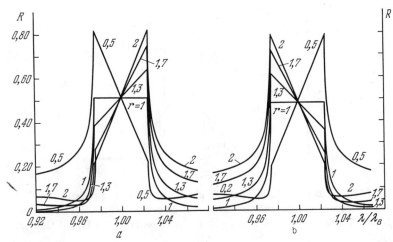

FIGURE 1.7. Calculated reflection spectra analogous to those in Fig. 1.6 for linear polarizations of the incident light. (a) Light polarization along director orientation; (b) light polarization normal to director orientation on the cholesteric surface [8].

(1.6) for ξ_2], it corresponds to the following physical picture: if there is an interface, the waves reflected from it interfere with those experiencing diffraction reflection. The phase of the wave experiencing diffraction reflection depends on frequency, whereas the phase of the wave reflected from the dielectric boundary is constant. Such interference is most prominent for plane-polarized light. (Recall that within the zeroth approximation with respect to δ, waves reflected from the dielectric boundary and those which experienced diffraction reflection have opposite circular polarizations and cannot interfere.)

As follows from (1.36), the reflection coefficient for the plane-polarized wave depends on the polarization orientation with respect to the director. This is similar to the case where $\bar{\epsilon} = \epsilon$ and this fact is used to determine the director orientation on the surface. In this case, however, polarization orientation at maximum and minimum reflection depends on light frequency (in the selective reflection band). The amplitude of modulation is determined not only by δ but, in the zeroth approximation in δ, also by $r - 1$, i.e., it is less sensitive to δ than in the case where $\bar{\epsilon} = \epsilon$ (i.e., $r = 1$) [8].

The following numerical calculations [8,10] (Figs. 1.6 and 1.7) illustrate this analysis: in the case of reflection from a half-space filled with a cholesteric, Figure 1.6 shows the computed reflection coefficients for light of diffracting R^+ and nondiffracting R^- circular polarizations versus frequency for different values of medium dielectric constant outside the cholesteric. In particular, Figure 1.6 shows how well equation (1.34) describes reflection from a cholesteric for small δ. For the value $\delta = 0.05$ used in the calculation, the deviation of R^+ from unity for $\bar{\epsilon} = \epsilon$ (i.e., $r = 1$) cannot be shown on the scale of the drawing in the selective reflection band since this

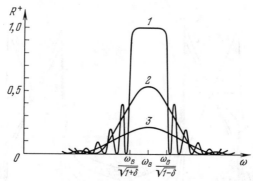

FIGURE 1.8. Reflection coefficient of light with diffracting polarization for normal incidence versus frequency in samples of different thickness. (1) Thick sample $(L\delta/p \gg 1)$; (2) sample of intermediate thickness $(L\delta/p \sim 1)$; (3) thin sample $(L\delta/p \ll 1)$.

difference is less than 2×10^{-4}. The corresponding calculated value of the reflection coefficient for nondiffracting polarization of all the wavelengths is also very small, less than 3×10^{-4}. Figure 1.6 also demonstrates that the constant reflection coefficient for circular polarizations in the selective reflection band is a good approximation for all reasonable values of r.

Figure 1.7 represents the calculated reflection coefficients of plane-polarized light versus frequency for two different orientations of the molecule on the cholesteric surface and shows that the constant reflection coefficient in the selective reflection band is a good approximation only if $\bar{\epsilon} = \epsilon$.

For a layer of finite thickness, the reflection coefficient R and transmission coefficient, $T = 1 - R$ are described by general formulae (1.21)–(1.23) and not discussed here, since they are cumbersome (see [4,7]). Although the general character and polarization properties of reflection are the same, reflection intensity decreases and Pendellösung beats and beats due to finite sample thickness appear in reflection (transmission) (Fig. 1.8).

1.3.6   Method of Characteristic Matrices

Another method of solving the boundary problem for a layered medium is to apply one procedure of characteristic matrices [1]. The method permits expression of reflection and transmission characteristics of multilayered system using the characteristics of one layer and this reveals how transmission and reflection depend on the number of layers and the properties inherent in each.

Although the structure of expressions derived within this approach is simple, generally, quantitative results are obtained only after numerical calculations. Therefore the analytical results obtained by this method apply to the simplest situations in periodic media with scalar dielectric permittivity

[1]. Nevertheless, this method applies to media with complicated structure [11] as well since it can be used in computer calculations, the accuracy of which is adjustable, in principle, since it is not limited by the method itself.

We briefly consider the fundamental idea of the method of characteristic matrices and refer the reader to the references in [11–13] for details.

If the electric (or magnetic) components of the transverse electromagnetic wave (TE or TM) at the entrance and exit surfaces of a plane-parallel layer are given by

$$Q = \begin{bmatrix} U(z) \\ V(z) \end{bmatrix}, \quad Q_0 = \begin{bmatrix} U_e \\ V_e \end{bmatrix} \tag{1.37}$$

then, without restriction to normal incidence, we obtain the following relationship between Q and Q_0 via the characteristic 2 by 2 matrix M for the layer:

$$Q_0 = MQ. \tag{1.38}$$

If a layered medium consists of N different layers (in the general case), it is also described by the characteristic matrix M_N related to the characteristic matrices of individual layers M_i as

$$Q_0 = M_N Q(z_n)$$
$$M_N = M_1 M_2 \ldots M_N \tag{1.39}$$

where $Q(Zn)$ determines the wave amplitude on the exit surface of the Nth layer. If all layers possess the same properties, the characteristic matrix has the form

$$M_N = (M_1)^N. \tag{1.40}$$

It is easy to extract the amplitude coefficients of reflection and transmission (for the TE-type waves) using elements m_{ik} of the characteristic matrix and the boundary conditions for the electric and magnetic component vectors of the wave

$$r = R/A = [m_{11} + m_{12}p_e)p_1 - (m_{21} + m_{22}p_e)][(m_{11} + m_{12}p_e)p_1$$
$$+ (m_{21} + m_{22}p_e)]^{-1}$$
$$= T/A = 2p_1[(m_{11} + m_{12}p_e) + (m_{21} + m_{22}p_e)]^{-1} \tag{1.41}$$

where A, R and T are the amplitudes of the incident, reflected and refracted waves

$$p_1 = \cos\theta_1 \sqrt{\epsilon_1/\mu_1}, \quad p_e = \cos\theta_e \sqrt{\epsilon_e/\mu_e}.$$

Here ϵ_1, μ_1, ϵ_e and μ_e are dielectric and magnetic permittivities of homogeneous media which limit the layered structure under consideration (see Fig. 1.9) and θ_1 and θ_e are the angles between the direction of stratification (z-axis) and direction of propagation of incident and transmitted waves, respectively.

Thus, the main idea of the approach is to seek the characteristic matrix of the layer, (1.38), and structure as a whole, (1.39). In case a, complicated

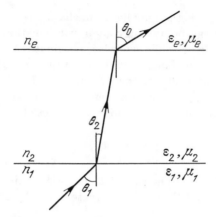

FIGURE 1.9. Schematic representation of light propagation through a layer with the jump in dielectric permittivity at its boundaries.

coordinate dependence of layer characteristics (even for scalar dielectric permittivity), the search for the characteristic matrix for the layer may be complicated. It is simplified, however, if a layered structure is considered as a sequence of different layers of dielectric properties. Thus, for a structure formed by an array of two-layered units with different properties, the characteristic matrix of the layered structure as a whole is expressed via the characteristics of individual layers and Chebyshev polynomials of order $N-1$ and $N-2$, which are functions of a dimensionless argument determined by the characteristics of the layers, frequency and incidence angle of the wave [1].

Since the simplified formulae are cumbersome, we give the exact expressions of the characteristic matrix for the structure built by N double layers and normal incidence which the condition that each of them is a quarter-wave layer, i.e. $h_1\sqrt{\epsilon_1}(\omega/c) = h_2\sqrt{\epsilon_2}(\omega/c) = \pi/2$ (the latter condition corresponds to the fulfillment of the Bragg condition):

$$M_N = \begin{vmatrix} (-\epsilon_2/\epsilon_1)^{N/2} & 0 \\ 0 & (-\epsilon_1/\epsilon_2)^{N/2} \end{vmatrix}. \tag{1.42}$$

Therefore, the reflection coefficients for the layered medium of N layers must be written assuming the above conditions, as

$$R_N = \left\{ [1 - \sqrt{\epsilon_{e2}/\epsilon_{e1}}(\epsilon_1/\epsilon_2)^N] \right.$$
$$\left. [1 + \sqrt{\epsilon_{e2}/\epsilon_{e1}}(\epsilon_1/\epsilon_2)^N]^{-1} \right\}^2. \tag{1.43}$$

Equation (1.43) demonstrates that reflection coefficient R_N with the number of layers caused by light diffraction increases and tends to unity for $N \to \infty$ even for an arbitrary small difference between ϵ_1 and ϵ_2. In our case, where absorption in the layers is absent, the transmission coefficient T is found from Eq. (1.41) or directly from the law of conservation of energy, $T = 1 - R$.

We stated the general principles of the method of characteristic matrices and illustrated its use with the simplest example. This method, combined with numerical computer calculations, solves the boundary problem in more complicated systems, such as optics of liquid crystals [13–15], X-ray optics of layered crystals [16,17], etc.

The method, although exact in principle, loses its clarity with complicated structure and makes anlaysis of the interaction between radiation and periodic media difficult. Therefore, the results obtained by this numerical method should be verified and compared with the results obtained by the approximate analytical solutions of corresponding problems. Moreover, if the accuracy of the approximate analytical method is sufficient, it is the method of choice due to its simplicity.

The examples given here almost exhaust the known exact analytical solutions of the Maxwell equations and the corresponding boundary problems for periodic media. The following chapters deal with more complicated structure, however, we always refer the reader to the simple results. On the other hand, in many complicated situations, the physics of the effect is studied using approximate methods since, in many important cases, the accuracy is sufficient and they provide the correct physical interpretation of the effects of radiation interaction with periodic media.

References

1. M. Born, E. Wolf: *Principals of Optics* (Oxford, Pergamon Press, 1965).

2. C. Elachi: Proc. IEEE **64**, 1666 (1976).

3. L. Brilluen, M. Parodi: *Propagation des Ondes dans les Milieux Periodicues* (Paris, 1956).

4. V.A. Belyakov, A.S. Sonin: *Optics of Cholesteric Liquid Crystals* (Moscow, Nauka, 1982) (in Russian).

5. H. De Vries: Acta Crystallogr. **4**, 219 (1951).

6. E.I. Kats: ZhTF **59**, 1854 (1970) [Sov. Phys. - JETP **32**, 104 (1971)].

7. V.A. Belyakov, V.E. Dmitrienko, V.P. Orlov: UFN **172**, 221 (1979) [Sov. Phys. - Usp. **22**, 63 (1979)].

8. M. Tur: Mol. Cryst. Liquid Cryst. **29**, 345 (1975).

9. V.A. Belyakov, V.E. Dmitrienko: Pis'ma v Zh. Tekh. Fiz. **7**, 19 (1981) (in Russian).

10. A.V. Tolmachev, A.S. Sonin: Fiz. Tverd. Tela **17**, 3096 (1975); Kristallografiya **21**, 794 (1976) (in Russian).

11. R.M.A. Azzam, N.M. Bashara: *Ellipsometry and Polarized Light* (Amsterdam, New York, Oxford, North Holland Publ. Comp. 1977).

12. W. Shercliff: *Polarized Light* (Cambridge, Harvard University Press, 1962).

13. S. Chandrasekhar: *Liquid Crystals* (Cambridge University Press, 1976).

14. D.W. Berreman, T.J. Scheffer: Phys. Rev. Lett. **25**, 577 (1970); Mol. Cryst. Liquid Cryst. **11**, 395 (1970); Phys. Rev. **5A**, 1397 (1971).

15. V.A. Belyakov, S.M. Osadchii, V.A. Korotkov: Kristallografiya **31**, 522 (1986) (in Russian).

16. A.V. Kolpakov, Yu.N. Belyaev: Vestn. MGU (Ser. 3, Fizika, Astronomiya **26**, 91 (1985) (in Russian).

17. M.A. Andreeva, K. Rocete, Yu.P. Khapachev: Phys. Stat. Sol (a) **88**, 455 (1985).

2

Approximate Description of Interaction of Radiation with Regular Media

In the previous chapter considering the interaction of an electromagnetic radiation with regular media, we have made no assumption that the modulation of their dielectric properties is small. Neither was made an assumption that for finite samples the intensity of the scattered radiation is low with respect to that of the incident radiation. The description of the interaction of an electromagnetic wave even with simplest periodic media is rather complicated and, as a rule, the problem stated in such a general form cannot be solved in an analytical form. But in many instances important for practice the problem has a small parameter both for electromagnetic and other radiations which provides a sufficiently accurate analytical solution of the problem. The physical conditions for a small parameter are: (i) weak interaction between radiation and individual scatterer or a subunit of the matter playing the role of such a scatterer. (ii) weak radiation scattering from a microscopic sample, i.e., small wave intensity relative to the scattered wave incident. If the first condition is met, the second one is necessarily fulfilled if the size of a scattering sample is sufficiently small. The first condition is related to the interaction between radiation and medium. Thus, a small parameter exists if X-rays interact with atoms [1], or neutrons with atoms and atomic nuclei [2] (scattering amplitudes for individual centers are much smaller than $(n\lambda)^{-1/2}$ where λ is radiation wavelength and n is scatterer density—in the continuum medium, amplitude of the spatial modulation of scatterer density. On the other hand, there is no such a small parameter, e.g., for low-energy electron scattering by atoms [3]. And, in resonance scattering of Mössbauer radiation by nuclei [4], the corresponding parameter may be sufficiently small to give the accuracy necessary to apply approximate methods.

A description of the interaction between radiation and regular media with complicated structure assuming the second condition (the so-called kinematical approximation of scattering theory) follows.

2.1   Kinematical Approximation

Let a plane monochromatic wave be incident onto a sample of regularly arranged scatterers of finite dimensions. Let scattered wave intensity I_s be much lower than that of incident wave I such that, in the first approximation, transmitted beam attenuation may be neglected relative to incident beam, i.e., it is possible to assume that incident wave intensity is constant within the sample.

For the time being, we consider only coherent elastic scattering of a wave from scatterers where $f_i(k_0 e_0, k_1 e_1)$ is the coherent scattering amplitude by an individual scatterer i, subscript i with discrete values and $k_0 e_0$ and $k_1 e_1$, wave and polarization vectors of incident and scattered waves, respectively. We reserve consideration of coherent scattering amplitude form until the sections which deal with specific regular media and varies types of radiations [1–4].

2.1.1   Scattering Cross Section

In the kinematical approximation, radiational scattering is usually described by cross sections $d\sigma/d\Omega$ defined as

$$d\sigma(\mathbf{k}_0, \mathbf{e}_0; \mathbf{k}_1, \mathbf{e}_1)/d\Omega_{\mathbf{k}_1} = \left| \sum_i f_i(\mathbf{k}_0, \mathbf{e}_0; \mathbf{k}_1, \mathbf{e}_1) e^{i(\mathbf{k}_0 - \mathbf{k}_1)\mathbf{r}_i} \right|^2 . \qquad (2.1)$$

Summation is over all sample scatterers and r_i is the ith scatterer coordinate. Expression (2.1) is transformed to the form

$$d\sigma(\mathbf{k}_0, \mathbf{e}_0; \mathbf{k}_1 \mathbf{e}_1/d\Omega_{\mathbf{k}_1} = \left| \sum_i f_i e^{i(\mathbf{k}_0 - \mathbf{k}_1)\mathbf{r}_i} \right|^2 \cdot \left| \sum_n e^{i(\mathbf{k}_0 - \mathbf{k}_1)\mathbf{r}_n} \right|^2 \qquad (2.2)$$

where summation is over all scatterers i within the unit cell of the periodic structure (in the first factor) and over all sample unit cells (in the second factor); r_n is the radius-vector of the nth unit cell. The first factor in (2.2),

$$F(\mathbf{k}_0, \mathbf{e}_0; \mathbf{k}_1, \mathbf{e}_1) = \sum_i f_i(\mathbf{k}_0, \mathbf{e}_0; \mathbf{k}_1, \mathbf{e}_1) e^{(\mathbf{k}_0 - \mathbf{k}_1)\mathbf{r}_i} \qquad (2.3)$$

is structure amplitude. It relates scattering and symmetry characteristics and scatterer type in the crystal unit cell. The second factor in (2.2) accounts for strong cross section dependence on $k_0 - k_1$ and leads to the Bragg condition for scattering maxima

$$\mathbf{k}_0 - \mathbf{k}_1 = \boldsymbol{\tau}, \quad |\mathbf{k}_0| = |\mathbf{k}_1| \qquad (2.4)$$

where τ is the structure reciprocal lattice vector.

In the limit of infinite scatterer number, scattering cross section (2.2) is related to one unit cell and takes the form known in the theory of X-ray diffraction

$$d\sigma/d\Omega = \lim_{N \to \infty} (|F|^2/N) \left| \sum_{n=1}^{N} e^{i(\mathbf{k}_0 - \mathbf{k}_1)\mathbf{r}_n} \right|^2$$

$$= (2\pi)^3/V|F|^2 \delta(\mathbf{k}_0 - \mathbf{k}_1 - \boldsymbol{\tau}) \qquad (2.5)$$

where V is unit cell volume.

Media scattering differs from X-ray diffraction in the complicated polarization and angular dependence of structure amplitude; however, as in X-ray and thermal-neutron diffraction, symmetry analysis of structure amplitude permits one to find reciprocal lattice vectors τ in relationships (2.2)–(2.5) for which F is zero, that is, diffraction reflection experiences extinction.

2.1.2 Polarization Characteristics

Thus, structure amplitude reflects features typical of periodic media with complicated structure. Structure complexity is associated microscopically with the vector nature of scattering from individual scatterers and the complex spatial structure of these scatterers in the medium.

The consequences of scatterer nonscalar nature are different extinction rules for such structures (see below) compared with those for structures formed by scalar scatterers and complicated scattering polarization characteristics.

We focus on the relation between scattering polarization characteristics and structure amplitude, give the general relationships between the polarization properties of both incident and scattered radiation and crystal structure and present the polarization dependent scattering cross sections for monochromatic radiation. Consider the differential scattering cross section of polarized radiation

$$d\sigma(\mathbf{k}_0, \mathbf{e}_1; \mathbf{k}_1, \mathbf{e}_1)/d\Omega_{\mathbf{k}_1} = (2\pi)^3/V|F_\tau(\mathbf{k}_0, \mathbf{e}_0; \mathbf{k}_1, \mathbf{e}_1)|^2 \delta(\mathbf{k}_0 - \mathbf{k}_1 - \boldsymbol{\tau}). \quad (2.6)$$

(Notation here is as earlier.) Expression (2.6), a function of e, reaches its maximum for values of $e_1 \equiv n_0'$. Vector $n_0'(e_0)$ is a polarization vector of the scattered radiation and corresponding cross section,

$$d\sigma(\mathbf{k}_0, \mathbf{e}_0; \mathbf{k}_1, \mathbf{n}_0)/d\Omega_{\mathbf{k}_1} \equiv d\sigma((\mathbf{k}_0, \mathbf{e}_0; \mathbf{k}_1)/d\Omega_{\mathbf{k}_1}$$

is the differential scattering cross section for a quantum with polarization vector e_0. Introducing a coherent scattering tensor T_{ik} for the unit cell such that

$$F(\mathbf{k}_0, \mathbf{e}_0; \mathbf{k}_1, \mathbf{e}_1) = \sum_{ik} e_{1i}^* T_{ik} e_{0k} = \mathbf{e}_1^* \hat{T} \mathbf{e}_0 \qquad (2.7)$$

yields the following formula for $\mathbf{n}_0'(\mathbf{e}_0)$

$$\mathbf{n}_0'(\mathbf{e}_0) = \hat{T}\mathbf{e}_0|\hat{T}\mathbf{e}_0|^{-1} \tag{2.8}$$

where $\hat{T}\mathbf{e}_0$ is a vector, the kth component of which equals $\sum_i T_{ki}e_{0i}$.

The explicit expressions for vector n_0 are given for some particular cases. The scattering cross section of unpolarized radiation is expressed by $d\sigma(\mathbf{k}_0, \mathbf{e}_0; \mathbf{k}_1)$ as

$$d\sigma/d\Omega_{\mathbf{k}_1} = 1/2 \sum_{i=1,2} d\sigma(\mathbf{k}_0, \mathbf{e}_{0i}; \mathbf{k}_1)/d\Omega_{\mathbf{k}_1}. \tag{2.9}$$

Summation is over two mutually orthogonal polarization vectors e_{0i}. The scattered radiation is partially polarized with polarization matrix

$$\rho = \sum_{i=1,2} \rho(\mathbf{n}_0'(\mathbf{e}_{0i}))(d\sigma(\mathbf{k}_0, \mathbf{e}_{0i}; \mathbf{k}_1)/d\Omega_{\mathbf{k}_1})$$

$$\times \left[\sum_i d\sigma(\mathbf{k}_0, \mathbf{e}_{0i}, \mathbf{k}_1)/d\Omega_{\mathbf{k}_1} \right]^{-1}. \tag{2.10}$$

In the case of electromagnetic radiation, $\rho(\mathbf{e})$ is the polarization density matrix for a photon with polarization vector e [5]. Its elements are given by

$$\rho_{ik} = e_i e_k^* \tag{2.11}$$

where e_i are the coefficients in the expansion of the polarization vector in polarization unit vectors.

The scattering cross sections of partially polarized radiation with polarization degree P has the form

$$(d\sigma/d\Omega_{\mathbf{k}_1})_P = (1-P)d\sigma/d\Omega_{\mathbf{k}_1} + Pd\sigma(\mathbf{k}_0, \mathbf{e}_0; \mathbf{k}_1/d\Omega_{\mathbf{k}_1}) \tag{2.12}$$

where $\mathbf{e}_0$ is the vector of polarization presented in the radiation. The polarization density matrix for the scattered radiation is equal to

$$\rho_P = \left[(1-P)(d\sigma/d\Omega_{\mathbf{k}_1})\rho + P\frac{d\sigma(\mathbf{k}_0, \mathbf{e}_0; \mathbf{k}_1)\rho(\mathbf{n}_0')}{d\Omega_{\mathbf{k}_1}} \right]$$

$$\times \left[(1-P)\frac{d\sigma}{d\Omega_{\mathbf{k}_1}} + P\frac{d\sigma(\mathbf{k}_0, \mathbf{e}_0; \mathbf{k}_1)}{d\Omega_{\mathbf{k}_1}} \right]^{-1}. \tag{2.13}$$

Note that the above expressions for polarization and scattering cross sections are obtained for monochromatic radiations.

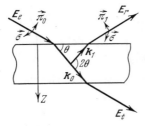

FIGURE 2.1. Geometry of diffraction scattering for a planar texture of a cholesteric.

2.1.3   Scattering of Light in Cholesterics

Consider selective (diffraction) light scattering by cholesterics as an illustration of the general formulae in the previous section. Unlike discrete scatterers, a cholesteric corresponds to continuous scatterer distribution (dielectric tensor).

Scattering Cross Section. Examine light transmission and reflection for a uniform plane parallel plate of a cholesteric with the optical axis normal to the surface (Fig. 2.1). To simplify the presentation and neglect light reflection and refraction from the boundary due to different refraction indices of the sample and the outer medium, we assume that the medium outside the crystal is homogeneous and its dielectric constant is equal to the mean dielectric constant $\bar{\epsilon}$ of the cholesteric. In our approximation, we neglect primary beam intensity variations and polarization characteristics in the sample. Thus, the problem reduces to the determination of scattered beam intensity and polarization characteristics as a function of the direction of primary beam propagation and polarization. These characteristics, for a monochromatic primary beam, are described by the differential (with respect to angle) scattering cross section.

In the kinematical approximation, we obtain from (2.1) and (2.2) that the scattering cross-section of light for a cholesteric sample is described by [6]

$$\frac{d\sigma(\mathbf{k}_0, \mathbf{e}_0; \mathbf{k}_1, \mathbf{e}_1)}{d\Omega_{\mathbf{k}_1}} = \left(\frac{\omega^2}{4\pi c^2}\right)^2 \left|\int [\mathbf{e}_1^*(\hat{\epsilon} - \bar{\epsilon})\mathbf{e}_0] \exp[i(\mathbf{k}_0 - \mathbf{k}_1)\mathbf{r}]d\mathbf{r}\right|^2 \quad (2.14)$$

where $\hat{\epsilon}(\mathbf{r})$ is the dielectric permittivity tensor of the cholesteric and $\mathbf{k}_0$, $\mathbf{k}_1$, $\mathbf{e}_0$, $\mathbf{e}_1$ are the wave and polarization vectors of the incident and scattered waves. Integration is over sample volume. Expression (2.14) may be transformed using the Fourier expansion of $\hat{\epsilon}(\mathbf{r})$

$$\hat{\epsilon}(\mathbf{r}) = \sum_{s=0,\pm 1} \hat{\epsilon}_s \exp[is\boldsymbol{\tau}\mathbf{r}] \quad (2.15)$$

where

$$\hat{\epsilon}_0 = \begin{pmatrix} \bar{\epsilon} & 0 & 0 \\ 0 & \bar{\epsilon} & 0 \\ 0 & 0 & \epsilon_3 \end{pmatrix}, \quad \hat{\epsilon}_1 = \epsilon_{-1}^* = \begin{pmatrix} 1 & \mp i & 0 \\ \mp i & -1 & 0 \\ 0 & 0 & 0 \end{pmatrix} \cdot \frac{\bar{\epsilon}\delta}{2}.$$

Here τ is the reciprocal lattice vector of the cholesteric. Substituting (2.15) into (2.14), produces

$$\frac{d\sigma(\mathbf{k}_0, \mathbf{e}_0; \mathbf{k}_1, \mathbf{e}_1)}{d\Omega_{\mathbf{k}_1}} = \left(\frac{\omega^2}{4\pi c^2}\right)^2 \left| \sum_s \mathbf{e}_1^* \hat{\epsilon}_s \mathbf{e}_0 \int \exp[i(\mathbf{k}_0 - \mathbf{k}_1 + s\tau]dr \right|^2.$$
(2.16)

In the limit of an infinite sample, the integral in (2.16) is proportional to the delta function $\delta(\mathbf{k}_0 - \mathbf{k}_1 + s\tau)$ [cf. (2.5)] and the scattering cross section acquires the form known in the theory of X-ray and neutron scattering [1,2]. In particular, scattering directions are determined by the kinematical Bragg condition (2.4) which gives no information about the intensity or polarization properties of scattering. This condition is also written in the form

$$\sin\theta = \frac{s\lambda}{p}$$
(2.17)

where 2θ is scattering angle (Fig. 2.1) and λ is light wavelength. If the Bragg condition is violated, the scattered wave has zero intensity—there is no scattering. The intensity and polarization characteristics of scattering depend on the details of sample structure and a structure factor, the coefficient before the the integral in expression (2.16) whereas X-ray scattering dependence on crystal structure is reflected in structure amplitude. Physically, structure amplitude refers to the amplitude of scattering by the crystal unit cell. For cholesterics, an analogue of the X-ray, structure amplitude is the quantity

$$F(\mathbf{k}_0, \mathbf{e}_0; \mathbf{k}_1, \mathbf{e}_1) = \mathbf{e}_1^* \hat{\epsilon}_s \mathbf{e}_0$$
(2.18)

which describes the scattering amplitude from a cholesteric layer with thickness equal to a half-pitch of the cholesteric helix.

The simple expressions given above are the basis for a series of qualitative features observed in optics of cholesterics [7]: Expression (2.16) represents cholesteric color dependence on the angle of observation and its change with the temperature; there are diffraction-reflected waves of different frequencies by different observation angles. In particular, for normal incidence, diffraction reflection is experienced by light with wavelength $\lambda = p$. Figure 2.2 (normal incidence) illustrates the situation where k_0 and k_1 are of opposite direction, that is, selective reflection of normally incident light is a particular case of diffraction reflection. The temperature related variations in cholesteric color explained by Eq. (2.16) correspond to the change in the wavelength of light diffracted from the cholesteric which, in turn, relates to the temperature dependence of cholesteric pitch.

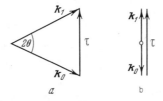

FIGURE 2.2. Illustration of the diffractive nature of selective reflection of light in a cholesteric. The Bragg condition for oblique (a) and normal (b) angles of incidence.

Polarization Characteristics. Expression (2.16) also delineates the polarization dependence of light reflection and transmission in a cholesteric. To see this, we analyze the dependence of structure amplitude $F(\mathbf{k}_0, \mathbf{e}_0, \mathbf{k}_1, \mathbf{e}_1)$ on polarization vectors $\mathbf{e}_0$ and $\mathbf{e}_1$. As follows from expression (2.15), $\hat{\epsilon}_s = 0$ for $|s| \geq 1$; thus, in the kinematical approximation, only first-order reflection occurs (it follows from (2.15) and (2.18) that for higher-order reflection $F \equiv 0$). Amplitude $F(\mathbf{k}_0, \mathbf{e}_0, \mathbf{k}_1, \mathbf{e}_1)$ depends on incident wave polarization vector $\mathbf{e}_0$ for a certain polarization $\mathbf{e}_0^\perp$ it goes to zero. This implies that polarization $\mathbf{e}_0^\perp$ does not experience diffraction scattering whereas polarization orthogonal to it $\mathbf{e}_0^s$ experiences the strongest diffraction scattering. Similarly, expression (2.18) for $F(\mathbf{k}_0, \mathbf{e}_0, \mathbf{k}_1, \mathbf{e}_1)$, depends on polarization vector $\mathbf{e}_1$. Amplitude, a function of $\mathbf{e}_1$, reaches its maximum for a certain polarization $\mathbf{e}_1^s$ and goes to zero for the orthogonal polarization $\mathbf{e}_1^s$ whereas the values of $\mathbf{e}_1^s$ and $\mathbf{e}_1^\perp$ are independent of incident wave polarization, $\mathbf{e}_0$. Therefore, scattered wave polarization $\mathbf{e}_1^s$ is independent of incident wave polarization.

To determine $\mathbf{e}_0^s$ and $\mathbf{e}_1^s$, expression (2.18) should be rewritten in explicit form. Let us choose as one of the polarization unit vectors σ in the direction normal to the scattering plane and write polarization vectors $\mathbf{e}_0$ and $\mathbf{e}_1$ in the form[1]

$$\mathbf{e}_0 = \sigma \cos \alpha_0 + i\pi_0 \sin \alpha_0; \quad \mathbf{e}_1 = \sigma \cos \alpha_1 + i\pi_1 \sin \alpha_1 \qquad (2.19)$$

where vectors $\sigma, \pi_0, \mathbf{k}_0(\sigma, \pi, \mathbf{k}_1)$ form the right-handed triad. Then expres-

---

[1]Representation of polarization vectors in form (2.19) implies that, in the most general form of the polarization vector, $\mathbf{e} = (\sigma \cos \alpha + \pi e^{i\beta} \sin \alpha)e^{i\eta}$, we assume that the phase, η, which is not important here, is zero and $\beta = \pi/2$. This corresponds to a certain choice of orientation for the axes of the polarization ellipse. The choice is determined by the symmetry of the problem; namely, one axis should be in the scattering plane while the other is normal to it.

sion (2.18) becomes

$$F(\mathbf{k}_0, \mathbf{e}_0; \mathbf{k}_1, \mathbf{e}_1) = -\frac{\bar{\epsilon}\delta}{2}(\cos\alpha_0 \pm \sin\alpha_0 \sin\theta)$$
$$\times (\cos\alpha_1 \pm \sin\alpha_1 \sin\theta). \qquad (2.20)$$

Minimizing expression (2.20) with respect to α_0, shows that the cholesteric provides the diffraction scattering for elliptically polarized light. Parameter $\alpha_0 = \alpha_0^s$ for that polarization [see (2.19)] depends on angle θ and is given by the expression

$$\alpha_0^s = \pm \operatorname{arctg}(\sin\theta). \qquad (2.21)$$

Light with orthogonal polarization, $\alpha_0^\perp = \mp \operatorname{arc tag}(\sin\theta)$ does not interact with the cholesteric. Scattered wave polarization also depends on the angle of incidence and corresponds to parameter $\alpha_1^s = \alpha_0^s$, i.e., is also determined by expression (2.21).

Thus, the kinematical approach accounts for selective reflection and transmission of light with different polarization and its dependence on the angle of incidence; in the special case of normal incidence, $(\theta = \pi/2)$, circularly polarized light is selectively scattered. The left-handed cholesteric helix produces scattered light with left-handed circular polarization, for the right-handed helix—right-hand circular polarization. In the limit of scattering angle tending to zero, light with linear σ polarization is scattered with linear polarization whereas that with π polarization is not scattered. In the general case, $0 < \theta < \pi/2$, elliptically polarized light experiences selective scattering and the scattered radiation is also elliptically polarized. In the kinematical approximation, polarization of scattered radiation is determined by angle θ and is independent of either the degree or type of primary beam polarization.

2.1.4   Kinematical Approximation Limitations

As stated, the kinematical approximation gives the clear physical picture of diffraction scattering and some useful relationships to describe the experimental results. However, the quantitative description is possible within this approximation, only in the situations where coherently scattering volumes are small to justify the neglect of radiation attenuation within coherently scattering volumes.

The limitations of the kinematical approximations may be illustrated on the case of a cholesteric. It is applicable either to very thin planar samples $(L\delta/p << 1$ where L is the sample thickness) or to mosaic samples with small perfect regions.

In thick crystals, multiple Bragg scattering becomes important and the kinematical approximation does not apply in quantitative analyses and even cannot be used for the explanation of some qualitative effects. Thus the kinematical approximation accounts for the rotation of polarization plane associated with the difference in interaction between left and right-handed polarization in cholesterics and the consequent differences in refraction index for each polarization. But the kinematical approach fails to explain the experimentally observed change in polarization plane rotation sign with frequency [7]. In addition, it does not rationalize higher-order diffraction reflection (for higher-order diffraction reflection the structure amplitude (2.18) goes to zero), frequency or angular width of the selective reflection bands for light in cholesterics and existence of frequency or angular range for light reflection of any polarization for the case of oblique incidence, etc. [7]. The above mentioned phenomena cannot be explained within the kinematical theory because it accounts for only single scattering and these phenomena require more rigorous consideration of the interaction of light with cholesterics [8,9]. The kinematical approximation also fails in the description of the interaction of other types of radiations (Mössbauer gamma radiation, neutrons, fast charged particles in crystals) with media. In each of these cases it is necessary to find the exact or numerical solutions of the Maxwell equations (see Ch. 1) or to find approximate analytical solutions based on the approach similar to the X-ray dynamical theory.

2.2   Dynamical Theory

Now consider the interaction of an electromagnetic wave with a sample of three-dimensional periodic structure in the case where the system does not meet the conditions necessary to apply the kinematical approximation. Assume that the dielectric permittivity tensor $\hat{\epsilon}(\mathbf{r})$ is a three-dimensional periodic function of coordinates and that the magnetic permittivity of the medium is unity, $\hat{\mu} \equiv 1$. As we seek the approximate analytical solutions of the Maxwell equations, we assume that all amplitudes $\hat{\epsilon}_{\boldsymbol{\tau}}$ for $\boldsymbol{\tau} \neq 0$ ($\boldsymbol{\tau}$ is the reciprocal lattice vector) in the Fourier expansion of the dielectric permittivity tensor

$$\hat{\epsilon}(\mathbf{r}) = \sum_{\boldsymbol{\tau}} \hat{\epsilon}_{\boldsymbol{\tau}} e^{i\boldsymbol{\tau}\mathbf{r}} \tag{2.22}$$

are small relative to amplitude $\hat{\epsilon}_0$. Such an assumption corresponds to many practical situations. The small parameter of the problem is $\sim 10^{-2}$ for optics of liquid crystals and $\sim 10^{-4} - 10^{-5}$ for Mössbauer and X-ray crystal optics. Thus, in most instances, the accuracy of this approximation satisfies the practical requirements of the experiment.

2.2.1   Set of Dynamical Equations

If the conditions of the previous section are met, it is possible to find from the Maxwell equations that an electric field vector for an arbitrary direction of light propagation in a periodic medium satisfies the following equation

$$\hat{\epsilon}\frac{\partial^2 \mathbf{E}}{\partial t^2} = -c^2 \text{rot rot } \mathbf{E} \tag{2.23}$$

where equation (2.22) defines $\hat{\epsilon}$. Since the medium is periodic, the solution of equation (2.23) is a Bloch wave [10]

$$\mathbf{E}(\mathbf{r}, t) = e^{i\mathbf{k}_0\mathbf{r} - i\omega t} \sum_{\boldsymbol{\tau}} \mathbf{E}_{\boldsymbol{\tau}} e^{i\boldsymbol{\tau}\mathbf{r}}. \tag{2.24}$$

Substituting (2.24) into (2.23), produces the set of homogeneous equations for $\mathbf{E}_{\boldsymbol{\tau}}$

$$-k_{\boldsymbol{\tau}}^2 \mathbf{E}_{\boldsymbol{\tau}} + \frac{\omega^2}{c^2} \sum_{\boldsymbol{\tau}'}{}' \hat{\epsilon}_{\boldsymbol{\tau}-\boldsymbol{\tau}'} \mathbf{E}_{\boldsymbol{\tau}'} + (\mathbf{k}_{\boldsymbol{\tau}} \mathbf{E}_{\boldsymbol{\tau}})\mathbf{k}_{\boldsymbol{\tau}} = 0 \tag{2.25}$$

where equation (2.22) defines $\mathbf{k}_{\boldsymbol{\tau}} = \mathbf{k}_0 + \boldsymbol{\tau}$ and $\hat{\epsilon}_{\boldsymbol{\tau}}$. The general properties of equation (2.23) are analyzed in numerous works (see e.g. [10, 11]). Using Floquet's theorem, it is shown [12] that, for fixed frequency and angle of incidence, there are four k_0 values in a crystal for which equation (2.23) has nontrivial solutions of form (2.24).

Numerical methods used for solving system (2.25) provide any desired accuracy. However, for solution analysis, it is preferable to have analytical expressions, even approximate ones. Thus, Eqs. (2.25) is solved below in the two-wave approximation of X-ray dynamical diffraction theory [4,8,9]. This approximation is based on the fact that if the Bragg condition (2.4) is met, the set of equations (2.25) has, as a rule, only two amplitudes $\mathbf{E}_0$ and $\mathbf{E}_{\boldsymbol{\tau}}$ whose values are as large as that of incident wave; all remaining amplitudes are smaller by a factor of at least $|\epsilon_{\boldsymbol{\tau}}|/|\epsilon_0|$.

The explicit form of the corresponding equations is given in the two-wave approximation and we isolate from system (2.25) two vector equations for waves $\mathbf{E}_1(\mathbf{E}_{\boldsymbol{\tau}})$ and $\mathbf{E}_0$ with wave vectors $\mathbf{k}_0$ and $\mathbf{k}_1 = \mathbf{k}_{\boldsymbol{\tau}} = \mathbf{k}_0 + \boldsymbol{\tau}$

$$\left(\hat{\epsilon}_0 - \frac{k_0^2 c^2}{\omega^2}\right)\mathbf{E}_0 + \hat{\epsilon}_{\boldsymbol{\tau}}\mathbf{E}_1 = 0$$

$$\hat{\epsilon}_{-\boldsymbol{\tau}}\mathbf{E}_0 + \left(\hat{\epsilon}_0 - \frac{k_1^2 c^2}{\omega^2}\right)\mathbf{E}_1 = 0. \tag{2.26}$$

The accuracy of the approximation permits neglect of the small (order of $\frac{|\epsilon_{\boldsymbol{\tau}}|}{|\epsilon_0|}$) nontransversality of the waves $\mathbf{E}_0$ and $\mathbf{E}_1$ in (2.26). Thus, it is possible to assume that they are orthogonal to $\mathbf{k}_0$ and $\mathbf{k}_1$, respectively.

We do not specify at this point, the form of equations (2.26) by writing the explicit form of Fourier-harmonics $\hat{\epsilon}(r)$ for the interaction of radiation with concrete periodic medium. Notice that equations (2.26) and the solutions of the corresponding boundary problems have been studied extensively (see e.g. [7,8]) for the simplest case of scalar dielectric permittivity $\epsilon(r)$ which corresponds, e.g., to X-ray diffraction from crystals. In this instance polarization separation occurs and vector equations (2.26) for E_0 and E_1 separate into two uncoupled sets of two equations for scalar amplitudes E_0 and E_1. In the case where the medium is characterized by tensor dielectric permittivity $\hat{\epsilon}(r)$, the solutions of (2.26) and the corresponding boundary conditions are complicated and have nontrivial polarization properties. This results in qualitatively new effects for media with structure complexity unobserved in diffraction optics of simple periodic media, e.g., the complicated polarization structure of forbidden bands [7] and some specific features of the Borrmann effects in such media [4], etc.

2.2.2 Dispersion Surfaces

We describe in general form the solution of the set of dynamical equations and boundary problems, in the two-wave approximation, of dynamical diffraction theory. The solution of the set of equations (2.26) determines the eigenwaves, 1.4., amplitudes $\mathbf{E}_0$ and $\mathbf{E}_1$ in expansion (2.24) and their dependence on the parameters of the problem—frequency ω, the direction of wave propagation, etc., and the dependence of wave vectors $\mathbf{k}_0$ and $\mathbf{k}_1$ of two plane waves in this expansion on those parameters.

Possible values of wave vectors $\mathbf{k}_0$ and $\mathbf{k}_1$ are determined from the solvability condition (2.26), i.e., the zero-value of its determinant. Different values of amplitudes $\mathbf{E}_0$ and $\mathbf{E}_1$ in the eigensolution correspond to different allowed values of $\mathbf{k}_0$ and $\mathbf{k}_1$. Since (2.26) is a set of four linear homogeneous equations, the condition of its solvability and the relation between $\mathbf{k}_0$ and $\mathbf{k}_1$ established by the Bragg condition (2.4), yields, in the general case, four different relationships between $\mathbf{k}_0$ and $\mathbf{k}_1$ which correspond to four different eigensolutions of form (2.24). In contrast to the kinematical approach, which under diffraction conditions [the Bragg condition (2.4) plus condition $|\mathbf{k}_0| = |\mathbf{k}_1|$] uniquely determines vectors $\mathbf{k}_0$ and $\mathbf{k}_1$, the dynamical approach permits some variations for $\mathbf{k}_0$ and $\mathbf{k}_1$ under diffraction conditions not yielding unique relations between k_0 and k_1 as in the kinematical approximation. The general solution of the set of equations (2.26) is a linear combination of all four eigensolutions of the system (2.26)

$$\mathbf{E}(\mathbf{r}, t)_j = (\mathbf{E}_{0j}e^{i\mathbf{k}_{0j}\mathbf{r}} + \mathbf{E}_{1j}e^{i\mathbf{k}_{1j}\mathbf{r}})e^{-i\omega t}$$
$$\mathbf{E}(\mathbf{r}, t) = \sum_{j=1} C_j \mathbf{E}(\mathbf{r}, t)_j \qquad (2.27)$$

where coefficients C_j and quantities $\mathbf{K}_{0j}$ and $\mathbf{k}_{1j}$ in each eigenwave are determined by the boundary conditions.

It is usually said that the values of $\mathbf{k}_0$ and $\mathbf{k}_1$ (consistent with the Bragg condition and Maxwell equations) determine the dispersion surfaces [8,9] which are the geometric loci of the ends of vectors $\mathbf{k}_0$ and $\mathbf{k}_1$ (Fig. 2.3). These surfaces are used to solve the following boundary problem. Dispersion surfaces are shown in Fig. 2.3 in the simplest case. There are four branches

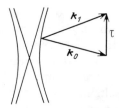

FIGURE 2.3. Dispersion surface.

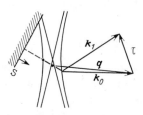

FIGURE 2.4. Dispersion surface and solution of the boundary problem (**S** is the surface normal, **q** is the wave vector outside the sample).

of the dispersion surface since the condition of zero-determinant for the set (2.26) reduces to an equation of degree four. However, two of the four branches are shown for clarity. If any point of the dispersion curves is connected with the ends of the reciprocal lattice vector τ (Fig. 2.3), the values of wave vectors $\mathbf{k}_0$ and $\mathbf{k}_1$ in expansion (2.24) satisfy both the Bragg condition and Maxwell equation.

Dispersion curves describe propagating modes. This means that wave vectors on the dispersion curve are real. In the space between the branches of the dispersion surfaces there are no propagating modes. The corresponding regions of k-space represent forbidden bands. In these regions, wave vectors $\mathbf{k}_0$ and $\mathbf{k}_1$ of the Maxwell equation solution, are complex quantities.

We determine wave vectors $\mathbf{k}_0$ and $\mathbf{k}_1$ in the eigensolution "excited" in a sample by an incident electromagnetic wave with dispersion equations and graphical construction. According to the wave vector boundary conditions, their tangential components on the sample surface are continuous. Thus, the difference between the wave vectors inside and outside the sample is directed along the surface normal. Hence, the following method is developed to seek eigensolutions "excited" in the sample for a given direction of wave propagation outside the sample (Fig. 2.4). Draw a straight line normal to sample surface S from the end of vector q corresponding to the wave vector outside the sample. The points of line intersection with the dispersion surfaces determine the values of wave vectors $\mathbf{k}_0$ and $\mathbf{k}_1$ of the eigensolutions "excited" in the sample.

In special cases (e.g., for surface normals to τ) it is possible that the

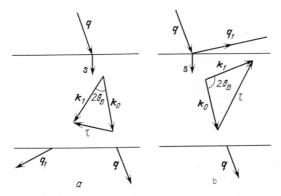

FIGURE 2.5. Diffraction geometry for the Bragg (b) and Laue (a) cases.

surface normal drawn through the end of vector q does not intersect the dispersion surface—the incident wave cannot "excite" propagating modes. This case corresponds to diffraction reflection of light from the sample. This is an illustration of the search for eigensolutions "excited" in a crystal with specific boundary conditions.

2.2.3 Solution of the Boundary Problem

In practice, an analytical method that does not use explicitly dispersion surfaces is more convenient when solving the boundary problem. This approach uses boundary conditions to establish relationships between wave vector $\mathbf{q}$ of the wave outside the sample and wave vectors $\mathbf{k}_0$ and $\mathbf{k}_1$ inside it. Then the solution of the dispersion equation is sought for a fixed value of q, on the condition that the determinant of system (2.26) is zero. The result describes the superposition of eigensolutions "excited" in the sample as a function of the parameters of the wave outside the sample, e.g. its angle of incidence or frequency (for the fixed angle of incidence).

Let a plane monochromatic wave $\mathbf{E}^e e^{i\mathbf{q}\mathbf{r}}$ be incident onto a crystal surface in the shape of a plate of finite thickness. For simplicity we neglect light refraction and reflection at the sample boundary associated with different refraction indices of two media and assume that the wave falls from a medium with dielectric permittivity that coincides with the mean dielectric permittivity $\bar{\epsilon}$ of the crystal.

Again, we assume that the sample is a plane-parallel plate. Accounting for the continuity of the tangential components of wave vectors k_0 and k_1 at the boundary, we write (see Fig. 2.5)

$$\mathbf{k}_0 = \mathbf{q} + q\xi s, \quad \mathbf{k}_1 = \mathbf{k}_0 + \boldsymbol{\tau} \tag{2.28}$$

where s is the inward surface normal and ξ is the small quantity of the order of $\epsilon_{\boldsymbol{\tau}}$. We determine the relation between the absolute values of wave vectors k_0 and k_1 and the angle of incidence $\gamma_0 = \hat{q, s} \approx \hat{k_0, s}$. We represent the vectors in the form

$$k_0 = q(1 + \eta_0), k_1 = q(1 + \eta_1). \tag{2.29}$$

From formulae (2.28), (2.29) and linear terms in small parameter ξ, we obtain the relation between η_1 and η_0 as

$$\eta_1 = \alpha/2 + \eta_0/b \tag{2.30}$$

where $b = \cos\gamma_0 / \cos\gamma_1, \gamma_1 = \hat{k_1 s}, \alpha = \tau(\tau + 2q)/q^2$ and parameter ξ obeys the equation $\eta_0 = \xi \cos\gamma_0$. Notice that parameter α is the measure of angular (frequency) deviation from the Bragg condition. Substituting equations (2.29) and (2.30) into the coefficients of system (2.26) and equating its determinant to zero, yields the dispersion equation which determines η_0 and therefore k_0 and k_1 as a function of parameter α. In the general case, [for an arbitrary dielectric permittivity tensor $\hat{\epsilon}(\mathbf{r})$], the dispersion equation is a fourth degree equation. For particular forms of $\hat{\epsilon}(\mathbf{r})$ [4,89,13] and specific scattering geometries [4,7] the equation becomes biquadratic. We present cases such as these below.

Upon finding the eigenvalues "excited" in the sample as functions of α, from the solutions of dispersion equations, solution of the boundary problem reduces to the determination of combined coefficients C_j in (2.27). They are determined from the continuity condition for tangential components of the electric and magnetic fields at the boundary (procedure similar to that used for solution of the boundary problem in Chapt. 1—corresponding solution in general form is given in Refs. [4,7]). The solution of the boundary problem determines amplitudes of diffracted, $\mathbf{E}^r$, and transmitted, $\mathbf{E}^t$, waves

$$\mathbf{E}^r = \sum_{j=1}^{4} C_j \mathbf{E}_{1j} e^{i\mathbf{k}_{1j}\mathbf{r}_e}, \quad \mathbf{E}^t = \sum_{j=1}^{4} C_j \mathbf{E}_{0j} e^{i\mathbf{k}_{0j}\mathbf{r}_t} \tag{2.31}$$

where $\mathbf{r}_e$ and $\mathbf{r}_t$ are radius-vectors on the entry and exit surfaces of the sample, respectively. In equation (2.31), only components of $\mathbf{k}_{1j}$ and $\mathbf{k}_{0j}$, normal to the surface, are essential in scalar products, $\mathbf{k}_{1j}\mathbf{r}_e$ and $\mathbf{k}_{0j}\mathbf{r}_t$, since their tangential components are equal for all eigenwaves (all j values).

The general scheme of the solution of the dynamical system (2.26) and boundary problem presented here is extended in the following section, to the description of the interaction between various types of radiation and specific media with complicated spatial structure (magnetically ordered structures, some liquid crystals, etc.). One approach to various problems demonstrates diffraction scattering property dependence (in previous media) on features of radiation interaction with the media, on the one hand, and the common physical nature of that interaction for different types of radiation, on the other.

References

1. R.W. James: *The Optical Principles of the Diffraction of X-rays* (London, Bell and Sons, 1967).

2. I.I. Gurevich, L.V. Tarasov: *Physics of Low-Energy Neutrons* (North-Holland, Amsterdam, 1968).

3. R.B. Hirsch, A. Howie, B.B. Nicholson, D.W. Pashley, M.J. Whelan: *Electron Microscopy of Thin Crystals* (London, Butterworth, 1965).

4. V.A. Belyakov: UFN **115**, 553 (1975) [Sov. Phys.-Usp. **18**, 267 (1975)].

5. V.B. Berestetskii, E.M. Lifshitz, L.P. Pitaevskii: *Relativistic Quantum Theory*, Part 1 (Moscow, Nauka, 1968) (in Russian).

6. L.D. Landau, E.M. Lifshitz: *Course of Theoretical Physics, vol. 8, Electrodynamics of Continuous Media* (Oxford, New York. Pergamon Press, 1960).

7. V.A. Belyakov, A.S. Sonin: *Optics of Cholesteric Liquid Crystals* (Moscow, Nauka, 1982) (in Russian).

8. Z.G. Pinsker: *Dynamical Scattering of X-rays in Crystals* (Berlin, Springer Verlag, 1978).

9. B.W. Batterman, H. Cole: Rev. Mod. Phys. **36**, 681 (1964).

10. L. Brilluen, M. Parodi: *Propagation des Ondes dans les Milieux Periodicues* (Paris, 1956).

11. C. Elachi, O. Yeh: J. Opt. Soc. Am. **63**, 840 (1973).

12. V.A. Belyakov, V.E. Dmitrienko: UFN **146**, 369 (1985) [Sov. Phys.-Usp. **28**, 535 (1985)]

3

Diffraction of Mössbauer Radiation in Magnetically Ordered Crystals

3.1 Diffraction of Mössbauer Radiation

We focus on specific periodic media characterized by complicated radiation diffraction interaction in the medium beginning with Mössbauer gamma radiation diffraction from magnetically ordered crystals.

Soon after the discovery of the Mössbauer effect [1,2], it was observed that coherent scattering of Mössbauer gamma quanta from crystals with Mössbauer nuclei was qualitatively different from X-ray scattering (see, e.g., [3–5] and references therein). It was revealed that this scattering would give not only information on nuclear properties and its interaction in the crystal but also crystal structure information unobtainable from other diffraction methods. Mössbauer radiation diffraction (hereinafter referred to as Mössbauer diffraction) is of particular interest for several reasons: its use in structure analysis shows promise and is known as Mössbauer diffraction structure analysis (analogous to X-ray and electron diffraction structure analyses). Mössbauer diffraction has unique features when compared with X-ray, neutron, and electron diffraction and provides crystal information such as X-ray structure, amplitude, phase which are usually beyond the scope of other diffraction methods. As an example remind that X-ray structure amplitude phase determination is essential for unambiguous structure determination. However, traditional diffraction methods fail to provide adequate phase information, especially for complicated compounds of unknown structures.

Determination of the structure amplitude phase follows a procedure based on a simple method of changing the phase and amplitude of resonance scattering of gamma quanta from a Mössbauer nucleus by changing the nuclear scattering resonance conditions using the Doppler effect, the amplitude of Rayleigh scattering (scattering by electrons) remains unchanged. Once the modulus of the total Rayleigh and nuclear amplitudes is measured experimentally for three values of the Doppler shift, it is possible to calculate the phase of Rayleigh amplitude from the known energy dependence of nuclear resonance amplitude.

Mössbauer diffraction permits the direct determination of crystal magnetic structure and is useful as a complement to magnetic neutron diffraction—the only known method to directly study magnetic structure. The physical grounds for this complimentarity is the relationship between Mössbauer radiation scattering amplitude dependence on magnetic field at the scattering nucleus and atomic magnitude and orientation of magnetic moment at a crystal lattice site. Mössbauer scattering amplitude depends on the electric field gradient (EFG) at the nucleus and this makes it possible to study the EFGs in a crystal. For example, Mössbauer diffraction data permits determination of EFG tensor principal axis orientation at equivalent sites of the unit cell. Another direction of Mössbauer diffraction studies is the collective interaction of nuclei in a crystal with gamma radiation which, if Bragg conditions are satisfied, changes the resonance energy of a gamma quantum and energy resonance line-width such that they do not match corresponding values of the Mössbauer level in an isolated nucleus. Thus, manifestation of the collective nature of the interaction between gamma quanta and nuclei is also the suppression of inelastic channels of nuclear reaction. This effect reveals itself in the following. Gamma radiation incident on a perfect crystal at the Bragg angle penetrates the crystal to a depth larger than that into which radiation penetrates imperfect or perfect crystals outside the Bragg condition—this is a nuclear analogue of the Borrmann effect known in X-ray diffraction. Mössbauer diffraction is interesting from the theoretical standpoint. Existing theories of various types of radiation diffraction from crystals, especially the dynamical theory of diffraction, are limited mainly to cases where scattering amplitude by an individual atom has the simplest form. Namely, theory uses, as a rule, scattering amplitude which corresponds to a scalar center and, if necessary, accounts for lattice thermal vibrations anisotropy using the Debye–Waller factor. The quantitative description of experiments of Mössbauer diffraction from crystals requires detailed theory for more complicated forms of scattering amplitudes for individual centers.

Thus, experimental studies of Mössbauer diffraction stimulate the development of detailed theory which accounts for the nonscalar nature of individual scatterers on the one hand and provides possible experimental verification of theoretical predictions on the other. The difference between the conclusions of such a theory and those known for scalar scatterers concerns mainly the polarization properties of eigenwaves in a crystal and related experimentally observed effects in diffraction studies.

Heretofore it was essential that a gamma radiation scattering crystal have Mössbauer nuclei at which gamma quanta resonance scattering occurred. However Mössbauer radiation scattering occurs in crystals without Mössbauer isotopes as well. At first glance, it seems that diffraction of gamma radiation in this case should proceed analogously to X-rays and cannot give any new information about the crystal in addition to that contained in X-ray data. But the detection of the scattered Mössbauer

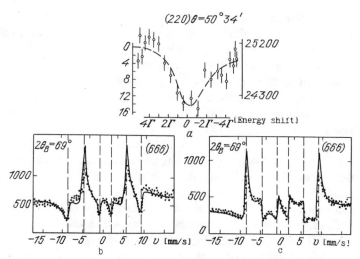

FIGURE 3.1. First observations of interference between nuclear and Rayleigh scattering for (200) reflection from an Fe single crystal [56] (a) and the interference in the Bragg (666) reflection in hematite for Zeeman splitting of a Mössbauer line (b,c). Antiferromagnetic axis is normal (b) to the scattering plane or in it (c) [57].

radiation yields new information unavailable for the X-ray technique. Extremely high resolution Mössbauer detectors permit separation of elastic and inelastic (diffuse) scattering in the region of diffraction maxima. Thus, it is possible to investigate the dynamics of a crystal lattice by studying, e.g., temperature variations of elastically scattered gamma quanta [4,5].

The first experiments on Mössbauer diffraction were performed by Moon, Black and co-workers of the Birmingham group [6,7]. The goal of their first experiments on polycrystalline samples was to demonstrate the coherence of nuclear elastic scattering. Experimental verification of this coherence was necessary because the duration of nuclear resonance scattering, determined by Mössbauer level lifetime is usually $10^{-7} - 10^{-8}s$, significantly greater than the reciprocal frequencies of excitation in a solid. Thus, the question arises whether phase relations are preserved in gamma quanta scattering by different nuclei of the crystal for such large scattering times. The coherence of nuclear resonance scattering was demonstrated experimentally by this group on Rayleigh and nuclear scattering interference in the diffraction maximum which appeared as an asymmetric dependence of scattering intensity on Mössbauer source velocity (Fig. 3.1).

Publication of the first experimental work was followed by a series of theoretical studies which suggested the use of Mössbauer radiation for diffraction experiments including determination of structure amplitude phase. For most Mössbauer transitions, the wavelength of gamma radiation is $\lambda \sim 10^{-8}$ cm, i.e., in the range convenient for diffraction experiments. This method is not used for experimental determination of structure amplitude

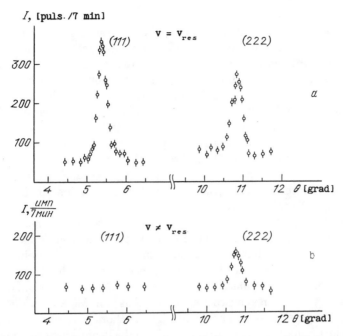

FIGURE 3.2. Magnetic diffraction maximum for Bragg scattering in hematite; the maximum is observed if there is resonance nuclear scattering (a), the maximum is absent if there is no nuclear scattering (b) [15].

phase in compounds of unknown structure, but Mössbauer and co-workers demonstrated the success [8,9] of such a determination using crystals of known structure.

The first experiments on Mössbauer diffraction from single crystals spurned further development of the theory. The first results on Mössbauer diffraction were generalized by Trammel and Hannon [9], Afanas'ev and Kagan [10], Belyakov and Aivazian, and Andreeva and Kuz'min [11].

As has already been noted, the dependence of the Mössbauer scattering amplitude on the magnetic field and EFG opens new possibilities of using Mössbauer diffraction for structure investigations in cases where X-ray diffraction fails or for studies of ordering of magnetic and electric fields in crystals.

The magnetic and quadrupole diffraction maxima not observed in X-ray diffraction were first predicted by Mössbauer diffraction theory [12–14] and confirmed experimentally [15,16,17]. Specific features of the coherent interaction between Mössbauer radiation and magnetically ordered crystals observed in these works stimulated the study of that interaction and its application [18–22]. A special term—magnetic Mössbauer diffraction analysis—was suggested which reflects the potential use of Mössbauer diffraction to delineate magnetic structure. At this writing, there has been one such determination [2,3].

There is considerable interest in work on Mössbauer optics of crystals under conditions where there is no Mössbauer diffraction. In the presence of resonance interaction between gamma quanta and nuclei, the optical characteristics of the crystal, with respect to gamma radiation, strongly depend on that interaction. For hyperfine magnetic or quadrupole splitting of the Mössbauer line in the crystal, the crystal shows birefringence and optical activity. The relative value of such effects ($\Delta n / |n - 1| \sim 1$) is much larger than the analogous value in the optical range. Similar to conventional optics, these phenomena can be used to extract information about magnetic and crystal structure. We restrict ourselves to Mössbauer diffraction, and refer readers interested in optics of Mössbauer radiation to the original literature [24,25].

In conclusion, we note that the experimental study of Mössbauer diffraction is intimately related to the problem of a Mössbauer source and, in some cases, to sample enrichment by a Mössbauer isotope. Both problems generate serious experimental difficulties, especially since Mössbauer sources decay and have a short lifetime. In connection with this, we should like to mention first works on synchrotron radiation used in Mössbauer diffraction experiments [26–29] (see Chapter 10). The authors have succeeded to obtain monochromatic synchrotron radiation and a beam which could be used as a source in Mössbauer diffraction experiments. The results obtained and successful solution of raised physical problems are very promising. They indicate that the application of synchrotron radiation provides well collimated beams of Mössbauer quanta with intensities higher by one or two orders of magnitude than those given by conventional sources. If to add that this is true for Mössbauer quanta of almost any energy (up to 100 keV) and to keep in mind that such a source does not decay, it is clear that the use of the synchrotron radiation opens new vistas for Mössbauer diffraction.

3.2   Amplitude of Coherent Mössbauer Scattering

Mössbauer radiation is gamma quanta with energies typically in the range from several to a hundred or more keV and wavelengths in the range from several angstroms to several tenths of an angstrom. Thus the energies and wavelengths of Mössbauer radiation are within the range of those in X-ray diffraction. Why then despite the fact that X-ray and Mössbauer gamma quanta are electromagnetic radiations of the same wavelength range their scattering by crystals may be qualitatively different. A general answer to this question is that the difference is caused by extremely narrow energy line width of Mössbauer radiation whose value is typically $\sim 10^{-8}$ eV (we recall that the corresponding value for X-ray characteristic radiation is of the order of 1 eV). As a consequence, in Mössbauer diffraction from crystals, along with scattering of quanta by atomic electrons (similar to the case of X-rays) an essential effect is also played by their resonance scattering at nuclei. Moreover, nuclear scattering of Mössbauer quanta often plays the key role, whereas for X-rays it is not important at all. Whence the consequence - a qualitative difference in scattering of these two types of radiation by crystals.

Of course the reader should bear in mind the conventional character of the above distinctions which are associated with the typical value of the energy linewidths for both radiations. Moreover, as has already been noted, powerful sources of synchrotron radiation in the X-ray wavelength range permit one to observe and study nuclear scattering of X-ray monochromatic radiation at Mössbauer nuclei and to obtain monochromatic X-ray beams with the linewidth close to that of the Mössbauer line [26–29] (see Chapter 10). In connection with the discussion of the energy linewidth, it should be noted that below we dwell on the interaction with crystals of strictly monochromatic radiation. Therefore in order to take into account the finite width and the shape of a line from the radiation source one should make corresponding energy averaging in all the relationships given below.

3.2.1   Amplitude of Elastic Scattering

Consider an elementary scattering event of resonance gamma quanta. First, consider the amplitude of elastic scattering of a resonance gamma quantum by an individual Mössbauer atom in a crystal. The scattering amplitude f is the sum of two terms—nuclear resonance term f^N and electron (Rayleigh) scattering term f^R

$$f(\mathbf{k}, \mathbf{e}; \mathbf{k}', \mathbf{e}') = f^N(\mathbf{k}, \mathbf{e}; \mathbf{k}', \mathbf{e}') + f^R(\mathbf{k}, \mathbf{e}; \mathbf{k}', \mathbf{e}') \qquad (3.1)$$

where $\mathbf{k}, \mathbf{e}$ and $\mathbf{k}', \mathbf{e}$ are wave and polarization vectors of the gamma quantum prior to and after scattering, respectively.

Amplitude F^R is identical to the scattering amplitude for X-rays and has the form $f^R(\mathbf{k}, \mathbf{e}, \mathbf{k}', \mathbf{e}') = Zr_e\mathbf{e}'^*\mathbf{e}f^a(\mathbf{k}' - \mathbf{k})$ where Z is the number of electrons per atom, r_e is the classical radius of electron and $f^a(\mathbf{k} - \mathbf{k}')$ is the atomic form-factor. The specific nature of Mössbauer diffraction is associated with nuclear resonance scattering. therefore we consider amplitude f^N. Since the lifetime of Mössbauer levels (characteristic value of the order of 10^{-7} s is rather large), it is assumed that the process of resonance elastic scattering occurs in two stages: (i) resonance absorption of a gamma quantum transferring the nucleus to a Mössbauer (excited) level and (ii) recoilless emission by an excited nucleus of a gamma quantum returning the nucleus back to the ground state.

Therefore, the scattering cross section is proportional to the product of the probability of a gamma quantum absorption without recoil and probability of recoilless emission and reflects the resonance dependence on the energy of scattered gamma quanta. Since, in the general case, a nucleus in a crystal is under the influence of electric and magnetic fields and its energy levels are split, the differences in the energies of the sublevels of the excited and ground states assume the role of resonance energies in scattering. Accordingly, the amplitude for resonance Mössbauer scattering of a gamma quantum via a definite sublevel of the ground m and excited m' states of the nucleus are represented in form [29]

$$f^N_{mm'} = \frac{\pi}{k}\Gamma_{mm'}f(\mathbf{k})f(\mathbf{k}')\frac{(\mathbf{en}^*_{mm'}(\mathbf{k}))(\mathbf{n}_{mm'}(\mathbf{k}')\mathbf{e}'^*)\sqrt{I^{mm'}(\mathbf{k})I^{mm'}(\mathbf{k}')}}{E - E_{mm'} + (i\Gamma/2)}$$

(3.2)

(we assume, for precision, that the nucleus is affected by magnetic field H and m and m' are the nuclear magnetic quantum numbers of the ground and excited states).

$$\Gamma_{mm'} = (2_{j'} + 1)\begin{pmatrix} j & L & j' \\ m & M & -m' \end{pmatrix}\Gamma_i$$

is the partial radiation width of the transition from level m' to level m; Γ and Γ_i are the total and radiation width of the Mössbauer level; L is the multipolarity of the transition (hereinafter, if not otherwise, we omit subscripts j and j' indicating the spin of the ground and excited state of the nucleus); $I^{mm'}(\mathbf{k})$ and $\mathbf{n}_{mm'}(\mathbf{k})$ are the normalized intensity ($\int I^{mm}(\mathbf{k})d\Omega_{\mathbf{k}} = 1$) and the polarization vector of the radiation emitted in direction $\mathbf{k}$ in transition $m' \to m$, $E_{mm'}$ is the difference in energies of levels m' and m and $f^2(k)$ is the Lamb–Mössbauer factor for emission (absorption) of a γ-quantum in direction $\mathbf{k}$. The general formulae for $I^{mm'}(\mathbf{k})$ and $\mathbf{n}_{mm'}(\mathbf{k})$ for nuclear transitions of pure and mixed multipolarities are given in [31–33]. We write the formulae for dipole and quadrupole transitions of pure multipolarity

$$I^M(\mathbf{k}) = \frac{1}{4\pi}[e_1^2(M, \mathbf{k}) + e_2^2(M, \mathbf{k})] \tag{3.3}$$

$$n_M(\mathbf{k}) = (\chi_2 \cos\alpha_M + i\sin\alpha_M\chi_1)e^{iM\varphi} \tag{3.4}$$

$$\chi_2 = \frac{[\mathbf{k}[\mathbf{Hk}]]}{|\,[\mathbf{k}[H\mathbf{k}]]\,|}, \chi_1 = \frac{[\mathbf{Hk}]}{|\,[H\mathbf{k}]\,|}, tg\alpha_M = \frac{e_2(M, \mathbf{k})}{e_1(M, \mathbf{k})} \tag{3.5}$$

where $M = m' - m$, θ is the angle between $\mathbf{k}$ and $\mathbf{H}$, φ is the azimuthal angle of vector k measured around $\mathbf{H}$ (Fig. 3.3) and e_1 and e_2 are listed in Table 3.1.

Note that in the numerator of equation (3.2), the product of k-dependent factors is proportional to the matrix element of gamma quantum absorption with wave vector $\mathbf{k}$ and polarization vector e by a nucleus in the $m \to m'$ transition. Similarly, the product of k'-dependent factors is proportional

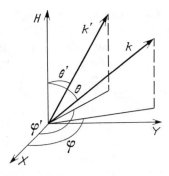

FIGURE 3.3. Definition of angles in the expression for scattering amplitude.

TABLE 3.1. Values of e_i for dipole and quadrupole transitions.

$e_i(M)$	$E1$	$M1$	$E2$	$M2$
$e_1(0)$	$-\sqrt{\frac{3}{2}}\sin\theta$	0	$\frac{1}{2}\sqrt{\frac{15}{2}}\sin 2\theta$	0
$e_1(\pm 1)$	$\pm\sqrt{\frac{3}{2}}\cos\theta$	$-\frac{\sqrt{3}}{2}$	$\frac{\sqrt{5}}{2}\cos 2\theta$	$\pm\frac{\sqrt{5}}{2}\cos\theta$
$e_1(\pm 2)$			$-\frac{\sqrt{5}}{4}\sin 2\theta$	$\pm\frac{\sqrt{5}}{2}\sin\theta$
$c_2(0)$	0	$\sqrt{\frac{3}{2}}\sin\theta$	0	$-\frac{1}{2}\sqrt{\frac{15}{2}}\sin 2\theta$
$e_2(\pm 1)$	$\frac{\sqrt{3}}{2}$	$\pm\frac{\sqrt{3}}{2}\cos\theta$	$-\frac{\sqrt{5}}{2}\cos\theta$	$-\frac{\sqrt{5}}{2}\cos 2\theta$
$e_2(\pm 2)$			$\pm\frac{\sqrt{5}}{2}\sin\theta$	$\frac{\sqrt{5}}{4}\sin 2\theta$

to the matrix element of a quantum emission with polarization vector $\mathbf{e}'$. Formula (3.2) describes resonance scattering if Zeeman splitting of nuclear levels markedly exceeds the Mössbauer level width. In the general case, the amplitude of resonance scattering for a nucleus in state m is obtained from expression (3.2) by summing over intermediate states m'.

As follows from Eq. (3.2), the explicit form of nuclear scattering amplitude as a function of scattering angle and quantum polarization may be complicated in the case of interaction between a Mössbauer nucleus and magnetic field or EFG. Neglecting details, we state that in scattering at a nucleus under the influence of a magnetic field, amplitude depends on field direction. This is evident from radiation intensities of the Zeeman transition $I^{mm'}(\mathbf{k})$ in (3.2) which depend on field direction, thus for dipole transition $I^{mm} \sim \sin^2 \theta$. In subsequent sections we consider, in more detail, this dependence for Zeeman splitting which exceeds the nuclear level width when resonance scattering proceeds through a definite Zeeman level of the excited nucleus and the nuclear scattering amplitude is directly described by expression (3.2).

If $\mathbf{e}$ and $\mathbf{e}'$ correspond to circular polarizations, amplitude (3.2) is conveniently expressed by elements $D^{(e)}_{mm'}$ of the finite-rotation matrix. Then formulae (3.2) take the form [33,34]

$$F^N(\mathbf{k}, \mu; \mathbf{k}', \mu') == (\mu\mu')^{L-\ell+1} \begin{pmatrix} j & L & j' \\ m & M & -m' \end{pmatrix} \cdot$$

$$\cdot \frac{f(\mathbf{k})f(\mathbf{k}')D^{(L)}_{\mu'M}(\mathbf{k}', \mathbf{H})D^{*(L)}_{\mu M}(\mathbf{k}, \mathbf{H}) \mid \chi(L,\ell) \mid^2}{E - E_{mm'} + (i\Gamma/2)} \tag{3.6}$$

where subscripts μ and μ' describe polarization of primary and scattered waves and assume two possible values $+1$ and -1 corresponding to right- and left-handed circular polarizations and $\chi(L, l)$ represent the reduced nuclear matrix elements of the transition. Similarly, it is possible to obtain the resonance scattering amplitude dependence on the orientation of the principal axes of the EFG tensor acting on the nucleus [36].

3.2.2   Isotope and Spin Incoherence

The description of Mössbauer diffraction involve coherent amplitude which requires averaging (3.2) over the crystal. Crystal atoms cause incoherent elastic scattering generally for two reasons (i) isotope incoherence and (ii) spin incoherence.

Isotope incoherence is associated with the fact that one definite isotope of a chemical element scatters Mössbauer radiation in resonance whereas the presence of another isotope at the crystal site is equivalent to absence of nuclear scattering. Accordingly the coherent amplitude is proportional to c, abundance of the Mössbauer isotope.

Spin incoherence reflects the dependence of resonance scattering amplitude on m, the projection of nuclear spin in the ground state. In this case coherent amplitude is the average over m, of the expression for the elastic scattering amplitude. Thus the final expression for the coherent amplitude has the form

$$f^N(\mathbf{k}, \mathbf{e}; \mathbf{k}'\mathbf{e}')_{coh} = c \sum_m a_m f_m(\mathbf{k}, \mathbf{e}; \mathbf{k}', \mathbf{e}') \tag{3.7}$$

where a_m is the occupancy number of the state with the projection m of the nuclear momentum and equation (3.2) determines f_m. If nuclear spins are polarized, dependence of a_m on m is important. This is associated with either crystal cooling [36] to temperatures at which nuclear spin ordering occurs or special external perturbations applied to the crystal. If nuclei are unpolarized (the present case), occupancy numbers are independent of m and $a_m \equiv (2j+1)^{-1}$.

3.2.3 Coherent Amplitude in Limiting Cases of Completely Split and Unsplit Lines

Generally, the dependence of f^N_{coh} on polarization and scattering angle is complex in explicit form. The corresponding expressions are substantially more simple in the limiting cases of Zeeman splitting, which markedly exceeds the width of the Mössbauer level and in the total absence of Zeeman splitting. In the former case, for fixed energy of a gamma quantum only one term is important in (3.7). It corresponds to resonance scattering through a definite Zeeman transition with coherent amplitude

$$f^N(\mathbf{k}, \mathbf{e}; \mathbf{k}', \mathbf{e}')_{coh} =$$
$$\frac{\pi c \Gamma_{mm'} f(\mathbf{k}) f(\mathbf{k}') (\mathbf{e}\mathbf{n}^*_{mm'}(\mathbf{K}')) (\mathbf{n}_{mm'}(\mathbf{K}')\mathbf{e}'^*) \sqrt{I^{mm'}(\mathbf{k}) I^{mm'}(\mathbf{k}')}}{k(2j+1)[E - E_{mm'} + (i\Gamma/2)]} \tag{3.8}$$

In the latter case, using expressions (3.2) and (3.6) where the energy containing denominator is now independent of m and m', it is possible to perform the summation in (3.7) in general form. Then, e.g., for dipole transitions the expression for the coherent amplitude takes the form

$$f^N(\mathbf{k}, \mathbf{e}; \mathbf{k}', \mathbf{e}')_{coh} = \frac{c(2j'+1)\Gamma_i f(\mathbf{k}) f(\mathbf{k}')}{4k(2j+1)[E - E_r + (i\Gamma/2)]} \begin{cases} (\mathbf{e}\mathbf{e}'^*) \\ [\mathbf{k}\mathbf{e}] \end{cases} \quad [\mathbf{k}'\mathbf{e}'^*] \tag{3.9}$$

where the upper line on the right side of the expression relates to electric dipole transitions, and the lower to magnetic dipole transitions. Using equations (3.8) and (3.9), one may readily obtain relative values of coherent scattering cross-sections, cross sections of gamma quantum absorption and Rayleigh scattering. Using the optical theorem, according to which $\sigma_t = (4\pi/k)I_m f(\mathbf{k}, \mathbf{e}; \mathbf{k}, \mathbf{e})$, equation (3.9) gives the total cross section of absorption per nucleus

$$\epsilon_t = \frac{c(2j'+1)\pi}{(2j+1)} f^2(\mathbf{k}) \frac{\Gamma\Gamma_i}{2k^2} \left| E - E_r + \frac{i\Gamma}{2} \right|^{-2} \tag{3.10}$$

and the cross section of coherent scattering per nucleus

$$\sigma_{coh} = \frac{\pi c^2 (2j'+1)^2}{6(2j+1)^2} \frac{f^2(\mathbf{k})f^2(\mathbf{k}')}{k^2} \left| \frac{\Gamma_i}{E - E_r + (i\Gamma/2)} \right|^2. \tag{3.11}$$

The radiation width Γ_i is related to Γ by the expression $\Gamma_i = \Gamma(1+\alpha)^{-1}$ where α is the gamma-ray internal conversion coefficient. Values from (3.9)–(3.10) for an Fe crystal give examples of the cross-section ratio for coherent nuclear scattering, absorption and Rayleigh scattering: the differential cross section of coherent nuclear scattering $(d\sigma/d\Omega)_N \sim 3p^2 \times 10^3$ barns at exact resonance; the cross section of coherent Rayleigh scattering $(d\sigma/d\Omega)_R$, as opposed to nuclear scattering, depends strongly on the angle, 40 barns for forward scattering and only 4 barns for scattering to $60°$; and, the section for nuclear absorption in resonance is $\sigma_t \sim p \cdot 10^6$ barns.

Since nuclear scattering is slow relative to the reciprocal frequencies of vibration in the crystal, the thermal factor enters the coherent amplitude as a product of two Lamb–Mössbauer factors $\exp(-k\langle x^2 \rangle_{\mathbf{k}}) \exp(-k^2 \langle x^2 \rangle_{\mathbf{k}})$ not in the form of the Debye–Waller factor $\exp(-|\mathbf{k} - \mathbf{k}'|\langle x^2 \rangle_{\mathbf{k}-\mathbf{k}'}$ as in the case of coherent scattering amplitude; for X-rays $(\langle x^2 \rangle_{\mathbf{k}}$ is the mean square of thermal vibrations in direction $\mathbf{k}$). This explains the smooth dependence of nuclear coherent scattering on scattering angle and the rapid drop of Rayleigh scattering cross section with this angle.

3.3   Kinematic Theory of Mössbauer Diffraction

The kinematical approximation of diffraction theory stated in Ch. 2 describes the specific features of coherent scattering of Mössbauer gamma quanta from crystals. This approximation is valid for gamma radiation scattering from small crystals when the scattered beam intensity is much lower than that of the primary. The approximation, with minor modifications, this approximation is also applicable to scattering from imperfect (mosaic) crystals of arbitrary dimensions when scattering from an individual crystallite (block) satisfies the condition above.

3.3.1   Structure Amplitude

As in X-ray diffraction, the main quantity which determines scattering is structure amplitude F, the sum of coherent scattering amplitudes (including positionally determined phase factors) of all unit cell atoms. A property of Mössbauer diffraction is that structure amplitude consists of two terms—one associated with electronic scattering of quanta equivalent to structure amplitude in the theory of X-rays, the other with nuclear scattering. Unlike X-rays, the latter term depends on the energy of gamma, quanta resonance and on the strength and structure of hyperfine (electric and magnetic) fields which affect Mössbauer nuclei, and, scattering cross section and polarization of scattered radiation depend on both the energy of Mössbauer quanta and the magnetic and electric structure of the crystal. Since X-ray structure amplitude is independent of energy (assume small changes in quantum energy close to resonance value) and the magnitude and phase of nuclear amplitude are essentially dependent on energy [see (3.2)], the scattering cross section, a function of energy, has an obvious interference term and is given as a sum of terms due to Rayleigh and nuclear scattering and interference. We consider, in the kinematical approximation, the energy and polarization characteristics of coherent Mössbauer scattering in paramagnetic and magnetically ordered crystals. The structure amplitude F_τ is given by the expression

$$F_\tau(\mathbf{k}, \mathbf{e}; \mathbf{k}', \mathbf{e}') = F_\tau^R(\mathbf{k}, \mathbf{e}; \mathbf{k}', \mathbf{e}') + F_\tau^N(\mathbf{k}, \mathbf{e}; \mathbf{k}', \mathbf{e}')$$

$$= \mathbf{e}\mathbf{e}'^* F_\tau^R + \sum_q f_q(\mathbf{k}, \mathbf{e}; \mathbf{k}'\mathbf{e}')_{coh} \exp[i(\mathbf{k} - \mathbf{k}')\mathbf{r}_q] \quad (3.12)$$

where summation in the last term is over Mössbauer nuclei in the unit cell and F^R is the X-ray structure amplitude. Consider diffraction of Mössbauer radiation by a crystal which yields an unsplit Mössbauer line. The structure amplitude may be conveniently written in terms of vectors $\mathbf{e}$ and $\mathbf{e}'$ which correspond to linear π (in the scattering plane) and σ (normal to that plane) polarizations. The structure amplitude for such polarizations are given in form

$$F_\tau = F_\tau^R P^R + F_\tau^N P^N \quad (3.13)$$

where P^R and P^N are Rayleigh and nuclear polarization factors, formula (3.12) determines F_τ^N, and (3.9) determines $f_{a\ coh}$ (omitting the product of polarization vectors). The form of P^N depends on multipolarity of the Mössbauer transition. Thus for $E1$ transition $P^N = P^R(P^R = \cos 2\theta$ for π polarization and $P^R = 1$ for σ polarization and 2θ is scattering angle.) For $M1$ transition $P^N = 1$ for π polarization and $P^N = \cos 2\theta$ for σ polarization.

A typical property of Mössbauer diffraction is the interference between nuclear and Rayleigh scattering. For an unpolarized incident beam, equation (3.14) describes the intensity of scattered radiation in the diffraction maximum

$$\sigma^R + \sigma^N + \sigma^{RN} = |F_\tau^R|^2 \frac{1 + \cos^2 2\theta}{2} + |\overline{F_\tau^N P^N}|^2$$

$$+ 2\mathrm{Re}\, F_\tau^N \overline{F_\tau^{R*} P^N} P^R \quad (3.14)$$

where bars above the second and third terms denote polarization averaging. For $E1$ nuclear transition and interference, the third term in equation (3.14) becomes $(1 + \cos^2 2\theta)\mathrm{Re}\, F_\tau^N F_\tau^R$; for $M1$ transition, $2\cos 2\theta \mathrm{Re}\, F_\tau^N F_\tau^R$. Since the magnitude and phase of nuclear scattering amplitude depend on the energy of gamma-quantum resonance [by factor $(E - E_r + i\Gamma/2)^{-1}$] and Rayleigh amplitude is constant for these energy changes, the intensity of scattered radiation, a function of E, has a distinct dispersion curve. The typical shape of that experimental curve, which clearly shows interference between nuclear and Rayleigh scattering, is shown in Fig. 3.1.

3.3.2 Phase Determination

Since the modulus and phase of nuclear amplitude depend on energy and are calculated for each value of gamma quantum energy, it is possible, with Mössbauer diffraction data, to determine the phase of X-ray structure amplitude F^R. Determination of structure amplitude phase from Mössbauer data is important in structure studies, especially in complex compounds with large numbers of atoms in the unit cell where the traditional methods of phase determination fail.

The method of phase determination from Mössbauer data is analogous to that of isomorphous substitution and anomalous dispersion used in X-ray diffraction and is illustrated by Fig. 3.4 taken from Ref. [8]. Determination of structure amplitude phase requires measurement of diffraction maximum intensity in the absence of nuclear scattering (large Doppler shifts of gamma quantum energy) and for two different values of E such that the amplitude of nuclear scattering is sufficiently large. The first measurement yields the magnitude of structure amplitude ($F_0(H)$ in Fig. 3.4), the two subsequent measurements, the modulus of the sum of the nuclear and structural amplitudes ($F_{0A}(H)$ and $\tilde{F}_{0A}(H)$ in Fig. 3.4) respectively. Thus, it is possible to uniquely determine structure amplitude phase φ_0 with calculated values of nuclear amplitudes n_{Fe} and $\tilde{n}_{Fe}$ for corresponding energies and geometrical construction (Fig. 3.4).

3.3.3 Diffraction at Magnetically Ordered Crystals

Consider Mössbauer diffraction in a magnetically ordered crystal where Mössbauer nuclei are influenced by a magnetic field. The form of the expression for a differential scattering cross section is the same as above (see (2.5,6)). However the corresponding τ is the reciprocal lattice vector of the crystal that accounts for magnetic structure. In this case the diffraction pattern is analogous to that of neutrons diffracted from magnetically ordered crystals. Therefore Mössbauer diffraction data, as magnetic neutron diffraction, make it possible to determine crystal magnetic structure.

While phase determination seems obvious (the Doppler effect can change the phase of nuclear scattering), magnetic structure determination is not

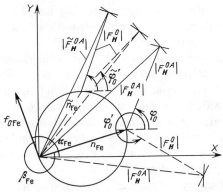

FIGURE 3.4. The principle of determination of the phase φ_0 of X-ray structure amplitude F^0 (F^{0A} and $\tilde{F}^{0A}$ are the total (for two energies) nuclear and Rayleigh scattering amplitudes per unit cell, $n_{Fe}(\tilde{n}_{Fe})$ is the amplitude of the nuclear gamma resonance scattering which depends on the energy of gamma quanta [8].

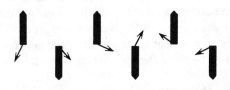

FIGURE 3.5. Illustrating the orientational disorder in nuclear moments (small arrows) in a magnetically ordered crystal (solid arrows indicate atomic magnetic moments).

obvious since, although atomic spins in magnetically ordered crystals are ordered, nuclear spins are disordered if the temperature is not very low (Fig. 3.5). Therefore it may seem surprising that diffraction of gamma radiation from the system of disordered nuclei may yield information on the magnetic order in the crystal. The situation changes, however, at very low temperatures when nuclear spins are ordered and diffraction pattern dependence on magnetic ordering raises no doubts.

However the possibility of magnetic structure determination by means of Mössbauer diffraction (even at elevated temperatures) does exist due to the narrow Mössbauer energy line width ($\Gamma \sim 10^{-8}$ev for Fe^{57}). Due to this sharpness of the line resonance scattering proceeds at nuclei whose spins have definite orientations relative to the magnetic field H; therefore coherent amplitude depends on H. Thus, for complete resolution of Zeeman splitting when the gamma-quanta energy coincides with that of one of the Zeeman transitions $E_{mm'}$, scattering proceeds through this transition alone. Hence, nuclei whose spin projection onto the magnetic field direction is equal to m participate in resonance scattering; nuclei with different values of spin projections effectively make no contribution. Since nuclear spins are randomly oriented, however, on the average, only one of $2j + 1$ Mössbauer nuclei participate in scattering. This diminishes the coherent amplitude by a factor of $2j + 1$ relative to the case of completely ordered nuclear spins, but leaves the qualitative character of the diffraction pattern unchanged. Strictly speaking, Mössbauer diffraction permits the direct determination of magnetic field ordering at Mössbauer nuclei. Since the type of magnetic field ordering at nuclei is determined by the type of atomic moment ordering in the crystal, however, we refer to it as magnetic structure determination from Mössbauer diffraction data.

The information on crystal magnetic structure appears in structure amplitude. For crystals, the magnetic space group of which differs from that of the crystal, the pattern of coherent Mössbauer radiation scattering is qualitatively different from X-ray diffraction—the former includes diffraction maxima due to magnetic structure ("magnetic maxima"). In terms of structure amplitude (3.12), which is the sum of terms associated with nuclear and Rayleigh scattering, this means that there are diffraction maxima for which $F_\tau^N \neq 0$ while $F_\tau^R \equiv 0$ for symmetry reasons. The nonzero value of F_τ^N is due solely to coherent nuclear amplitude dependence on magnetic field and is explained by the difference between crystal magnetic and lattice symmetries.

The information about magnetic structure may be extracted not only from observed magnetic maxima and their intensities but also from the polarization properties of these maxima. Expression (3.12) describes both diffraction maxima determined by the unit cell of the crystal ("crystal diffraction maxima") and those determined by the magnetic unit cell ("magnetic diffraction maxima"). We neglect magnetic Rayleigh scattering effects since Rayleigh scattering amplitude dependence on atomic magnetic moment orientation is very weak[2] [37]; and, magnetic diffraction maxima contain the contribution from nuclear scattering alone with no interference effects form nuclear and Rayleigh scattering. In contrast, crystal diffraction maxima contain contributions from both Rayleigh and nuclear scattering and their interference is manifest in those maxima.

Scattering polarization characteristics are described, in the kinematic approximation, by the general formulae (2-6)–(2.12). The specific feature of Mössbauer scattering, however, is that the resonance property of nuclear scattering results in strong scattering polarization property dependence on energy E of a gamma quantum. Since the interference of Rayleigh and nuclear scattering is strongly dependent on E, crystal maxima polarization

---

[2]Hopefully with the widespread use of synchrotron radiation the observed [60–62] magnetic dependence of X-ray scattering amplitude ("magnetic reflections" [59,67]) becomes a working tool for the study of magnetic structure by X-ray diffraction methods [63–66] (see Chapt.9)

[in particular equation (2.13)] is a function of energy E of gamma quanta]. Therefore even if incident radiation is completely polarized, averaging over the Mössbauer energy line width of the source results in the polarization density matrix $\bar{\rho}$ which corresponds to a partially polarized beam [averaging equation (2.13) yields partially polarized scattering radiation in the crystal maximum]. In the nuclear maxima where ρ is independent of E, a polarized incident radiation will yield $\bar{\rho}$ which corresponds to a completely polarized scattered radiation.

The general form of the scattering cross section for unpolarized radiation incident onto a magnetically ordered crystal and its dependence on energy are given by (3.14). The character of the corresponding interference determined by the third term in (3.14) depends, however, on the type of magnetic ordering in the crystal, the Zeeman transition through which scattering proceeds and the orientation of magnetic fields at Mössbauer nuclei. This directly follows from the corresponding equations of the coherent amplitude equation (3.8).

3.3.4 Resolved Zeeman Splitting of a Mössbauer Line

Consider the explicit form of the general expressions above applied to some types of magnetic ordering in crystals. We assume, for simplicity, that Zeeman splitting in a crystal is much larger than the incident radiation line width and the Mössbauer line in the crystal. That is, the situation is one in which nuclear scattering proceeds through a definite Zeeman level of the ground and excited states of the nucleus. The differential cross sections of unpolarized gamma quanta for electric $E1$ and magnetic $M1$ dipole nuclear transitions are represented by the following expressions.

(i) *Ferromagnets.* The magnetic and crystal unit cells coincide and there are no magmatic diffraction maxima. From (3.12), we obtain the following expression for nuclear and interference components of the cross section. For nuclear component

$$\sigma_f^N = \frac{C^2(\mathbf{k},\mathbf{k}')\Gamma^2|\tilde{F}_T^N|^2}{2k^2(1+M^2)^2|\Delta E|^2}[1-(-1)^M\cos^2\theta][1-(-1)^M\cos^2\theta']. \quad (3.15)$$

For the interference term σ^{RN}
(a) $E1$ transition

$$\sigma_f^{RN} = \frac{r_e}{k}\frac{C(\mathbf{k},\mathbf{k}')\Gamma}{(1+M^2)|\Delta E|^2}\left\{\left[\frac{1}{4}\sin 2\theta \sin 2\theta' \cos\phi + \sin^2\theta \sin^2\theta'\right.\right.$$

$$+M^2(\cos^2\theta+\cos^2\theta'-\sin^2\theta\sin^2\theta'\sin^2\phi]Re(\Delta E\tilde{F}_T^{N*}F_T^R)$$

$$-M\left(\frac{1}{4}\sin 2\theta \sin 2\theta' \sin\phi + \frac{1}{2}\sin^2\theta\sin^2\theta'\sin 2\phi\right)I_m(\Delta EF_T^{N*}F_T^R)\right\}$$

$$(3.16a)$$

(b) $M1$ transition

$$\sigma_f^{RN} = \frac{r_e}{k} \frac{C(\mathbf{k}, \mathbf{k}')\Gamma}{(1 + M^2)|\Delta E|^2} [\sin \theta \sin \theta' \cos \phi \, Re(\Delta E \tilde{F}_T^{N*} F_T^R)$$

$$+ 2M^2 \cos \theta \cos \theta' Re(\Delta E \tilde{F}_T^{N*} F_T^R) - M \sin \theta \sin \theta' \sin \phi I_m (\Delta E \tilde{F}_T^{N*} F_T^R)]$$
$$(3.16b)$$

where

$$C(\mathbf{k}, \mathbf{k}') = \frac{3c\Gamma_{mm'} f(\mathbf{k}) f(\mathbf{k}')}{8\Gamma(2j + 1)}, \quad \tilde{F}_T^N = \sum_q e^{i(\mathbf{k} - \mathbf{k}')\mathbf{r}_q}.$$

The summation is over Mössbauer nuclei of the unit cell, $\Delta E = E - E_{mm'} + (i\Gamma/2)$, $\phi = \varphi - \varphi'$ is the difference between the azimuthal angles of vectors $\mathbf{k}$ and $\mathbf{k}'$ [axis Z is directed along the magnetic field (Fig. 3.3)] and the remaining symbols are as stated above.

Formulae (3.16) show that the explicit form of the interference term depends on crystal structure and is different for different lines of the Mössbauer spectrum. The interference term has distinct forms for the electric and magnetic dipole nuclear transitions. This principle is explained by: Rayleigh scattering of electric dipole one and its interference with nuclear scattering yield different results for $E1$ and $M1$ transitions. The interference term dependence on transition multipolarity is, in principle, used to determine nuclear transition type.

(ii) *Antiferromagnets.* The magnetic field at nuclei assumes two values, H and $-H$. In the magnetic diffraction maximum in Eq. (3.12) only the nuclear term is nonzero

$$\sigma_{af}^N = \frac{M^2 C^2(\mathbf{k}, \mathbf{k}')\Gamma^2}{2k^2|\Delta E|^2} |\tilde{F}_{(H)}^N|^2 (\cos^2 \theta + \cos^2 \theta' + \sin^2 \theta \sin^2 \theta' \sin^2 \phi) \quad (3.17)$$

where $\tilde{F}_H^N = \sum_q e^{i(\mathbf{k} - \mathbf{k}')\mathbf{r}_q}$ and the summation is over nuclei in field H within the unit cell.

The nuclear component for crystal diffraction maxima is

$$\sigma_{af}^N = \frac{C^2(\mathbf{k}, \mathbf{k}')}{2k^2(1 + M^2)^2|\Delta E|^2} \{[1 - (-1)^M \cos^2 \theta][1 - (-1)^M \cos^2 \theta']$$

$$\times \left(|\tilde{F}_{(H)}^N|^2 + |\tilde{F}_{(-H)}^N|^2 \right) + 2 \sin^2 \theta \sin^2 \theta' Re(\tilde{F}_{(H)}^N \tilde{F}_{(-H)}^{N*} e^{i2M\phi}) \} . \quad (3.18)$$

The interference term for $E1$ and $M1$ transitions is given by equations (3.16a) an (3.16b) respectively, if the expression after Re is substituted by $\Delta E F^R (F_H^{N*} + \tilde{F}_{-H}^{N*})$ and that after Im by $\Delta E F^R (\tilde{F}_{(H)}^{N*} - \tilde{F}_{(-H)}^{N*})$.

Notice that, as shown by Eq. (3.17), there are no magnetic diffraction maxima for $M = 0$. This result is independent of transition multipolarity and is associated with the fact that the amplitude of Mössbauer scattering for transitions with $M = 0$ remains unchanged if $\mathbf{H}$ is substituted by $-\mathbf{H}$.

As follows from the general consideration, polarization characteristics of radiation scattered into the diffraction maxima depend on the interference of nuclear and Rayleigh scattering and, in particular, the ratio of corresponding structure amplitudes and the shape of the Mössbauer line. Therefore, we give only the explicit form of the polarization density matrix for nuclear diffraction maxima alone, for which Rayleigh structure amplitude vanishes and averaging over energies does not change the form of the polarization density matrix.

Consider the polarization characteristics of the radiation at magnetic diffraction maxima for an antiferromagnet. As follows from the above, the corresponding polarization density matrix is the same across the width of the Mössbauer line. If the primary radiation has polarization vector e, the vector of the scattered radiation is obtained from equations (3.8), (2.7), and (2.8) as

$$\mathbf{n}_0' = \mathbf{N}/|\mathbf{N}|, \quad \mathbf{N} = \mathbf{n}_t(\mathbf{k}')(\mathbf{e}\mathbf{n}_t^*(\mathbf{k})) - \mathbf{n}_t^*(\mathbf{k}')(\mathbf{e}\mathbf{n}_t(\mathbf{k})) \tag{3.19}$$

where $\mathbf{n}_t$ is the transition polarization vector (see Eq. (3.4)). For unpolarized incident radiation, the polarization matrix is determined by (2.10). For $M1$ transitions (2.10) leads to the following Stokes parameters (in the basis of unit vectors given by Eq. (3.5) for direction $\mathbf{k}'$)

$$\xi_1 = \rho_{12} + \rho_{21} = \frac{\sin^2 \theta \cos \theta' \sin 2\phi}{\cos^2 \theta + \cos^2 \theta' + \sin^2 \theta \sin^2 \theta' \sin^2 \phi}$$

$$\xi_2 = i(\rho_{12} - \rho_{21}) = 0$$

$$\xi_3 = \rho_{11} - \rho_{22} = \frac{\cos^2 \theta' - \cos^2 \theta - \sin^2 \theta(1 + \cos^2 \theta') \sin^2 \phi}{\cos^2 \theta + \cos^2 \theta' + \sin^2 \theta \sin^2 \theta' \sin^2 \phi} \tag{3.20}$$

and the degree of scattered radiation polarizatin is $P = [\sum_{i=1}^{3} \xi_i^2]^{1/2}$.

The corresponding expressions for $E1$ transitions follow if we change the signs of ξ_1 and ξ_2. The density matrix (3.20) is real, physically, this means that the scattered radiation is partially linearly polarized. The corresponding polarization vector makes angle α with the unit vector $\boldsymbol{\chi}_q'$ determined by the relationship $tang\, 2\alpha = \xi_1/\xi_2$. For $\xi_1 = 0$, the density matrix is diagonal in the basis of the unit vectors above and the scattered radiation is partially polarized along one of these unit vectors (for $\xi_3 > 0$ along $\boldsymbol{\chi}_1'$, for $\xi_3 < 0$ along $\boldsymbol{\chi}_2'$). The scattered radiation is polarized along one of the unit vectors $\boldsymbol{\chi}_1', \boldsymbol{\chi}_2'$ if the magnetic field direction is in the scattering plane or normal to it. In the former case the degree of polarization is $P = |\cos^2 \theta - \cos^2 \theta'|(\cos^2 \theta + \cos^2 \theta')^{-1}$ and in the latter case the radiation is completely polarized. For $\xi_1 = \xi_3 = 0$ the radiation is unpolarized.

We note that in this case, for an antiferromagnet, the radiation polarization density matrix for radiation at the magnetic diffraction maxima is independent of crystal structure and is determined by magnetic structure alone. Therefore, the polarization density matrix (3.20) at the magnetic maximum uniquely gives the orientation of the antiferromagnetic axis in the crystal. The radiation polarization in the crystal maxima depends on both magnetic and crystal structures of the antiferromagnet. Therefore, polarization measurements yield the information on crystal structure [38]. Methods for accounting for absorption [39] and analogous consideration of diffraction from more complicated magnetic structures and polycrystals are found in Refs. [30,40,41,55].

3.3.5   Crystals with Mössbauer Nuclei at Sites with Inhomogeneous Electric Field (EFG)

Consider diffraction from crystals containing Mössbauer nuclei at sites with an inhomogeneous electric field. Coherent nuclear amplitude depends on the EFG and therefore the intensity of diffraction maxima and radiation polarization which provide the information on the EFG and may be used in structure studies [42]. Neglecting a detailed description, we focus on diffraction from crystals which give quadrupole diffraction maxima not observed in scattering of other types of radiation [13]. Scattering amplitude dependence on magnetic field results in the appearance of magnetic diffraction maxima. Similarly, Mössbauer scattering amplitude dependence on the orientation of the EFG tensor principal axes results in the appearance of quadrupole diffraction maxima forbidden by the crystal space group. Quadrupole diffraction maxima are seen in crystals where Mössbauer nuclei occupy crystallographically equivalent sites and are affected by an EFG with principal axes oriented differently. Mössbauer amplitude dependence on the orientation of an EFG tensor axes makes Mössbauer nuclei which occupy equivalent crystallographic positions in the crystal lattice nonequivalent in scattering (Fig. 3.6). The result is the appearance of diffraction maxima forbidden by the crystal space group. In terms of structure amplitudes (3.12), this means that there are reciprocal lattice vectors τ for which $F_\tau^N \neq 0$ and $F_\tau^R = 0$, in accordance with symmetry considerations. A structure for which quadrupole maxima were first observed experimentally is sodium nitroprusside [16]. The anisotropy of thermal lattice vibrations may cause difference in scattering by atoms in equivalent positions which produce "dynamical diffraction maxima" forbidden by the crystal space group [43].

Note also that in addition to the magnetic and quadrupole maxima, Mössbauer scattering may also produce the experimentally unobserved up to now combined maxima [38]. Such maxima should appear in structures where Mössbauer nuclei experience the action of both magnetic field and EFG. They may appear in situations where crystallographically equivalent lattice sites are nonequivalent while under the simultaneous influence of magnetic field and EFG, although equivalent when these forces act separately.

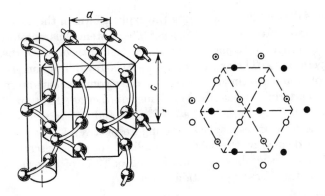

FIGURE 3.6. The structure of the tellurium unit cell (at the projection of the unit cell onto the plane normal to the screw axis). Crystallographically nonequivalent sites are designated by different symbols in accordance with three main orientations of the EFG axes at these sites [17].

3.4 Dynamical Theory of Mössbauer Diffraction

In the previous section, Mössbauer diffraction was considered in the kinematical approximation. Quantitative description within this approximation is possible for thin crystals with thickness $L < V/\sigma_{coh}$ where σ_{coh} is defined by Eq. (3.10) and V is unit cell volume. Thus for a 14.4 keV transition in Fe^{57} and 100% Mössbauer isotope content, the corresponding crystal thickness $L \sim 10^{-4}$cm at exact resonance. This crystal thickness increases as $C^{-2}|(E - E_R)/\Gamma + i/2|^2$ with deviation from exact resonance and at lower concentration of Mössbauer isotope; for natural abundance of Mössbauer isotope ($C = 0.025$), L is of the order of 10^{-2} cm at exact resonance.

Quantitative descriptions of diffraction patterns should include multiple scattering and crystal absorption of gamma quanta with propagation in the crystal. These factors may be taken into consideration within the framework of the dynamical theory of diffraction (Chapt. 2) which not only quantitatively describes diffraction from thick perfect crystals but also produces some new qualitative results relative to the kinematical approximation: suppression of inelastic channels of nuclear reactions in a crystal under the Bragg condition (the Kagan–Afanas'ev effect) [45,46], and Pendelösung beats of gamma quanta intensity transmitted through a crystal, etc. These effects have direct analogues in X-ray diffraction and some specific properties associated with the interaction between Mössbauer gamma quanta and nuclei. Thus, e.g., suppression of inelastic channels of nuclear reactions is independent of crystal temperature whereas the X-ray analogue—anomalous transmission of X-rays (the Borrmann effect)–manifests a strong temperature dependence. The equation for the dynamical theory of Mössbauer diffraction is derived by substituting the explicit form of the dielectric permittivity tensor into $\hat{\epsilon}_T$, Eq. (2.26) [3]. The system thus obtained differs from the corresponding equations of the dynamical theory of X-rays by the form of operator $\hat{\epsilon}_T$. The expression for $\hat{\epsilon}_T$ in X-ray diffraction [47,48] that describes photon scattering by electrons needs a complemented term related to nuclear scattering to describe Mössbauer diffraction. Thus, in (2.26) the operator acquires the form $\hat{\epsilon}_T = \hat{\epsilon}_T^R + \hat{\epsilon}_T^N$ ($\hat{\epsilon}_T^R$ and $\hat{\epsilon}_T^N$ are terms related to Rayleigh and nuclear resonance scattering, respectively). The nuclear component $\hat{\epsilon}_T^N$ is defined by the coherent amplitude of nuclear scattering (3.7) and is related to the above introduced tensor (2.7) of unit cell coherent scattering by the following equation

$$\hat{\epsilon}_T^N = \frac{4\pi}{Vk^2} \hat{T}_T^N \tag{3.21}$$

We begin the analysis of the set of dynamical equations assuming of the simplest form of tensor $\hat{T}_T^N$ which represents an unsplit Mössbauer line.

3.4.1   Unsplit Mössbauer Line

The operator $\hat{\epsilon}_T^N$, in the basis of polarization unit vectors π and σ, is diagonal with respect to the polarization subscripts π and σ.[4] Using equations (3.7) and (3.21) it is presented as

$$\epsilon_0^N = \frac{4\pi}{Vk^2} F_0^N, \quad \hat{\epsilon}_{\pm T}^N = \frac{4\pi}{Vk^2} \hat{P}^N F_{\pm T}^N \tag{3.22}$$

where F_T^N is the nuclear structure amplitude determined by (3.12). The explicit form of the polarization factor P^N for dipole nuclear transitions follows from (3.9).

For an unsplit line, the solutions of (2.26) are analogous to those for X-rays. Specifically, the π and σ polarizations are eigenpolarizations and the set of vector equations is divided into two uncoupled systems of scalar equations which determine k_0, k_1 and E_0/E_1 for σ and π polarizations

$$\left(\epsilon_0^R + \frac{4\pi}{Vk^2} F_0^N - \frac{k_0^2 c^2}{\omega^2} \right) E_0^\mu + \left(\epsilon_T^R P^R + \frac{4\pi}{Vk^2} P^N F_T^N \right) E_1^\mu = 0$$

$$\left(\epsilon_{-T}^R P^R + \frac{4\pi}{Vk^2} P^N F_{-T}^N \right) E_0^\mu + \left(\epsilon_0^R + \frac{4\pi}{Vk^2} F_0^N - \frac{k_1^2 c^2}{\omega^2} \right) E_1^\mu = 0 \tag{3.23}$$

where superscript μ takes two values corresponding to σ and π polarizations and ϵ_T is the sum of Rayleigh [47,48] and nuclear (3.21) terms.

The solvability condition of system (3.23) (the zero value of the determinant) for each eigensolution determines its own region of k_1 and k_0 values

---

[4] Unit vectors with π polarization are in the scattering plane whereas those with σ polarization are normal to the plane.

that satisfy the Bragg condition (2.4) and the Maxwell equations. These regions in k-space, as stated in Chapt. 2, form four dispersion surfaces (in accordance with four eigensolutions). Each eigensolution has a corresponding point on the dispersion surface. In our case two eigenvalues of system (3.23) correspond to σ and π polarization, i.e., two dispersion surfaces (Figs. 2.3 and 2.4). Thus, in the dynamical theory, a diffracted wave exists over a small range of $\Delta\theta$ angles close to the Bragg condition as opposed to the kinematical approximation where it exists for one value of the angle between $\mathbf{k}_0$ and $\mathbf{k}_1$. The value of this range is determined by parameters that characterize the interaction between a quantum and crystal, i.e., ϵ_τ.

Equations (2.28)–(2.30) yield the eigensolutions of system (3.23) "excited" in the sample by a plane monochromatic wave in the shape of a plane parallel plate.

Substituting equations (2.28)–(2.30) into (3.23), we find from the solvability condition, the following expression for η:

$$
\begin{aligned}
\eta_0 &= \frac{1}{4}(\epsilon_0^0 + b\epsilon_0^1 - b\alpha) \pm \frac{1}{4}\left\{(\epsilon_0^0 + b\epsilon_0^1 - b\alpha)^2 + 4f[\epsilon_0^0\alpha \right. \\
&\quad \left. - (\epsilon_0^0\epsilon_0^1 - \epsilon_\tau^\mu\epsilon_{-\tau}^\mu)]\right\}^{1/2}
\end{aligned}
\tag{3.24}
$$

$$\eta_1 = \alpha/2 + \eta_0/b, \quad \epsilon_0^{0,1} = \hat{\epsilon}_0 - 1$$

for the wave propagating along $\mathbf{k}_0$ and $\mathbf{k}_1$, respectively.

Parameter α is the measure of deviation from the Bragg condition. The values found for η_0 correspond to the following amplitude ratio in the eigensolutions

$$R = B^2 = |E_1|^2/|E_0|^2 = |(2\eta_0 - \epsilon_0^0)/\epsilon_\tau^\mu|^2. \tag{3.25}$$

It follows from equations (3.24) and (3.25) it follows that $|E_1|^2/|E_0|^2 \sim 1$ in the angular range $\Delta\theta \sim| \epsilon_\tau^\mu |$.

The solution of system (3.23) above is analogous to the solution of the corresponding system for X-rays; and, it permits description of all the details of Mössbauer diffraction connected with resonant nuclear scattering for unsplit Mössbauer lines.

Now consider diffraction from a crystal for two cases: the Bragg case, where primary and diffracted beams are on one side of the entrance surface (Fig. 2.5a) and the Laue case, where primary and diffracted beams are on different sides of the crystal surface (Fig. 2.5b). We assume that the crystal surface is either parallel (Bragg case) or normal (Laue case) to the reflecting planes of the crystal to simplify the formulae. These cases are usually referred to as symmetric cases of diffraction with boundary conditions of form

$$\sum_j C_j \mathbf{n}'_j E_1^j = 0, \quad E_0 = \sum_j C_j \mathbf{n}_j E_0^j \tag{3.26}$$

in the Laue case and

$$\sum_j C_j \mathbf{n}_j E_0^j = E_0, \quad \sum_j C_j \mathbf{n}'_j E_1^j \exp(i\mathbf{k}_{1_j}\mathbf{s}L) = 0 \qquad (3.27)$$

in the Bragg case, where E_0 is the amplitude of the incident wave and L is crystal thickness.

3.4.2   Suppression of Inelastic Channels of Nuclear Reactions

We omit the explicit form of coefficients C_j in expansion (2.27), which are found from boundary conditions (3.26), and analyze the spatial attenuation of its terms in the sample for the symmetric Laue case ($b = 1$).

Far from the Bragg condition, $|\alpha| >> 1$, the amplitude of wave E_1 is negligibly small what corresponds to propagating wave E_0. The corresponding wave vector k_0 is determined by expression (3.24) with a plus sign before the braces, i.e., $k_0 = q(1 + \epsilon_0/2)$. The wave corresponding to the minus sign in (3.24) is not excited in the crystal [45,47]. The wave attenuation is determined by the imaginary part of wave vector $\mathbf{k}_0$; far from the Bragg condition it is equal to $0.5qIm\epsilon_0$. Neglecting attenuation due to the interaction of gamma quanta with electrons, the expression for wave intensity as a function of penetration depth, into the crystal becomes

$$I \sim \exp(-\mu x), \quad \mu = qIm\epsilon_0 = \frac{\sigma_t}{V} \qquad (3.28)$$

where σ_t is the total resonance absorption cross section for gamma quanta in the crystal given by (3.10).

Under the Bragg condition ($|\alpha| << |\epsilon_{\boldsymbol{\tau}}^{\mu}|$), expansion (3.27) contains the four eigensolutions, however different eigenwaves may attenuate differently in the crystal. For example, the attenuation of two eigenwaves, under Bragg conditions may be much weaker than outside Bragg conditions, and, under definite conditions, the attenuation of one may be zero. As an illustration, we consider wave propagation in a crystal for the value of parameter Δ small relative to $|\epsilon_{\boldsymbol{\tau}}^N|$:

$$\Delta = \epsilon_0^{N_0}\epsilon_0^{N_1} - \epsilon_{\boldsymbol{\tau}}^N \epsilon_{-\boldsymbol{\tau}}^N. \qquad (3.29)$$

Under this assumption the upper relationship in (3.24) is written in approximate form as

$$\eta_0^- = -\frac{1}{2}\frac{\epsilon_0^0 \alpha - \Delta}{\epsilon_0^0 + \epsilon_0^1}, \quad \eta_0^+ = \frac{1}{2}\left(\epsilon_0^0 + \epsilon_0^1 - \alpha + \frac{\epsilon_0^0 \alpha - \Delta}{\epsilon_0^0 + \epsilon_0^1}\right) \qquad (3.30)$$

where the superscript of η_0 corresponds to the sign before the braces in (3.24). It follows from (3.30), that the attenuation coefficients of the eigensolutions (and of the corresponding terms in (2.27) which correspond to η_0^+

are of the same order as in expression (3.28). The second pair of eigensolutions, which correspond to η_0^-, have weaker attenuation and the absorption coefficient is equal to

$$\mu^- = 2Imq\eta_0^- \tag{3.31}$$

This means that under the above conditions the gamma radiation may penetrate into the crystal much deeper than in the case of large deviations from the Bragg condition. This phenomenon is known in X-ray diffraction as the Borrmann effect; in Mössbauer diffraction it is known as the suppression of inelastic channels of nuclear reactions or the Kagan–Afanas'ev effect. Both phenomena are of analogous physical nature. Under diffraction conditions the incident wave is transformed so that the electromagnetic field in a crystal becomes a coherent superposition of two plane waves. The phase relationships of these waves are such that, for two eigensolutions, the intensity minima of electric or magnetic field (depending on the multipolarity of scattering) correspond to crystal lattice sites. The result is the reduction of inelastic-scattering cross sections, i.e., photoeffect at atoms and resonance absorption of gamma quanta at nuclei, and an explanation of the observed decrease in the absorption coefficient.

Since the characteristic times for Rayleigh and nuclear resonance scattering are essentially different, the manifestations of these effects are qualitatively different. The time for resonance scattering at nuclei $\tau_N \sim 10^{-7}s$ is much larger than the reciprocal frequencies of lattice vibrations $\tau_p \sim 10^{-12}s$ and the time for Rayleigh scattering by electrons is much smaller than τ_p. Thus, the complete suppression of inelastic channels of nuclear reactions may be attained for the Kagan–Afanas'ev effect whereas, in principle, the complete suppression of inelastic processes cannot be attained for the Borrmann effect. The condition for the complete suppression of inelastic processes is that μ^- for π and σ polarizations vanish simultaneously. It follows from (3.24), that this is possible only when $\alpha = 0$ and $\Delta = 0$.

For X-rays (i.e., scattering at electrons) Δ cannot be zero since the relationship between the characteristic time of lattice vibrations and scattering time suggests that amplitudes F_T^R include Debye–Waller factors which, for F_0^R (scattering at zero-angle), are identically equal to unity and, for F_T^R, are always less than unity. For Mössbauer scattering, however, the process is slow and F_T^N includes (instead of Debye–Waller factors) Lamb–Mössbauer factors in the form $\exp(-k_i^2\langle x^2\rangle_{\mathbf{k}_i})\exp(-k_p^2\langle x^2\rangle_{\mathbf{k}_p})$ (where $i,p = 0,1$); the products are equal for both terms in (3.29) and Δ may be zero. Thus in the case of an unsplit Mössbauer line, the complete suppression of inelastic channels is possible (if Rayleigh scattering is neglected for σ polarization, the Mössbauer transition is an electric dipole transition and, for π polarization, if it is a magnetic dipole transition [45]).

The same relationship between scattering times and lattice vibration reciprocal frequencies explains the fact that the Kagan–Afanas'ev effect is independent of temperature whereas the Borrmann effect exhibits strong temperature dependence (decrease of the anomalous transmission with temperature rise). In Rayleigh scattering, Δ increases with temperature since the first term in (3.29) is constant and the second decreases with the temperature. In Mössbauer scattering, temperature variation does not change the zero values of Δ since thermal factors in (3.29) are equal for both terms.

Suppression of inelastic channels of nuclear reactions reveals itself in the angular range close to the Bragg condition $\Delta\theta \leq |\epsilon_T|$ which is of several angular seconds. Therefore, this effect should be studied experimentally on crystals with high degrees of perfection using beams of gamma radiation collimated up to several angular seconds.

References to the early experimental studies of the Kagan–Afanas'ev effect are found in Ref. [3] and subsequent achievements are reviewed in [67,68,71–74].

3.4.3 Bragg Reflection from Crystals

We now analyze the solution of the problem of diffraction reflection in the symmetric Bragg case ($b = -1$). In expansion (2.27) for crystals of arbitrary thickness [49], we limit ourselves, for simplicity, to thick crystals for which $|LqIm(\eta^+ - \eta^-)| >> 1$. Similar to the case of X-rays, two eigenwaves with wave vectors determined by η_0^+ in (3.24) are present in (2.27). Thus, if an incident wave is π (or σ) polarized, the reflection coefficient R is defined by equation (3.25) for eigensolutions with π (or σ-) polarization corresponding to η_0^+. Therefore, if, for some eigenpolarization, $\Delta = 0$ [see (3.29)] $R = |E_1|^2/|E_0|^2 = 1$, for this polarization and certain values of parameter α, i.e., total reflection takes place for a certain angle of incidence. This is a sequence of complete suppression of inelastic channels. For X-rays, when we account for inelastic processes the ratio (3.25) is smaller than unity because inelastic processes cannot be suppressed completely and reflection is accompanied by intensity losses. But, in contrast to X-ray scattering where absorption is low and reflection coefficient almost constant in the region of diffraction reflection (almost zero outside it); absorption is essential in Mössbauer scattering and the reflection curve has a different shape (Fig. 3.7). Since the sum of nuclear and Rayleigh amplitudes and absorption (the imaginary part of the coherent amplitude) are dependent essentially on the energy of a gamma quantum the shape of reflection curve (Fig. 3.7) depends also on the energy deviation from the exact resonance condition.

Since the angular dimensions of the diffraction reflection region are small (about ten angular seconds) and the beams of gamma quanta used in experiments are essentially divergent, the measure of integrated reflected intensity is

$$R_T = \int \frac{|E_1|^2}{|E_0|^2} d\theta = \frac{1}{2 \sin 2\theta_B} \int_{-\infty}^{+\infty} \frac{|E_1|^2}{|E_0|^2} d\alpha. \qquad (3.32)$$

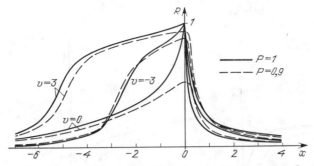

FIGURE 3.7. The calculated angular dependence of the Bragg reflection from a perfect crystal for three different energies of gamma quanta (the energy parameter $\nu = 2(E - E_r)/\Gamma$, the angular variable $x = (\epsilon_0)^{-1}[\alpha - 2(Re\,\epsilon_0 - Re\,\epsilon_r p^\mu)]$ [49].

In the right hand part of Eq. (3.32), integration over angles is replaced by integration over parameter α using the relation $d\theta = -d\alpha/2 \sin 2\theta_B$ where $2\theta_B$ is the scattering angle for the exact Bragg position. Integration over α in (3.32) noting the small value of Δ, Eq. (3.29), results in the expression analogous to the formula known in X-ray diffraction [47]

$$R_{T\mu} = \frac{8|\epsilon_T^\mu|}{3 \sin 2\theta_B} \mathfrak{F}^\mu \qquad (3.33)$$

where $\mathfrak{F}^\mu$ is close to unity.

Consider now characteristic properties of Mössbauer diffraction due to strong amplitude F_T^N dependence on the energy of gamma quanta. For pure nuclear scattering, it follows from (3.33), that for large deviations from resonance $\nu = (E - E_r)/\Gamma$, the reflection coefficient $R \sim |\nu|^{-1}$. Consequently, the integrated reflection dependence on ν, for a thick crystal, deviates from the Lorentz curve and the energy width of the corresponding curve is much broader than that for a thin crystal when $R \sim |\nu|^{-2}$ for $|\nu| >> 1$. The reflection curve in (Fig. 3.7) and the integrated reflection coefficient also reflect the typical interference between nuclear and Rayleigh scattering. This interference is seen most clearly in the integrated intensity dependence (Fig. 3.8) on deviation from resonance energy. The above formulae and illustrations relate to diffraction from an ideal crystal.

We note that the energy dependence of scattering in a mosaic crystal is different [49] from that in an ideal crystal and will not perform analysis of mosaic crystals within the framework of the kinematical approximation [50]. Figure 3.8 also shows the differences in the interference patterns of ideal versus mosaic crystals.

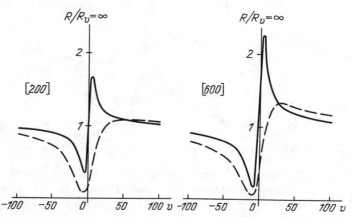

FIGURE 3.8. Integrated (with respect to angle) reflection coefficient from perfect (solid line) and imperfect (dashed line) tin crystals calculated for (200) (a) and (600) (b) reflections at 77 K [49].

3.4.4   Solution of the Dynamical Equations for Hyperfine Splitting of Mössbauer Line

If there is hyperfine splitting of a Mössbauer line for definiteness, we consider here magnetically ordered crystals, the set of dynamical equations (2.26) does not reduce, in the general case, to the set of two scalar equations for amplitudes with known eigenpolarizations. Now, each of the four solutions of system (2.26) has different eigenpolarizations n_j and n'_j [51,52]. Eigenpolarizations (vectors n_j and n'_j) depend on magnetic structure and vary, within the diffraction region, with deviation from the Bragg condition. If the Bragg condition is violated, they pass to eigenpolarizations for direct transmission [2,5]. Eigenpolarization depends not only on the angle of incidence of gamma quanta but also on their energy (in the presence of Rayleigh scattering). In the limit of large gamma quanta energy deviations from resonance values, when nuclear scattering may be neglected, eigenpolarizations become conventional π and σ polarizations known for X-rays and Mössbauer diffraction from magnetically ordered crystals looses its specific nature.

Thus the character of the diffraction pattern from magnetically ordered crystals is complicated compared with the cases above where there is no magnetic ordering. New qualitative effects appear which were absent in diffraction for the case of an unsplit Mössbauer line. For example, it becomes possible to completely suppress inelastic nuclear processes [53,54]. The complexities are associated mainly with the polarization properties of the eigensolutions of the set of equations (2.26).

In addition to the scattered radiation polarization dependence on crystal magnetic structure, different polarization properties of eigensolutions manifest themselves in Pendellösung beats. In a magnetically ordered crystal, Pendellösung generally yields six periods of intensity beats of the primary and secondary waves with crystal thickness [52] due to nonorthogonality of eigenpolarizations compared with only two periods observed for X-rays [47,48].

The results above follow from the general structure of the diffraction problem solution represented by Eq. (2.27). The explicit form of this expansion is obtained by setting boundary conditions assuming the crystal is a plane-parallel plate. Similar to the case of an unsplit line, using the relation between $\mathbf{k}_0$ and $\mathbf{k}_1$ (2.28)–(2.30) imposed by the boundary conditions, it is possible to uniquely determine from (2.26) all the parameters of the four eigenwaves excited in a crystal for the given angle of incidence of the primary wave. The four values of parameter η_0 which determine the wavevectors of the eigensolutions are set by the roots of the equation which follow from the condition that the determinant of the system (2.26) be zero.

Since, generally, the corresponding equation is an equation of fourth order relative to η_0 and its roots cannot be analytically expressed by the parameters of the problem, we omit the detailed solution of (2.26) and the boundary problem (see [3]). We present, however, the qualitative results related to the general case. From the properties of the eigensolutions—their nonorthogonality, the dependence of their polarization properties of the incidence angle even in the small region of strong diffraction scattering and their strong dependence on the energy of gamma quanta—it follows that the diffraction pattern from not too thick crystals ($q|E_T|L \sim 1$) is complicated. The intensity and polarization characteristics of the direct and diffracted beams experience beats (up to six periods) with the variation of both sample thickness and gamma quanta energy. Therefore, the qualitative description of experimental results normally requires some numerical calculations. Nevertheless, general analysis of the solution permits delineation of the typical qualitative characteristics of dynamical diffraction of Mössbauer radiation from magnetically ordered crystals. Such characteristics are best revealed analysis of the limiting case of very thick crystals ($q|\epsilon_T|L >> 1$).

As mentioned previously, only two of four eigensolutions are "excited" in a semi-infinite crystal and attenuate while propagating in the crystal. This result, obtained earlier for negligibly absorbing media, is valid in Mössbauer diffraction where absorption is essential and normally not negligible. In this situation the four eigensolutions of (2.26) may correspond to attenuating waves in the crystal; of these only two, for which attenuation is maximal, are "excited" in the crystal [3]. Therefore, in a thick sample (formally semi-infinite crystal), these eigensolutions describe diffraction scattering and its polarization dependence. The weight with which an incident wave "excites" each solution is defined by the coefficients of the polarization expansion for the incident wave in eigenpolarizations $\mathbf{n}_1$ and $\mathbf{n}_2$ of the $\mathbf{E}_0$ wave. In this case the reflection curves of the waves whose polarization coincides with eigenpolarizations are described by the angular dependence of functions B_1 and B_2 in (3.25) where $B_i = E_{1i}/E_{0i}$ is the amplitude ratio in the ith eigensolution. Notice that eigenpolarizations generally change along the reflection curve. Since eigenpolarizations $\mathbf{n}'_1$ and $\mathbf{n}'_2$ are not orthogonal (the prime denotes polarizations of wave E_1 in the eigensolution), the reflection curve for unpolarized radiation is not a half-sum of curves for $\mathbf{n}'_1$ and $\mathbf{n}'_2$ but also contains interference addition. The reflection coefficient thus takes the form

$$R = \frac{1}{2}(1 - |\mathbf{n}_1\mathbf{n}_2^*|)^{-1}[B_1^2 + B_2^2 - 2Re(\mathbf{n}'_1\mathbf{n}_2'^*)(\mathbf{n}_1^*\mathbf{n}_2)B_1B_2]. \qquad (3.34)$$

The polarization density matrix of the scattered radiation is described as

$$\bar{\rho} = A\left\{B_1^2\rho(\mathbf{n}'_1) + B_2^2\rho(\mathbf{n}'_2) - B_1B_2[(\mathbf{n}_1^*\mathbf{n}_2)\rho(12) + (\mathbf{n}_2^*\mathbf{n}_1)\rho(21)]\right\} \quad (3.35)$$

where $\rho(\mathbf{n})$ is the density matrix corresponding to the polarization vector $\mathbf{n}$; the matrices $\rho(ik)$ are set by relationships $\rho(ik)_{pq} = (\mathbf{n}'_i)_p(\mathbf{n}'^*_k)_q$ and A is the normalization factor. The degree of polarization of the scattered radiation corresponding to matrix (3.35) is

$$P = 2[(f^-)^2 + |D|^2]^{1/2}\left\{B_1^2 + B_2^2 - 2B_1B_2Re[(\mathbf{n}_1^*\mathbf{n}_2)(\mathbf{n}'_1\mathbf{n}_2'^*)]\right\}^{-1} \quad (3.36)$$

where

$$f^- = \frac{1}{2}\left\{B_1^2(|n'_{11}|^2 - |n'_{12}|^2) + B_2^2(|n'_{21}|^2 - |n'_{22}|^2)\right.$$
$$\left. - 2B_1B_2Re(\mathbf{n}_1^*\mathbf{n}_2)[n'_{11}n_{21}'^* - n'_{12}n_{22}'^*]\right\},$$
$$D = B_1^2 n'_{11}n_{12}'^* + B_2^2 n'_{21}n_{22}'^* - B_1B_2[n'_{11}n_{22}'^*(\mathbf{n}_1^*\mathbf{n}_2) + n'_{21}n_{12}'^*(\mathbf{n}_2^*\mathbf{n}_1)]$$

and n_{pq} is the projection of vector $\mathbf{n}_p$ onto the qth polarization unit vector. The vector of partial polarization in the scattered radiation $\mathbf{n}' = \cos\alpha\chi'_1 + e^{i\beta}\sin\alpha\chi'_2$ is determined by parameters α and β which satisfy relationship

$$tag\,\alpha e^{i\beta} = D\left\{[(f^-)^2 + |D|^2]^{1/2} - f^-\right\}^{-1} \qquad (3.37)$$

As in the procedure of the previous section, the integrated characteristics with respect to incident angles and the gamma quanta energy are obtained from equations (3.34) and (3.35).

Using these expressions, we analyze the general character of reflection radiation intensity and its polarization dependence on small variations in incidence angle close to the Bragg reflection. We consider the symmetric Bragg case ($b = -1$) for precision. Consider the case of a Hermitian matrix for system (2.26), i.e., neglect absorption. Then, it follows from the solution properties, that $B_i \equiv 1$ if η_{0i}, the root of the secular equation, is a complex quantity. That is, in the range of angles $\Delta\theta$ (or the corresponding values of parameter α) where the four roots of the secular equation are

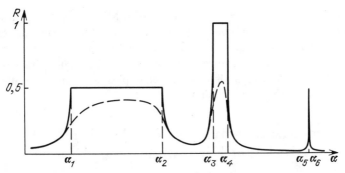

FIGURE 3.9. Qualitative reflection curve of an unpolarized beam for a perfect magnetically ordered crystal. Solid line indicates neglect of absorption, dashed line indicates the case with absorption. In the absence of absorption, angular interval $\alpha_3 < \alpha < \alpha_4$ corresponds to total reflection of any polarization.

complex quantities, the reflection coefficient of a wave with any polarization is unity. In the angular range where two roots are real and two are complex quantities, the reflection coefficient reaches unity for the eigenpolarization determined by the solution which attenuates in the crystal depth. In the range where four roots are real, the reflection coefficient for a wave of any polarization is less than unity. For large deviation angles from the Bragg condition the four roots η_{0i} are real and reflection coefficient R tends to zero. Thus, the typical reflection curve of unpolarized radiation has the form shown in Fig. 3.9. The ranges of α values $\alpha_1 < \alpha < \alpha_2$ and $\alpha_5 < \alpha < \alpha_6$ correspond to two real and two complex roots of the secular equation. In these ranges reflection coefficient $R \cong 1/2$. The reflected radiation is completely polarized and the polarization vector coincides with polarization vector $\mathbf{n}'$ for the eigensolution which attenuates in the crystal depth. In the range $\alpha_3 < \alpha < \alpha_4$, $R = 1$, and the reflected radiation is unpolarized. Figure 3.9 depicts the typical situation where the region of the total reflection of any polarization is separated from that of polarization selective reflection. Depending on the concrete form of matrix ϵ_T^N, positions may change for regions of complete and selective reflection; specifically, one or both regions of selective reflection may directly abut the region of total reflection. The angular dimensions of the strong reflection region $(\alpha_6 - \alpha_1)$ depend not only on the scattering amplitudes at a certain angle but also on the forward scattering amplitudes and are of the order of $|\alpha_6 - \alpha_1| \sim \max\{|\epsilon_T^N|, |\epsilon_0^0 - \epsilon_0^1|\}$.

The general shape of the reflection curve (three maxima) is preserved if we account for absorption. Everywhere, however, except probably some points, the reflection coefficient is smaller than in cases where absorption is absent (qualitative behavior of the reflection coefficient is shown by the dashed line in Fig. 3.9).

The present section was devoted to the general analysis of the diffraction problem for magnetic hyperfine splitting of a Mössbauer line in a crystal and quadrupole and combined hyperfine splitting (neglecting the explicit form of operator ϵ_T^N). Analysis of the obtained expressions for the general case is difficult; thus, we analyze the qualitative aspects of diffraction from magnetically ordered crystals on some examples admitting the analytical solution of the diffraction problem.

3.4.5   Polarization-Independent Amplitude of Scattering at Zero Angle

If directions $\mathbf{k}_0$ and $\mathbf{k}_1$ coincide with the high-symmetry crystallographic axes (e.g., three-fold or higher symmetry rotational axes), the scattering amplitude at zero angle in these directions is independent of polarization and the polarization of a wave scattered at zero angle coincides with the initial polarization.[5] That is, for the direct propagation of a wave in highly symmetric directions, any polarization is the eigenpolarization. In this case the eigenpolarization of the solutions of (2.26) is determined by the scattering amplitude from direction 1 to direction 2 and from direction 2 to direction 1 and is readily obtained from the form of these amplitudes. However, for a polarization-independent forward scattering amplitude, the set of four scalar equations (2.26) separate into two independent systems of two equations, if we use those unit vectors which simultaneously diagonalize operators $\hat{\epsilon}_T$ and $\hat{\epsilon}_{-T}$ as polarization unit vectors. Thus the search for eigenpolarizations reduces to the simultaneous diagonalization of two second-rank matrices describing 1–2 and 2–1 scattering. The matrices [see (3.21)]

$$\hat{T}_{\pm\tau} = \begin{pmatrix} T_{\pm\tau}^{11} & T_{\pm\tau}^{12} \\ \\ T_{\pm\tau}^{21} & T_{\pm\tau}^{22} \end{pmatrix}$$

are diagonalized in the polarization unit vectors determined by relationship

$$\mathbf{n} = \frac{\hat{S}\mathbf{n}}{|\hat{S}\mathbf{n}|}, \quad \mathbf{n}' = \frac{\hat{S}'\mathbf{n}'}{|\hat{S}\mathbf{n}|}, \quad \hat{S} = \hat{T}_\tau \hat{T}_{-\tau}, \quad \hat{S}' = \hat{T}_{-\tau} \hat{T}_\tau. \tag{3.38}$$

Equation (3.38) yields, e.g., the following expression for the eigenpolarization vector n in terms of the initial unit vectors of Eq. (3.21)

$$\mathbf{n}_{1,2} = [S_{12}\boldsymbol{\chi}_1 + (\xi_{1,2} - S_{11})\boldsymbol{\chi}_2][|S_{12}|^2 + |\xi_{1,2} - S_{11}|^2]^{-1/2} \tag{3.39}$$

where

$$\xi_{1,2} = (S_{11} + S_{22})/2 \pm [(S_{11} + S_{22})^2/4 + S_{12}S_{21} - S_{11}S_{22}]^{1/2}.$$

---

[5]In a cubic crystal with a center of inversion the forward scattering amplitude in any direction is independent of polarization.

The solution of the diffraction problem proceeds analogously to the case of an unsplit line. Note that in the general case matrix $\hat{S}$ is not Hermitian; thus the eigenpolarization vectors $\mathbf{n}_1$ and $\mathbf{n}_2$ defined by Eq. (3.39) are not orthogonal. Similar to that for an unsplit line, the eigenpolarizations do not vary in the region of diffraction reflection. This simplifies the analysis of polarization properties.

Now let the eigenpolarizations be orthogonal. Then, in the Bragg case of unpolarized radiational scattering from a thick crystal, the density matrix (3.35) averaged over the diffraction reflection region corresponds to partially polarized radiation along vectors $\mathbf{n}_1$ or $\mathbf{n}_2$, depending on which of the two scattering amplitudes from direction 1 to 2 is larger—that for polarization $\mathbf{n}_1$ or $\mathbf{n}_2$. The degree of polarization P is expressed in terms of matrix elements $T^{ii}_{\mathcal{T}} \equiv T^i_{\mathcal{T}}$ written in the basis of eigenpolarizations by the following formula

$$P = \frac{\left| |T^1_{\mathcal{T}}|\mathcal{F}^1 - |T^2_{\mathcal{T}}|\mathcal{F}^2 \right|}{|T^1_{\mathcal{T}}|\mathcal{F}^1 + |T^2_{\mathcal{T}}|\mathcal{F}^2} \qquad (3.40)$$

where $\mathcal{F}^j$ are the same quantities as in (3.33).

Expression (3.40) differs from the results obtained in the kinematical approximation [see (2.10)] where the radiation is partially polarized along the same vector but the degree of polarization is $(|T^1_{\mathcal{T}}|^2 - |T^2_{\mathcal{T}}|^2)/(|T^1_{\mathcal{T}}|^2 + |T^2_{\mathcal{T}}|^2)$. As follows from (3.35), for nonorthogonal eigenpolarizations the vector which describes partial polarization differs from vectors $\mathbf{n}'_1$ and $\mathbf{n}'_2$. Integrated polarization characteristics are obtained from expressions (3.35) and (3.36). The system of dynamical equations is solvable also if forward scattering amplitudes depend on polarization and polarization vectors $\mathbf{n}_j$ and $\mathbf{n}'_j$ determined from (3.38) coincide with eigenpolarizations vectors for direct transmission. Such a situation may occur with special mutual orientations of $\mathbf{k}_0$ and $\mathbf{k}_1$ and directions characterizing crystal magnetic structure; for magnetic maxima in an antiferromagnetic crystal this situation may occur if an antiferromagnetic axis is in the plane $\mathbf{k}_0$, $\mathbf{k}_1$, normal to it or the difference between the azimuthal angles of wave vectors $\mathbf{k}_0$ and $\mathbf{k}_1$ measured around the antiferromagnetic axis is $\pi/2$. In the two former cases eigenpolarizations are π and σ polarizations, whereas in the latter eigenpolarizations are linear but given by unit vectors (3.5).

3.4.6 Completely Resolved Zeeman Splitting of Mössbauer Line

Analysis of the general expressions given in Sect. 3.4.4 and Sect. 3.4.5 is simplified if Mössbauer scattering proceeds through one or two Zeeman transitions. The transitions are considered different not only for different energies but also for equal energies of Zeeman transitions for degenerate levels (e.g., for quadrupole splitting). If the unit cell of a crystal has several

Mössbauer nuclei, the transitions in nuclei of different positions in the unit cell are also considered different transitions independent of their energies. Thus, in an antiferromagnetic with two magnetic atoms per unit cell and completely resolved Zeeman splitting, each line of the Mössbauer spectrum has contributions from the transitions of nuclei at sites $\mathbf{H}$ and $-\mathbf{H}$, i.e., two different transitions per line.

Consider now the effect of the suppression of inelastic channels [53]. For one Zeeman transition the matrix element in system (2.26) (see also (3.21)) is $\epsilon_T^{pq} \sim \mathbf{n}_t(\mathbf{k}_n)_q \mathbf{n}_t(\mathbf{k}_m)_p e^{i(\mathbf{k}_n - \mathbf{k}_m)\mathbf{r}}$ where r determines nuclear position which makes the columns of matrix $\hat{T}$ linearly dependent on one another since they differ by a common factor. The consequence of such a linear dependence is that the secular equation at $\alpha = 0$ has nonzero coefficients before η_0^4 and η_0^3 i.e. three roots are equal to zero. This means that for $\alpha = 0$ the three corresponding eigensolutions do not attenuate and inelastic channels are suppressed completely. If scattering proceeds through two different transitions, each element of matrix T includes two terms of the type above. The linear dependence of columns is such that for $\alpha = 0$ the secular equation has, in addition to η_0^4 and η_0^3, also η_0^2. Consequently, two roots of the equation are zero and two eigensolutions do not attenuate. The two zero roots are due to the fact that either gamma quanta of some polarization does not interact with nuclei or, the complete suppression of inelastic channels, $\alpha = 0$, takes place independently of nuclear transition multipolarity. A similar analysis shows that for three transitions at $\alpha = 0$ only one root is zero and the suppression occurs only for the polarization corresponding to the solution with zero eigenvalue. For four transitions the suppression effect should not occur in the general case, i.e., there is no such superposition of waves E_0 and E_1 for which the amplitude of the excited nucleus formation is zero in the transitions considered.

Notice that for complete Zeeman splitting of a Mössbauer line the suppression effect, as opposed to the case of an unsplit line, may take place not only for such a field configuration in a crystal whose strength at the nuclear site goes to zero, and for a configuration with nonzero field at the nucleus which does not interact with the nucleus due to polarization selection rules. The analogous suppression of absorption (Borrmann effect) is known for cholesterics (Chapt. 4).

For a split line and partially overlapping components, the interesting interference effects may be observed in the interaction between different kinds of scattering in the crystal unit cell. In addition to the interference between Rayleigh and nuclear scattering there may be interference between nuclear scattering proceeding through different Zeeman transitions [22] (Fig. 3.10). Omitting specific structures, we state that for resolved hyperfine splitting, the diffraction problem is solved analytically for several scattering geometries in an antiferromagnet [3]. Moreover, in some interesting physical situations the secular equation reduces to biquadratic with respect to parameter η_0 [3]. That case, not examined here since it is analogous to light diffraction

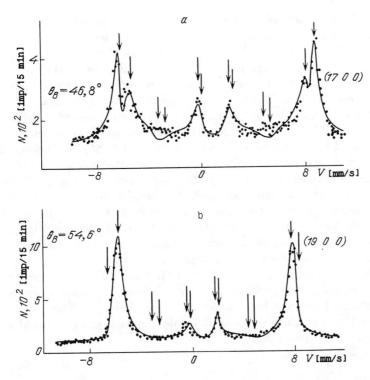

FIGURE 3.10. Spectra of nuclear (magnetic) reflection of Fe_3BO_6 crystals for (a) constructive (reflection (1700)) and (b) destructive ((1900) reflection) interference of nuclear scattering at Fe^{57} nuclei occupying nonequivalent crystallographic sites [22].

in a cholesteric with oblique incidence (see Chapt. 4), and the corresponding results are readily applied to Mössbauer diffraction.

In conclusion, we note that there has been a significant improvement in the state of experimental art Mössbauer diffraction studies in recent years. Thus, it is possible now to investigate not only qualitative theoretical predictions [18–23, 67–75] but also very fine (including dynamical) effects of the coherent interaction between Mössbauer radiation and crystals [76–96].

The magnetic structure of Fe_3BO_6 and also its EFG structure were investigated by the means of Mössbauer diffraction [85–89]. Detailed theoretical and experimental studies have been initiated to examine coherent inelastic Mössbauer scattering [90–96]. Mössbauer scattering, particularly, coherent forward scattering, was used to investigate relaxation phenomena in magnetically ordered compounds [93] and to measure the line width of long-living isomer nuclear levels [94]. A review of experimental investigations of coherent collective nuclear excitation in a crystal performed with convention of Mössbauer sources is presented in [96]. We note, however, that the most promising investigations in this direction and in some other Mössbauer diffraction problems are those in which Mössbauer experiments are performed with synchrotron radiation (see Chapt. 10). The corresponding investigations are in the state of fast development and one should await new interesting results here.

References

1. R.L. Mössbauer: Z. Phys. **151**, 124 (1985).

2. R.L. Mössbauer: Naturwissenchaften **45**, 538 (1958).

3. V.A. Belyakov: UFN **115**, 553 (1975) [Sov. Phys. -Usp. **18**, 267 (1975)].

4. G Albanese, A. Deru: Riv. Nuovo Cimento Serie 3, **2**, No. 9, 1 (1979).

5. D.C. Champeney: Rep. Prog. Phys. **42**, 1017 (1979).

6. P.J. Black, P.B. Moon: Nature **188**, 481 (1960).

7. P.J. Black, D.E. Evans, D.A. O'Connor: Proc. Roy. Soc. **A270**, 168 (1962).

8. F. Parak, R. Mössbauer, U. Biebl: Z. Phys. **244**, 456 (1971).

9. R. Mössbauer, F. Parak, W. Hoppe in: *Mössbauer Spectroscopy II*, ed. U. Gonser (Berlin, Heidelberg, Springer-Verlag, 1981).

10. Yu. Kagan, A.M. Afanas'ev, in: *Mössbauer Spectroscopy and Its Application* (Vienna, IAEA, 1972) p. 143.

11. M.A. Andreeva, R.N. Kuz'min: *Mössbauer Gamma Optics* (Izd-vo MGU, 1982) (in Russian).

12. V.A. Belyakov, Yu.M. Aivazyan: Pis'ma v ZhETF **7**, 477 (1968) [JETP Lett. **7**, 368 (1968)].

13. V.A. Belyakov, Yu.M. Aivazyan: Pis'ma v ZhETF **9**, 637 (1969) [JETP Lett. **9**, 393 (1969)].

14. M.A. Andreeva, R.N. Kuz'min: DAN SSSR. **185**, 1282 (1969) [Sov. Phys. - Doklady **14**, 298 (1969)].

15. G.V. Smirnov, V.V. Sklyarevskii, R.A. Voskanyan, A.N. Artem'ev: Pis'ma v ZhETF **9**, 123 (1969) [JETP Lett. **9**, 70 (1969)].

16. R.M. Mirzababaev, G.V. Smirnov, V.V. Sklyarevskii, A.N. Artem'ev, A.N. Izrailenko, A.V. Bobkov: Phys. Lett. **A37**, 441 (1971).

17. V.S. Zasimov, R.N. Kuz'min, A.Yu. Aleksandrov, A.I. Firov: Pis'ma v ZhETF **15**, 394 (1972) [JETP Lett. **15**, 277 (1972)].

18. H.F. Maurus, Van Bürck, G.V. Smirnov, R.L. Mössbauer: J. Phys. C: Solid State Phys. **17**, 1991 (1984).

19. V. Van Bürck, H.F. Maurus, G.V. Smirnov, R.L. Mössbauer: J. Phys. C: Solid State Phys. **17**, 2003 (1984).

20. G.V. Smirnov, Yu.V. Shvyd'ko, E. Realo: Pis'ma v ZhETF **39**, 33 (1984) [JETP Lett. **39**, 41 (1984)].

21. P.P. Kovalenko, V.G. Labushkin, A.K. Ovsepyan, E.R. Sarkisov, E.V. Smirnov: Pis'ma v ZhETF **39**, 471 (1984) [JETP Lett. **39**, 573 (1984)].

22. P.P. Kovalenko, V.G. Labushkin, A.K. Ovsepyan, E.R. Sarkisov, E.V. Smirnov, I.G. Tolpekin, ZhETF **88**, 1336 (1985) [Sov. Phys. - JETP **61**, 793 (1985)].

23. P.P. Kovalenko, V.G. Labushkin, A.K. Ovsepyan, E.R. Sarkisov, E.V. Smironv, A.R. Propkopov, V.N. Seleznev: Fiz. Tverd. Tela **26**, 3068 (1984).

24. *Mössbauer Spectroscopy II* ed. by U. Gonser (Berlin, Heldelberg, Springer-Verlag 1981).

25. Yu.M. Aivazyan, V.A. Belyakov: FTT **13**, 968 (1971) [Sov. Phys. - Solid State **13**, 808 (1971)].

26. A.N. Artem'ev, V.A. Kabanik, Yu.N. Kazakov, G.N. Kulipanov, V.A. Moleshko, V.V. Sklyarevskii, A.N. Skrinsky, E.P. Stepanov, V.B. Khlestov, A.I. Chechin: Nucl. Instrum. Methods **152**, 235 (1978).

27. A.I. Chechin, N.V. Andronov, M.V. Zelepukhin, A.N. Artem'ev, E.P. Stepanov: Pis'ma v ZhETF **37**, 531 (1983) [JETP Lett. **37**, 633 (1983)]; G.V. Smirnov, Yu.V. Svyd'ko: ibid **44**, 431 (1986) [JETP Lett. **44**, 556 (1986)].

28. E. Gerdau, R. Rüffer, H. Winkler, W. Tolksdorf, C.P. Klages, J.P. Hannon: PHys. Rev. Lett. **54**, 835 (1985).

29. V.A. Belyakov, R.Ch. Bokun: Izv.AN SSSR, Ser. Fiz. **36**, 1476 (1972) (in Russian).

30. V.A. Belyakov, Yu.M. Aivazyan: Phys. Rev. B. **1**, 1903 (1970).

31. V.A. Belyakov: ZhETF **54**, 1366 (1968) [Sov. Phys. - JETP **27**, 622 (1968)].

32. V.A. Belyakov, V.P. Orlov: ZhETF **56**, 1366 (1969) [Sov. Phys. JETP **29**, 733 (1969)].

33. G.T. Tremmel: Phys. Rev. **126**, 1045 (1962).

34. M.K.F. Wong: Phys. Rev. **149**, 378 (1966).

35. V.A. Belyakov, in: Proc. 12th Int. cong. on Low Temp. Physics, Kyoto, Japan, 1970, p. 727.

36. Yu.M. Aivazian, V.A. Belyakov: ZhETF **56**, 346 (1969) [Sov. Phys.- JETP **29**, 191 (1969)].

37. D.T. Keating: Phys. Rev. **178**, 732 (1969).

38. M.A. Andreeva, R.N. Kuz'min: Kristallograpfiya **18**, 407 (1973) [Sov. Phys. - Crystallogr. **18**, 645 (1973)].

39. I.P. Perstnev, F.N. Chukhovskii: FTT **16**, 3011 (1974) [Sov. Phys. - Solid State, **16**, 1946 (1975)].

40. A.G. Grygoryan, dV.A. Belyakov. Vestnik MGU Ser. 3 (Physics, Astronomy) No. 6, 668 (1971) (in Russian); H. Winkler, R. Eeisberg, E. Alp et al.: Z. Phys. **49**, 331 (1983).

41. V.A. Belyakov, Yu.M. Aivazyan, Trudy VNIIFTRI vyp. 15(45) Moscow (1972), p. 34 (in Russian).

42. Yu.M. Aivazyan, V.A. Belyakov: ZhETF **56**, 346 (1969) [Sov. Phys. - JETP **29**, 191 (1969)].

43. V.A. Belyakov: FTT **13**, 2170 (1971) [Sov. Phys. - Solid State **13**, 1824 (1972)].

44. A.V. Kolpakov, E.N. Ovchinnikova, R.N. Kuz'min: Phys. Stat. Sol. 93, 511 (1979).

45. A.M. Afanas'ev, Yu.M. Kagan: ZhETF 48, 327 (1965) [Sov. Phys. - JETP 21, 215 (1965)].

46. Yu.M. Kagan, A.M. Afanas'ev: ZhETF 49, 1504 (1965) [Sov. Phys. - JETP 22, 1032 (1966)].

47. B.W. Batterman, H. Cole: Rev. Mod. Phys. 36, 681 (1964).

48. Z.G. Pinsker: *Dynamical Scattering of X-rays in Crystals* (Berlin, Springer-Verlag, 1978).

49. Yu.M. Kagan, A.M. Afanas'ev, I.P. Perstnev: ShETF 54, 1530 (1968) [Sov. Phys. - JETP 27, 819 (1968)].

50. V.E. Dmitrienko, V.A. Belyakov: ZhETF 73, 681 (1977) [Sov. Phys. - JETP 46, 356 (1977)].

51. V.A. Belyakov: FTT 13, 3320 (1971) [Sov. Phys. - Solid Stae 13, 2789 (1971)].

52. V.A. Belyakov, E.V. Smirnov: ZhETF 68, 608 (1975) [Sov. Phys. - JETP 41, 301 (1975)].

53. A.M. Afanas'ev, Yu. Kagan: ZhETF 64, 1958 (1973) [Sov. Phys. - JETP 37, 987 (1973)].

54. A.M. Afanas'ev, I.P. Perstnev: ZhETF 65, 1271 (1973) [Sov. Phys. - JETP 38, 630 (1974)].

55. V.A. Belyakov, R.Ch. Bokun: Acta Crystallogr. A31, 737 (1975).

56. P.J. Black, G. Longworth, D.A. O'Connor: Proc. Phys. Soc. 83, 925 (1964).

57. An.N. Artem'ev, I.P. Persnev, V.V. Sklyarevskii, G.V. Smirnov, E.P. Stepanov: ZhETF 64, 261 (1973) [JETP Lett. 15, 226 (1972)].

58. E.V. Zolotoyabko, E.M. Iolin: *Coherent Rayleigh Scattering of Mössbauer Radiation* (Riga, Zinatne, 1986) (in Russina).

59. O.L. Zhizhimov, I.V. Khriplovich: ZhETF 87, 547 (1984) [Sov. Phys. - JETP 60, 313 (1984)].

60. F. De Bergevin, M. Brunel: Phys. Lett. A39, 141 (1972); Acta Crystallogr. A37, 314 (1981).

61. N.F. Fallev, A.A. Lomov, V.G. Iabushkin: Acta Crystallogr. 37A, Suppl. p. C-374 (1981).

62. M. Brunel, F. De Bergevin: Acta Crystallogr. **37**, 324 (1981); Acta Crystallogr. **39**, 84 (1983).

63. D. Gibbs, D.E. Moncton, K.L.D'Amico, J. Bohr, B.H. Grier: Phys. Rev. Lett. **55**, 234 (1985).

64. D. Gibbs, D.E. Moncton, K.L. D'Amico: J. Appl. Phys. **57** (No. 8, pt. 2A) 3619 (1985).

65. M. Blum: J. Appl. Phys. **57**, (No. 8, pt. 2A) 3615 (1985).

66. P.N. Platzman: J. Appl. Phys. **57**, (No. 8, pt. 2) 3623 (1985).

67. R.N. Platzmann, N. Tzoar: Phys. Rev. B. **2**, 3556 (1970).

68. Van Bürck, G.V. Smirnov, Mössbauer et al.: J. Phys. C: Solid State Phys. **11**, 2305 (1978).

69. G.V. Smirnov, Yu.V. Shvyd'ko: Pis'ma v ZhETF **34**, 409 (1982) (in Russian).

70. G.V. Smirnov and A.I. Chumakov: ZhETF **89**, 1169 (1985). [Sov. Phys. - JETP **62**, 673 (1985)].

71. I.G. Tolpekin, P.P. Kovalenko, V.G. Labushkin et al.: Pis'ma v ZhETF **43**, 474 (1986) (in Russian).

72. G.V. Smirnov, Yu.V. Shvyd'ko, U. Van Bürck et al.: Phys. Stat. Sol. (b) **134**, 465 (1986).

73. U. Van Bürck, G.V. Smirnov, H.J. Maurus: J. Phys. C: Solid Sate Phys. **19**, 2557 (1986).

74. U. Van Bürck, G.V. Smirnov, R.L. Mössbauer: J. Phys. C: Solid state Phys. **19**, 2567 (1986).

75. G.V. Smirnov: Hyp. Int. **27**, 203 (1986).

76. Yu. Kagan, A.M. Afanas'ev, V.G. Kohn: Phys. Lett. **68A**, 339 (1978).

77. G.T. Trammel, J. Hannon: Phys. Rev. B. **18**, 165 (1978).

78. Yu. Kagan, A.M. Afanas'ev, V.G. Kohn: J. Phys. C: Solid State Phys. **12**, 615 (1979).

79. V.A. Belyakov, R.Ch. Bokun: Acta Crystallogr. **A31**, 737 (1975).

80. G.P. Hannon, N.V. Hung, G.T. Trammel: Phys. Rev. B. **32**, 5068, 4081 (1985).

81. E.V. Smirnov, V.A. Belyakov: ZhETF **79**, 883 (1980) (in Russian).

82. U. Van Bürck: Hyp. Int. **27**, 219 (1986).

83. V.A. Belyakov, Yu.M. Aivazyan: Trans. of 6th All-Union Conf. on the Use of Synchrotron Radiation (SI-84), Novosibirsk, Inst. of Nuclear Physics of the Siberian Branch of hte USSR Acad. Sci., 318 (1984) (in Russian).

84. V.A. Belyakov: UFN **151**, 7699 (1987) [Sov. Phys. - Usp. **30**, 331 (1987)].

85. I.G Tolpekin, P.P. Kovalenko, V.G. Labvushkin, E.R. Sarkisov: *Interaction of Mössbauer Radiation with the Matter*, ed. by R.N. Kuz'min (Moscow, Moscow University, 1987) p. 133.

86. I.G. Tolpekin, P.P. Kovalenko, V.G. Labushkin, E.N. Ovchinnikova, E.R. Sarkisov, E.V. Smirnov: Sov. Phys. - JETP **67**, 404 (1988).

87. I.G. Tolpekin, V.G. Labushkin, E.N. Ovchinnikova, E.V. Smirnov, Ya. A. Sornikov: Phys. Lett. **A147**, 323 (1990).

88. P.P. Kovalenko, V.G. Labvushkin, E.R. Sarkisov, I.G. Tolpekin: Fiz. Tverd. Tela **29**, 593 (1987).

89. I.G. Tolpekin, V.G. Labushkin, E.N. Ovchinnikova, E.V. Smirnov, Pisma v Zh. Tekh. Fiz. **14**, 2024 (1988).

90. V.A. Belyakov, Yu. M. Aivazian: Nuc. Instrum. Methods A **282**, 628 (1989).

91. V.A. Belyakov: Abstr. Latin American Conf. Appl. Mössbauer Effect, Havana, Cuba, p. 8.3 (1990).

92. S.L. Popov, G.V. Smirnov, Yu.V. Shvyd'ko: Hyp. Int. **58**, 2463 (1990).

93. N.V. Zelepukhin, V.E. Sedov, G.V. Smirnov, V.N. Dubinin, O.N. Razumov: Pisma v JETP **49**, 143 (1989); Hyp. Int. **56**, 1507 (1990).

94. Yu.V. Shvyd'ko, G.V. Smirnov: Nuc. Instrum. Methods in Phys. Res. **B51**, 452 (1990).

95. Yu. V. Shvyd'ko, G.V. Smirnov, S.L. Popov, and T. Herfrich: Pisma v JETP **53**, 63 (1991).

96. Yu. V. Shvyd'ko, G.V. Smirnov: J. Phys.: Condens. Matter **1** 10563 (1989).

4

Optics of Chiral Liquid Crystals

In Chapt. 3 we considered the interaction between Mössbauer (short wave) radiation and crystals within the framework of diffraction theory. Here we demonstrate that the same approach describes the interaction between optical (long wave) radiation and periodic media of complex structure, specifically, chiral liquid crystals.

Different types of liquid crystals have very interesting and useful applications which exploit their optical properties [4–5]. The most unusual and sophisticated optical properties are found in chiral liquid crystals [4,5] the main representatives of which are cholesteric liquid crystals (or cholesterics), chiral smectic liquid crystals and the blue phases of liquid crystals.

Although these types of liquid crystals have different structures and properties, all consist of mirror-asymmetric (chiral) molecules and exhibit spatial periodicity with periods generally in the range of visible light wavelengths. This explains the observation that the periodicity of these liquid crystals manifests itself most prominently in their optical properties, typically those due to light diffraction.

4.1   Optics of Cholesteric Liquid Crystals (CLC)

In Chpts. 1,2 we presented the optical properties of CLCs in the general discussion of periodic media optics. Those properties were considered in the context of the exact solution of the Maxwell equations for light propagating along the optical axis and within the kinematical approximation for an arbitrary propagation direction.

Since the purpose of those considerations was to emphasize the general features of periodic media optics, only some of the interesting properties of CLC optics were mentioned. Here we present a detailed consideration of the problem and begin with a description of the salient facts of CLC optics.

Cholesteric liquid crystals selectively reflect light of definite polarization and wavelength. For example, when a light beam is incident on a CLC along its optical axis, the beam component with some circular polarization is reflected in a relatively narrow frequency band width while the other (orthogonal) polarization penetrates into the crystal without reflection. The color of a CLC depends on the angle at which one looks at the crystal. The optical pecularities of CLCs are connected with their structure shown in Fig. 1.3 for a single crystal sample; the long axes of cholesteric molecules are shown as oblong marks.

It was observed that the light polarization planes rotate by several thousand degrees per millimeter in thin layers; this is far above the natural optical activity of molecules. The wavelength at which these anomalous events occur changes in the presence of an external field (electric or magnetic) or temperature variance.

In any plane perpendicular to the optical axis Z, all molecules are similarly oriented, however, but there is no long-range order at the molecular centers of gravity. The orientation of the molecule's axes varies with coordinate Z as

$$\varphi(z) = \frac{2\pi z}{P} \qquad (4.1)$$

where $\varphi(z)$ is the angle of molecular rotation around the Z axis. The quantity P is the period or pitch of a cholesteric helix—at that distance the molecular axis rotates by 2π. The value of P is usually several thousand angstreams.

The structure shown in Fig. 1.3 of cholesteric crystals is idealized . Specifically, for a fixed value of Z, the orientations of individual molecules may slightly differ from that determined by (4.1), which gives the Z dependence of the mean orientation of molecules. This direction is usually described by unit vector $\mathbf{n}$, the director. The degree of orientational ordering of molecules is characterized by the order parameter [1].

$$S = (3\langle\cos^2\theta\rangle - 1)/2 \qquad (4.2)$$

where $\langle\cos^2\theta\rangle$ is the mean-square cosine of the angle of deviation of the molecular long axis from the director. Hence, equation (4.1) describes the spatial variation of the director's orientation which, at $S \neq 1$, coincides with average molecular orientation.

The principal values of the CLC dielectric permittivity tensor $\hat{\epsilon}(\mathbf{r})$ (1.6) are determined by molecular dielectric anisotropy and depend on the order parameter, so that in an isotropic liquid with $S = 0$ the parameter δ (dielectric anisotropy) vanishes [6]. Development of a quantitative theory of CLC dielectric permittivity, and other types of liquid crystals encounters difficulties [6]. Usually it is assumed that the anisotropy δ is proportional to S and molecular anisotropy, while the proportionality factor remains without precise determination [1].

4.1.1   The Fundamental Equations

Consider light propagation in a CLC at an arbitrary angle to the CLC optical axis. This case is difficult for analysis because there is no exact solution to the Maxwell equations. Thus, in a series of papers, the solutions for light propagation in an arbitrary direction with respect to cholesteric axis were analyzed in general, whereas the solutions were obtained by numerical calculations for specific values of the parameters (see references in [4]). However, there is a small parameter, the anisotropy of dielectric properties δ, which makes it possible to use the simple and exact theory described in Chapter 2.

Oblique incidence in CLC optics qualitatively differ from that of normal incidence (propagation along the optical axis); there are higher order reflections at frequencies which are multiples of the Bragg frequency; also, first-order reflection differs from that of normal incidence: a region appears in which any polarization is reflected and the polarization properties of solutions become more complex. The physical cause for this is that when light propagates at an angle to the helical axis we must account for both the diffractive scattering of light with the polarization given by (2.21) and birefringence which affects the polarization properties of the solutions. Hence, the results of kinematic theory, according to which diffraction scattering occurs for light whose polarization is defined by (2.21), becomes invalid for thick crystals and the diffraction pattern of light obliquely incident on a CLC becomes more complicated.

The more exact description of CLC light diffraction in the range of the first-order reflexes, (considered within the kinematic theory in Chapter 2), requires that we use the system of dynamic equations in the two-wave approximation (2.26) and substitute into it the explicit form of the CLC dielectric tensor (1.6). To solve the system (2.26) it is convenient to reduce it to a form which is used routinely in the dynamic theory of diffraction:

$$[1 - k_0^2/q^2 + (\delta/2)\cos^2\theta]E_0^\sigma - (\delta/2)\,E_1^\sigma + (i\delta/2)\sin\theta E_1^\pi = 0$$

$$[1 - k_0^2/q^2 - (\delta/2)\cos^2\theta]E_0^\pi - (i\delta/2)\sin\theta E_1^\sigma - (\delta/2)\sin^2\theta \cdot E_1^\pi = 0$$

$$- (\delta/2)\,E_0^\sigma + (i\delta/2)\sin\theta \cdot E_0^\pi + [1 - k_1^2/q^2 + (\delta/2)\cos^2\theta]\cdot E_1^\sigma = 0$$

$$-(i\delta/2)\sin\theta \cdot E_0^\sigma - (\delta/2)\sin^2\theta \cdot E_0^\pi + [1 - k_1^2/q^2 - (\delta/2)\cos^2\theta]\cdot E_1^\pi = 0 \quad (4.3)$$

where E^σ and E^π are the σ and π components of amplitudes $\mathbf{E}_0$ and $\mathbf{E}_1$, respectively, and $q = (\omega/c)\,[\bar\epsilon(1 - \frac{\delta}{2}\cos^2\theta)]^{1/2}$ is the mean value of the vector of the wave propagating at an angle $\frac{\pi}{2} - \theta$ to the optical axis.

Equating the determinant of (4.3) to zero, yields the conditions for the system to be solvable in a form similar to (1.14):

$$t^4 - (2\Delta^2 + 2m^2 - 1)t^2 + \Delta^4 - (2m^2 + 1)\Delta^2 + 2m^2\Delta = 0 \quad (4.4)$$

where we assume

$$t = (k_1^2 - k_0^2)[q^2\delta(1 + \sin^2\theta)]^{-1}, \quad m = \cos^2\theta(1 + \sin^2\theta)^{-1}$$
$$\Delta = (2q^2 - k_0^2 - k_1^2)[q^2\delta(1 + \sin^2\theta)]^{-1}. \quad (4.5)$$

4.1.2   Eigensolutions

We use the method described in Sec. 2.3 of Chapter 2 to solve the boundary problem and we consider the sample a plane-parallel layer as above. When the optic axis is perpendicular to the CLC surface, expressions (2.28–2.30) show that the parameter in (4.4) coincides within α, to within a factor, and is associated with the deviation of the incidence angle (or wavelength) from the value given by the Bragg condition (2.4):

$$\Delta = 2\sin\theta(2q\sin\theta - \tau)[\delta q(1 + \sin^2\theta)]^{-1}. \qquad (4.6)$$

The quantity Δ is a convenient parameter to describe the boundary problem and the CLC optical properties. We must find the dependence of all optical parameters, particularly eigenvalue dependence on Δ. The dependence of Δ on t is given by (4.4) as

$$t_j = \pm\left\{\Delta^2 + m^2 - 1/2 \pm [(\Delta^2 + m^2 - 1/2)^2\right.$$
$$\left. + (2m^2 + 1)\cdot\Delta^2 - 2m^2\Delta - \Delta^4]^{1/2}\right\}^{1/2} \qquad (4.7)$$

Solutions t_j are enumerated as in the case of normal incidence (see (1.15)), that is, if the sign in the square brackets is "plus", then $j = 1, 4$ respectively for signs "plus" and "minus" in the braces; if the sign in the square brackets is "minus", then $j = 2, 3$ respectively for signs "plus" and "minus" in the braces. The value of t given by (4.7) is similar to η of Eq. (1.14).

The eigensolutions of (4.3) are expressed by t_j

$$\mathbf{E}_j(\mathbf{r}, t) = e^{-i\omega t}(\mathbf{E}_{0j}e^{i\mathbf{k}_{0j}\mathbf{r}} + \mathbf{E}_{1j}e^{i\mathbf{k}_{1j}\mathbf{r}}) \qquad (4.8)$$

where $\mathbf{E}_{0j} = E_{0j}^{\sigma}\cdot\boldsymbol{\sigma} + E_{0j}^{\pi}\boldsymbol{\pi}$, $\mathbf{E}_{1j} = E_{1j}^{\sigma}\boldsymbol{\sigma} + E_{1j}^{\pi}\boldsymbol{\pi}_1$,

$$E_{0j}^{\sigma} = a_{1j} = (\Delta - m + t_j)[(\Delta - t_j)^2 - m^2],$$
$$E_{0j}^{\pi} = a_{2j} = -i\sin\theta(\Delta + m + t_j)[(\Delta - t_i)^2 - m^2] \qquad (4.9)$$
$$E_{1j}^{\sigma} = a_{3j} = (\Delta - m - t_j)(\Delta + t_j - m^2)$$
$$E_{1j}^{\pi} = a_{4j} = -i\sin\theta(\Delta + m - t_j)(\Delta + t_j - m^2)$$
$$\mathbf{k}_{0j} = \mathbf{q} + qs\delta(\Delta + m + t_j)[2(1 + m)\sin\theta]^{-1}, \quad \mathbf{k}_{1j} = \mathbf{k}_{0j} + \boldsymbol{\tau}. \quad (4.10)$$

We note that equation (4.10) was obtained using the boundary conditions for wave vectors (the continuity of their tangential components for the case when the CLC surface is perpendicular to the CLC optical axis). Here $\mathbf{q}$ is the incident wave vector, and $\mathbf{s}$ is the vector normal to the crystal surface.

The eigensolutions of (4.8) are the superpositions of two waves with wave vectors $\mathbf{k}_{0j}$ and $\mathbf{k}_{1j}$, the polarization of which are generally elliptical.

The frequency dependence of polarization parameters for waves 1 and 2 are presented in Fig. 4.1. In contrast to normal incidence in which eigenpolarizations are circular, eigenpolarizations in oblique incidence are elliptical and vary noticeably within the selective reflection band range [8].

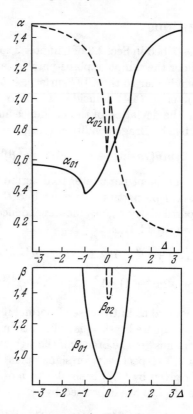

FIGURE 4.1. Frequency (angular) dependences of eigensolution polarization parameters for the case of oblique light propagation with respect to the cholesteric axis. The parameters α_j and β_j describe the eigenpolarization vectors in the representation $n_j = \overline{\sigma} \cos \alpha_j + \overline{\pi} \sin \alpha_j \exp(i\beta_j)$. The orientation of the polarization ellipse with respect to $\overline{\sigma}$ is given by ψ, and the ratio of its axes is given by $b = tg\,\alpha'$ where $tg\,2\psi_j = tg\,2\alpha_j \cos \beta_i$; $\sin 2\alpha' = \sin 2\alpha_j \sin \beta_j$. The calculations were done for the following values of the parameters: $\theta = \pi/4$, $\overline{\epsilon} = 2.445$, $\delta = 0.05$, $p = 0.78\,\mu$m.

The rocking curves (angular dependence of the reflection coefficient) for linear polarizations have plateau-like shapes as in X-ray diffraction, the asymmetry is a manifestation of eigenpolarization angular (frequency dependence) which exhibits maximal diffractive reflection (see also the analogous measured frequency dependences at Fig. 4.11).

4.1.3   The Solution of the Boundary Problem

The general solution to (4.3) is the superposition of eigensolutions of type (2.27). If the dielectric constant of the medium adjacent to the CLC boundary is equal to the CLC mean dielectric constant, the problem is solved as above.

First, we find the amplitudes of the reflected and transmitted waves. Let a wave of amplitude E^e and polarization $\mathbf{e}$ be incident from outside the sample surface located at $z = z_1$ and

$$\mathbf{E}^e = \mathbf{e} E^e e^{i(\mathbf{q r} - \omega t)} \tag{4.11}$$

where

$$\mathbf{e} = \boldsymbol{\sigma} \cos\alpha + \boldsymbol{\pi}_0 \sin\alpha e^{i\beta}. \tag{4.12}$$

We seek the amplitudes of the reflected and transmitted waves in the form $\mathbf{E}^r = E^r_\sigma \boldsymbol{\sigma} + E^r_\pi \boldsymbol{\pi}_1$ and $\mathbf{E}^t = E^t_\sigma + E^t_\pi \boldsymbol{\pi}_0$ respectively, while the field within the sample has form (2.27). The continuity condition of components of $\mathbf{E}$ and $\mathbf{H}$ on the crystal's surface produces a system of equations for E^r_σ, E^r_π, E^t_σ and E^t_π. The boundary conditions for the fields in oblique incidence are as Eq. (1.17), except that only tangential components must be equated in this case. After simple and cumbersome calculations we obtain, from (1.17) the following system of equations:

$$\sum_j a_{1j} C'_j = E^e_\sigma, \quad \sum_j a_{2j} C'_j = E^e_\pi$$

$$\sum_j a_{3j} C'_j e^{i\Delta k_j L} = 0, \quad \sum_j a_{4j} C'_j e^{i\Delta k_j L} = 0$$

$$E^r_\sigma = \sum_j a_{3j} C'_j e^{i\tau z_1}, \quad E^r_\pi = \sum_j a_{4j} C'_j e^{i\tau z},$$

$$E^t_\sigma = e^{iqsL} \sum_j a_{1j} C'_j e^{i\Delta k_j L};$$

$$E^t_\pi = \sum_j e^{iqsL} a_{2j} C'_j e^{i\Delta k_j L} \tag{4.13}$$

where $C'_j = C_j e^{ik_{0j}}$ s and z_1 of equal size and Δk_j is the diffractive correction to the wave vector as defined by (4.10) and a_{tq} are given by (4.9). The following expressions for the amplitudes of reflected and transmitted waves are obtained from Eq. (4.13).

$$E^r_\sigma = (E^e/D) D^r_\sigma = (E^e/D) \begin{vmatrix} a_{11} & a_{12} & a_{13} & a_{14} & \cos\alpha \\ a_{21} & a_{22} & a_{23} & a_{24} & e^{i\beta}\sin\alpha \\ \gamma_1 a_{31} & \gamma_2 a_{32} & \gamma_3 a_{33} & \gamma_4 a_{34} & 0 \\ \gamma_1 a_{41} & \gamma_2 a_{42} & \gamma_3 a_{43} & \gamma_4 a_{44} & 0 \\ a_{31} & a_{32} & a_{33} & a_{34} & 0 \end{vmatrix}$$

$$(4.14)$$

$$D = \begin{vmatrix} a_{11} & a_{12} & a_{13} & a_{14} \\ a_{21} & a_{22} & a_{23} & a_{24} \\ \gamma_1 a_{31} & \gamma_2 a_{32} & \gamma_3 a_{33} & \gamma_4 a_{34} \\ \gamma_1 a_{41} & \gamma_2 a_{42} & \gamma_3 a_{43} & \gamma_4 a_{44} \end{vmatrix}$$

$$\gamma_j = e^{i\ell(\Delta + t_j + m)}, \quad \ell = \delta q L[2(1+m)\sin\theta_B]^{-1} \qquad (4.15)$$

where L is crystal length.

The other components of $\mathbf{E}^t$ and $\mathbf{E}^r$ are determined by (4.14) if we substitute the first four elements of the last row in matrix D^r_σ by a_{4j} for D^r_π, $\gamma_j a_{1j}$ for D^t_σ and $\gamma_j a_{2j}$ for D^t_π where $j = 1, 2, 3, 4$. The parameters α^t, β^t and α^r, β^r which determine, as in (4.12), $\mathbf{e}^t$ and $\mathbf{e}^r$, the polarization vectors of the transmitted and reflected waves, are given by

$$tg\alpha^t e^{i\beta^t} = D^t_\pi/D^t_\sigma; \quad tg\alpha^r e^{i\beta^r} = D^r_\pi/D^r_\sigma \qquad (4.16)$$

Hence, Eqs. (4.14) and (4.16) yield a solution of the boundary problem and describe light transmission and reflection for arbitrary incidence angles and sample thickness.

Before analyzing expression (4.14) in the general case, we introduce some special cases which result in simpler expressions than those produced in the analysis of expression (4.14).

Thin Crystals. Consider scattering by thin crystals that satisfies condition $\ell << 1$ (see (4.15)). The determinants of (4.14) show that wave polarization with maximal reflectivity (4.12) is determined by

$$tg\alpha e^{i\beta} = -i\sin\theta_B. \qquad (4.17)$$

The wave reflected from a thin crystal is always completely polarized and its parameters α and β are given by (4.17) no matter what the polarization of the indicent wave. Concerning the first approximation, the wave whose polarization is orthogonal to that defined by (4.17) is not reflected at all. This result agrees with the kinematic consideration of Chapter 2.

Thick Crystals. For thick crystals, $\ell >> 1$, the solution to the boundary problem is much simpler than that for $\ell \sim 1$ because only eigensolutions with nonzero coefficients in (2.27) are those which correspond to wave damping inside the crystal depth. The damping of eigensolutions for real $\hat{\epsilon}$ is due to diffraction scattering and the fact that parameters $\mathbf{k}_0$ and $\mathbf{k}_1$ in (4.10) have nonzero imaginary parts in a certain range of Δ, that is, there are zones in k-space where wave propagation is forbidden. As a result, diffraction reflection bands are present [9–12] which boundaries are determined from (4.4).

From the solution of (4.4), three different cases of light reflection are distinguished in the dependence of parameter Δ (see (4.5)), i.e., on deviation of the incidence angle (or frequency) from that given by the Bragg condition (2.4). If Δ is such that all the solutions t_j of Eq. (4.4) are real, the wave vectors in (4.10) are also real for all solutions of (4.8). Such waves pass through a crystal without attenuation and the reflectivity in this range is small. The other specific case is when Δ is such that two solutions, $t_{2,3}$, are imaginary (with different signs) and the two others, $t_{1,4}$, real. There, one of the eigenwaves (say $\mathbf{E}_2$) is attenuated exponentially inside the crystal, the other, $\mathbf{E}_3$, increases. Then, the incident wave with elliptical polarization corresponding to $\mathbf{E}_{02}$ is completely reflected, while the orthogonally polarized wave excites nondamping waves $\mathbf{E}_1$ and $\mathbf{E}_4$ in the crystal and only slightly contributes to reflection. In that region of Δ, reflectivity for unpolarized light is practically $1/2$ (the region Δ_p). There is one more region of Δ in which the four roots t_j are complex and complex-conjugated in pairs. There, light with any polarization is almost completely reflected and the reflectivity for unpolarized light is unity (the region Δ_T).

Eq. (4.4) determines boundaries of the regions Δ_p and Δ_T abut for different values of parameter $m = \cos^2\theta(1 + \sin^2\theta)^{-1}$ For instance, at $0 < m < (1 + \sqrt{13})/18$,

$$-(1 + \sqrt{1 + 8m^2})/2 < \Delta_p < 0 < \Delta_T < (\sqrt{1 + 8m^2} - 1)/2 \qquad (4.18)$$

$$(\sqrt{1 + 8m^2} - 1)/2 < \Delta_p < 1.$$

The variations of Δ at reflection band boundaries over the range of θ are presented in Fig. 4.2.

Consider now the results of dynamic diffraction theory in different regions. When the roots are real, the formulae obtained from (4.14) are simple only for incidence angles far from the exact Bragg condition at $|\Delta| \gg 1$

$$E_\sigma^r = \left\{ -\sin\theta(1 + m)(1 - e^{2i\tilde{t}_1\ell})\cos\alpha - i(1 - m)[1 - e^{i(\tilde{t}_1 + \tilde{t}_2)\ell}] \times \right.$$
$$\left. \times e^{i\beta}\sin\alpha \right\}(2\Delta\sin\theta)^{-1}$$

$$E_\pi^r = (1/2\Delta)\left\{ -(1 - m)(1 - e^{2i\tilde{t}_2\ell})e^{i\beta}\sin\alpha + i(1 + m)\sin\theta \times \right.$$
$$\left. \times [1 - e^{i(\tilde{t}_1 + \tilde{t}_2)\ell}]\cos\alpha \right\} \qquad (4.19)$$

where $\tilde{t}_{1,2} = (\Delta/|\Delta|)t_{1,2}$ and $t_{1,2}$ is given by (4.7).

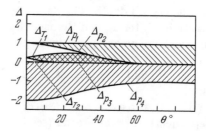

FIGURE 4.2. The diffraction scattering band dependence on the incidence angle of light in a cholesteric liquid crystal. The parameter Δ characterizes deviation from the Bragg condition. The selective reflection band of an individual polarization are hatched and the reflection bands of any polarization are double-hatched.

Equation (4.19) shows that the intensity of reflected waves decreases as Δ^{-2} and the polarization, determined by E^r_π/E^r_σ, depends on Δ ad L in a complicated way. In the region Δ_T light with any polarization is reflected and all t_j are complex and complex-conjugated in pairs. For $\ell \gg 1$, two parameters $\gamma_j = \exp[i\ell(\Delta + m + t_j)]$ in determinant (4.14) (for example, at $j = 1, 2$) are exponentially small, the two others exponentially large. As follows from (4.14), $E^t_\sigma = 0$ and $E^t_\pi = 0$ because the columns in D^T_σ and D^t_π are linearly dependent but $D \neq 0$. Thus, in region Δ_T the intensity of the transmitted wave is zero; hence, a wave with any polarization is completely reflected since there is no absorption in CLCs.

Expression (4.16) determines the polarization of the reflected wave in Δ_T and is simplified as compared with the general case:

$$tg\alpha^r e^{i\beta} = \begin{vmatrix} \cos\alpha & a_{11} & a_{12} \\ e^{i\beta}\sin\alpha & a_{21} & a_{22} \\ 0 & a_{41} & a_{42} \end{vmatrix} \times \begin{vmatrix} \cos\alpha & a_{11} & a_{12} \\ e^{i\beta}\sin\alpha & a_{21} & a_{22} \\ 0 & a_{31} & a_{32} \end{vmatrix}^{-1} \qquad (4.20)$$

where a_{ij} are given by (4.9) for complex t_1 and t_2 which correspond to the solutions that damp within the crystal. Figure 4.3 shows the dependence of $\Delta\theta_T$ (the angular width of the total reflectin band Δ_T) on incidence angle. The frequency width of this band is given by

$$\Delta\omega_T/\omega = \Delta\theta_T ctg\theta. \qquad (4.21)$$

Now consider the intermediate region in which t_1 and t_2 are imaginary and t_3, t_4 real. The amplitude of transmitted light is, neglecting exponentially small corrections,

$$E^t_\sigma = D^{-1}(A_{21}\cos\alpha - a_{11}e^{i\beta}\sin\alpha) \begin{vmatrix} a_{32} & a_{33} & a_{34} \\ a_{42} & a_{43} & a_{44} \\ a_{12} & a_{13} & a_{14} \end{vmatrix}$$

$$E^t_\pi = D^{-1}(a_{21}\cos\alpha - a_{11}e^{i\beta}\sin\alpha) \begin{vmatrix} a_{32} & a_{33} & a_{34} \\ a_{42} & a_{43} & a_{44} \\ a_{22} & a_{23} & a_{24} \end{vmatrix} \qquad (4.22)$$

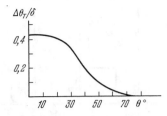

FIGURE 4.3. Total reflection band angular width $\Delta\theta_T$ dependence on Bragg's angle θ [12].

$$D = \begin{vmatrix} a_{11} & 0 & a_{13} & a_{14} \\ a_{21} & 0 & a_{23} & a_{24} \\ 0 & a_{32} & \gamma_3 a_{33} & \gamma_4 a_{34} \\ 0 & a_{42} & \gamma_3 a_{43} & \gamma_4 a_{44} \end{vmatrix}.$$

Here t_1 corresponds to the solution which attenuates inside the crystal. Equation (4.22) shows that transmitted wave polarization is determined by E_1^t/E_σ^t and depends neither on incident polarization nor on crystal thickness. It also follows from the formulae that there is always a polarization which is reflected completely, for this polarization $tg\alpha_1 \exp(i\beta_1) = a_{21}/a_{11}$. The polarization of the reflected wave is, in this case, given by

$$tg\alpha_1^r \exp(i\beta_1^r) = a_{41}/a_{31}. \tag{4.23}$$

Also, (4.22) shows that the transmission is maximal if the incident wave is polarized so that $tg\alpha_2 \exp(i\beta_2) = \frac{a_{21}^*}{a_{11}^*}$, i.e., if it is orthogonal to that of the wave which experiences total reflection.

The results of this section permit analysis of the interaction between light and a cholesteric liquid crystal for arbitrary incidence angles. The optical properties become anomalous in the vicinity of the angle which satisfies the Bragg condition, the corresponding region is of the order of δ, i.e., determined by the local anisotropy of the dielectric constant.

Although the solutions corresponding to the diffraction reflection band in CLCs are similar to those for X-rays diffraction, they are more complicated than those of X-ray diffraction theory but more simple than Mössbauer diffraction for the general structures of Chapter 3. However, polarization properties of eigensolutions are more complex, therefore, optical polarization properties of cholesteric crystals are unusual. The most interesting results are those of dynamic theory, related to oblique incidence of light on a sample, because exact solutions are still unknown for this case; and, the corresponding light scattering properties differ considerably from those in the case of light propagating along the axis.

When light is incident obliquely, in addition to the selective reflection band, there is a band of total light reflection with any polarization near the region where the Bragg condition holds. The polarization of the selectively reflected light and total reflection band depend on incidence angle. In

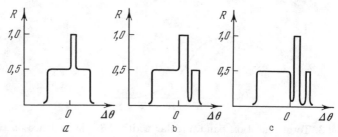

FIGURE 4.4. The qualitative behavior of the reflection coefficient of unpolarized light as a function of deviation from the Bragg condition. Depending on the incidence angle θ there are either one ($90° > \theta \geq 32°$) (a), two ($32° \geq \theta \geq 25°$) (b), or three ($25° \geq \theta \geq 0°$) (c) selective reflection bands.

normal incidence, the angular and frequency width of the reflection band is zero and the totally reflected polarization is circular. As seen in Fig. 4.3, the total reflection band Δ_T is at a maximum when incidence angle $\theta \lesssim 25°$; in this region, $\Delta\theta_T \approx 0.4\delta$, or, at $\delta = 0.01$, it is $14'$. Therefore, the dimension of the diffraction region is such that the above relation is verifiable experimentally.

According to a detailed analysis for small incidence angles $\theta \leq 30°$, weak reflection occurs far from the Bragg condition as well as between the total and selective reflection bands. Thus, in the region of diffraction reflection of obliquely incident light, there is a complicated and interesting dependence of reflected light polarization properties and reflection coefficient on incidence angle at a fixed frequency or on frequency at a fixed angle (Fig. 4.4).

It was shown in references [9–12] that the angular (frequency) width of the bands of polarization selective reflection and total reflection may be either adjacent to or separated from the regions where there is no diffraction reflection. The behavior of unpolarized light reflection as a function of frequency (angle) is shown in Fig. 4.5. The value, $1/2$, the reflection coefficient corresponds to the selective reflection band for a definite polarization which varies along the reflection curve and the value 1 corresponds to total reflection band. The total and selective reflection bands for small θ may not be adjacent but separated by intervals of weak reflection. This fact follows from Fig. 4.4 which shows the structure of reflection regions as a function of wavelength and incidence angle as obtained from (4.4). It is seen that for normal incidence ($\theta = 90°$) the width of the polarization-independent band becomes zero in agreement with the results of the exact solution.

4.1.4 High-Order Reflection for Oblique Incidence

General Considerations. Consider one more specific feature of the case of oblique incidence, namely, the existence of higher-order reflection corre-

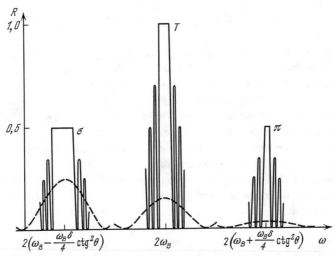

FIGURE 4.5. Calculated for unpolarized light reflection coefficient dependence on frequency in the second order [12]: σ means that σ-polarization is reflected; T corresponds to reflection of any polarization, π means that π-polarization is reflected; solid line is for a thick sample; dashed line is for a thin sample.

sponding to the case of diffraction reflection from CLCs at $|s| > 1$ in (2.17). That is, diffraction reflection exists at a fixed incidence angle at frequency ω_B and at frequencies $s\omega_B$ where s is reflection order.

Higher-order reflection was first discovered theoretically while numerically solving the Maxwell equations [10,13] and was later observed experimentally [13]. As mentioned above there is no reflection at multiple frequencies $s\omega_0$ in the case of normal incidence. In addition, according to the results of kinematic theory for oblique incidence, no direct scattering occurs with changes in the wave vector by $s\tau(|s| > 1)$ because the Fourier expansion (1.6) of tensor $\hat{\epsilon}(\mathbf{r})$ contains no corresponding harmonics. However, using dynamic diffraction theory, particularly (2.25), we demonstrate the existence of higher-order refections for oblique incidence and analyze their s dependence. It was found that diffraction reflection intensity decreases rapidly as reflection order s increases.

Although there is no direct scattering of waves E_0 to E_s, such scattering is possible through intermediate waves $E_1, E_2 \ldots, E_{s-1}$ [12]. Description of these processes demands that we retain at least $S + 1$ equations in (2.25) which contain the amplitudes of intermediate waves. These amplitudes are at least a factor of δ less then E_0 and E_s and may be excluded from the system by expressing them through E_0 and E_s. Thus, we obtain for E_0 and E_s the system of equations, similar to (2.26), with the exception that $\hat{\epsilon}_{\pm\tau}$ is substituted by $F_s\hat{\epsilon}_{\pm\tau}$, where

$$F_s = \left(\frac{\delta}{8}\right)^{s-1} \frac{s^{2s}}{(s!)^2} ctg^{2s-2}\theta. \tag{4.24}$$

As δ in the case of first-order reflection, the value δF_s describes characteristic angular (or frequency) diffraction reflection bands, whose widths are of the order of δ^s, i.e., they decrease as s increases. We note that to observe the s-order reflection we must have a rather thick crystal with $L \sim p/\delta F_s$ and reflection is observed when s is not very large. Figure 4.5 shows the qualitative dependence of reflection coefficient on the frequency for second-order reflection [12]; two lateral maxima are located at the distance $\pm 2\omega_s \delta ctg^2\theta/4$ from the central one.

The width of the σ-maximum is $\Delta\omega_\sigma = 2\omega_B \delta^2 ctg^2\theta/4 \cdot \sin^2\theta$. In this maximum, light with linear σ-polarization is reflected as σ-polarized, while in the π-maximum, whose width is $\Delta\omega_\pi 2\omega_B \delta^2 ctg^2\theta/4$, π-polarized light is reflected as π-polarized. In the central T-maximum σ-polarized light becomes π-polarized after reflection and vice versa, i.e., reflection occurs for any polarization. The width of this maximum is $\Delta\omega_T = 2\omega_B \delta^2 ctg^2\theta/4 \sin\theta$. If crystal thickness is small, the maxima in Fig. 4.5 broaden (dashed line) and become much less prominent.

A similar picture of three maxima is observed with incidence angle variance at a fixed frequency. Second-order reflection takes place near the Bragg angle, $\theta_B = \arcsin(\tau/q)$ and expression (4.25) defines angular widths of the σ-, π- and T-maxima respectively:

$$\frac{\Delta\theta_\sigma}{\Delta\omega_\sigma} = \frac{\Delta\theta_\pi}{\Delta\omega_\pi} = \frac{\Delta\theta_T}{\Delta\omega_T} = \frac{1}{\omega}tg\theta. \tag{4.25}$$

Studies of higher-order reflection (particularly second-order) are useful for CLCs with large pitch whose first-order reflection occurs in the infrared. Notice that for these crystals Bragg angles are small and hence more easily observed, high reflection orders. This is demonstrated in the formulae above where reflection band frequency and angular width decrease and crystal thickness (necessary for reliable observation) increases.

Second-Order Selective Reflection. We will use the general results for higher-order reflection in CLCs presented above for the case of second-order reflection within the framework of multiwave dynamic diffraction theory. Second-order reflections are due to multiple light scattering events and are, to a certain extent, analogous to the Reninger reflections in X-ray scattering [14]. No such reflections appear in kinematic diffraction theory since the corresponding scattering amplitudes are strictly zero. The phenomenon in question is more complicated in the case of CLCs because of the specific light polarization properties in CLC.

Assume that the direction of a monochromatic wave incident on a CLC is close to or coincides with the Bragg direction for second-order reflection, i.e., $s = 2$ in (2.17) (Fig. 4.6). This means that the selectively reflected light

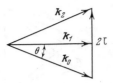

FIGURE 4.6. Bragg condition in vector form for second order selective reflection.

frequency is $2\omega_B$, if the corresponding first-order frequency for the same incidence angle is ω_B.

We consider second-order Bragg reflexes using expressions (2.25) and (2.26). Since the two-wave approximation is not sufficient to obtain second-order reflection, we retain in the infinite system (2.25) the equations for three waves: E_0, E_1 and E_2. Neglecting higher-order terms in δ (usually, δ is about 10^2), we obtain

$$\left(\hat{\epsilon}_0/\bar{\epsilon} - \frac{k_0^2}{q^2}\right)\mathbf{E}_0 + (\hat{\epsilon}_{-\tau}/\bar{\epsilon})\,\mathbf{E}_1 = 0$$

$$(\hat{\epsilon}_\tau/\bar{\epsilon})\,\mathbf{E}_0 + \left(1 - k_1^2/q^2\right)\mathbf{E}_1 + (\mathbf{k}_1\mathbf{E}_1)\mathbf{E}_1/q^2 + (\hat{\epsilon}_{-\tau}/\bar{\epsilon})\mathbf{E}_2 = 0$$

$$(\hat{\epsilon}_\tau/\bar{\epsilon})\,\mathbf{E}_1 + \left(\hat{\epsilon}_0/\bar{\epsilon} - k_2^2/q^2\right)\mathbf{E}_2 = 0. \qquad (4.26)$$

From the second equation we find

$$\mathbf{E}_1 = \left[q^2\hat{\epsilon}_\tau\mathbf{E}_0 - \mathbf{k}_1(\mathbf{k}_1\hat{\epsilon}_\tau\mathbf{E}_0) + q^2\hat{\epsilon}_{-\tau}\mathbf{E}_2 - \right.$$
$$\left. - \mathbf{k}_1(\mathbf{k}_1\hat{\epsilon}_{-\tau}\mathbf{E}_2)\right]\left[(k_1^2 - q^2)2\bar{\epsilon}\right]^{-1} \qquad (4.27)$$

and substitute $\mathbf{E}_1$ into the third and first equations. When retaining only terms with the lowest orders in δ, we obtain

$$(\hat{\epsilon}_0/\bar{\epsilon} - k_0^2/q^2)\mathbf{E}_0 + \frac{\delta}{2}ctg^2\theta\hat{\epsilon}_{-\tau}\mathbf{E}_2 = 0$$

$$\frac{\delta}{2}ctg^2\theta\hat{\epsilon}_\tau\mathbf{E}_0 + (\hat{\epsilon}_0/\bar{\epsilon} - k_2^2/q^2)\mathbf{E}_2 = 0. \qquad (4.28)$$

It was taken into account that the component of $\mathbf{k}_1$ along the Z-axis was zero, while that perpendicular to Z was $q\cos\theta$ (see Fig. 4.6). Note that the coefficients of $\hat{\epsilon}_{\pm\tau}$ in (4.28) are nonzero because intermediate wave $\mathbf{E}_1$ is not transversal.

Let us analyze the solutions for system (4.28). This system differs from system (2.26,4.3) in the coefficients of $\hat{\epsilon}_{\pm\tau}$; however, its solution is less complex than that of (4.3) because (4.28) is reduced to two equations which separately describe π- and σ-polarized light scattering. Physically, this separation is due to CLC birefringence. For second and higher order diffraction scattering, the difference between wave vectors of eigenwaves is proportional to birefringence values, i.e., to δ, and greater than the diffraction corrections to wave vectors; these corrections are proportional to δ^s. This means that

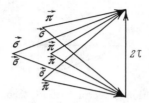

FIGURE 4.7. Bragg condition in vector form for a CLC with strong birefringence.

polarization properties of eigenwaves in the presence of diffraction are the same, to within the order of δ, as those in the case of direct transmission in the absence of diffraction. That is, higher-order diffractive scattering is too weak to appreciably change the polarization properties and wave vectors of eigenwaves. This means that, in the presence of diffraction, the polarizations of $\mathbf{E}_0$ and $\mathbf{E}_2$ practically coincide with σ- and π- polarizations, the eigenpolarizations for direct transmission without diffraction. Thus, there are four possible polarization combinations: $\mathbf{E}_0$ and $\mathbf{E}_2$ are either both σ- or π-polarized, π- and σ- or σ- and π-polarized respectively. Since the bire-fringence is strong compared with δ^2, the angular (or frequency) regions in which the Bragg condition is satisfied for the combinations above (Fig. 4.7). Thus, in corresponding angular (or frequency) intervals, whose widths are of the order of δ^2, one wave with eigenpolarization is subject to diffractive scattering, while that of the other eigenpolarization does not satisfy the Bragg condition and experiences negligible diffractive scattering. This is the physical reason for simplification of (4.28) which is reduced to a system of two equations. Naturally, the same result is obtained by solving the complete system and neglecting small polarization corrections of the order of δ. Thus, system (4.28) is reduced to the system of scalar equations:

$$(1 - k_0^2/q_p^2)E_0^p + F_{pq}E_2^q = 0$$
$$F_{qp}E_p^r + (1 - k_2^2/q_q^2)E_2^q = 0 \qquad (4.29)$$

where indices p and q at the scalar amplitudes E_0 and E_2 and q indicate corresponding waves polarizations (these indices may be π or σ as well)

$$q_q = \frac{\omega\sqrt{\epsilon_q}}{c}; \quad \epsilon_\pi = \bar{\epsilon}\left(1 - \frac{\delta}{2}\cos^2\theta\right), \quad \epsilon_\sigma = \bar{\epsilon}\left(1 + \frac{\delta}{2}\cos^2\theta\right)$$
$$F_{\pi\sigma} = F_{\sigma\pi} = \frac{\delta^2}{4}\frac{\cos^2\theta}{\sin\theta}; \quad F_{\pi\pi} = \frac{\delta^2}{4}\cos^2\theta, \quad F_{\sigma\sigma} = \frac{\delta^2}{4}ctg^2\theta.$$

System (4.29) is identical with the system of equations from X-ray diffraction theory [15,16].

The Bragg angle θ_B corresponding to the polarization combination above is:

– for scattering $\sigma - \pi$ and $\pi - \sigma$: $\theta_B^{\pi\sigma} = \theta_B^{\sigma\pi} = \theta_B = \arcsin\frac{\tau c}{\omega\sqrt{\bar{\epsilon}}}$,

– for scattering $\sigma - \sigma$ and $\pi - \pi$: $\theta_B^{qq} = \theta_B + \Delta\theta^{qq}$,

where $\Delta\theta^{\sigma\sigma} = -\Delta\theta^{\pi\pi} = \delta ctg\theta_B$. If we assume that the incidence angle is close to some of the θ_B^{pq} angles and solve the boundary problem (as for first-order diffraction reflection) or use the results of X-ray diffraction theory [15,16], we obtain the reflection coefficient R and transmission coefficient T dependence on incidence angle and thickness L of a planar cholesteric texture [4,12].

We assumed above that, in an experiment the incidence angle is varied at fixed frequency ω. Conversely, if angle θ is fixed and frequency varied, diffraction scattering occurs in the vicinity of Bragg's frequencies ω_B^{pq}. For polarization-changing scattering $\omega_B^{\sigma\pi} = \omega_B^{\pi\sigma} = \omega_B = c\tau/(\sqrt{\bar{\epsilon}}\sin\theta)$ and polarization-conserving scattering $\omega_B^{qq} = \omega_B + \Delta\omega^{qq}$, $\frac{\Delta\omega^{\pi\pi}}{\omega_B} = \frac{-\Delta\omega^{\sigma\sigma}}{\omega_B} = -\delta ctg\theta/4$.

We note that the mentioned angular (or frequency) positions of diffraction maxima (Fig. 4.5) are given, within the approximate approximation, to the accuracy of δ and widths, to the accuracy of δ^2, i.e., to a first nonvanishing approximation, a more exact solution requires that we use a more accurate approximation. For example, positions of maxima can be found to the accuracy of δ^2 by retaining, in (4.28), higher-order terms in δ on the basis of (4.26).

Thus, higher-order diffraction scattering and corresponding polarization properties are simpler than those of first-order. This analysis indicates that the complex polarization properties of first-order diffraction scattering at oblique incidence are due to birefringence and diffraction scattering of the same strength.

4.1.5   A Planar Layer

The explicit expressions related to light propagation in CLCS are given for an unlimited sample or a halfspace filled with a CLC in Chapter 1 and in this chapter as well. These expressions are used occasionally to describe experiments, but the most typical and interesting experimental cases are those on samples that are not thick $(\delta L/p \sim 1)$.

Notice that, in many cases, experiments involved normal or near normal incidence of light on a planar texture. The analytical solution for this problem may be used for any layer's thickness. In the general case, however, it is complicated and it is necessary to invoke numerical methods. Thus, we analyze the optical properties of a layer of arbitrary thickness L for normal incidence assuming no reflection at dielectric boundaries. The analytic description of this case is simple; for a layer of finite thickness, we study the polarization and frequency properties of both transmitted and reflected light. Thus, we neglect light reflection at CLC boundaries and assume that anisotropy δ is small and $\bar{\epsilon} = \epsilon$ or $r = (\bar{\epsilon}/\epsilon)^{1/2}$ differs from unity by less than δ. Under these conditions, three eigensolutions are excited in a sample; two correspond to diffracting polarization, the third, to the non-diffracting wave propagating along the direction of incidence. Systems (2.18) and (2.19), which describe the solution to the boundary problem in a general case, become more simple and take the form (1.24) except that $r = 1$ and index j assumes values 1, 2 and 3. Solutions of the corresponding systems are described by expressions of form (1.21) if the fourth row and the fourth column are canceled in the corresponding matrices. Thus the problem reduces to solving a system of three linear equations.

The assumption of no boundary reflection permits other simplifications which reduce the problem to a system of two linear equations. This is so because, in this case, the polarizations in system (1.20) become separated since the incident wave with diffracting circular polarization excites only solution No. 2 and No. 3, whereas the wave with non-diffracting polarization excites only solution No. 1. This means that the amplitude and phase of the eigenwave at the CLC surface is determined by E_e^-, i.e., it is identical with corresponding values for the non-diffracting component of the incident wave.

Accounting for the approximation, i.e., neglect of boundary reflection, which introduces an error of the order of δ, we obtain the equation for the amplitudes of diffracting eigensolutions E_2 and E_3 retaining only the terms of lowest order in δ. We do this by assuming that in system (1.20) $\eta_2^+ = \eta_3^+ = 1$, $\eta_2^- = \eta_3^- = -1$. This yields

$$E_2^+ + E_3^+ = E_e^+, \quad e^{iK_2^- L}\xi_2 E_2^+ + e^{iK_3^- L}\xi_3 E_3^+ = 0 \qquad (4.30)$$

From the expression for the amplitudes of reflected and transmitted waves with diffracting polarization (1.19) we find:

$$E_r^+ = \sum_{j=2,3} \xi_j E_j^+, \quad E_t^+ = \sum_{j=2,3} e^{iK_j^+ L} E_j^+. \qquad (4.31)$$

And from (4.30) and (4.31), or from the corresponding simplified form of (1.21), we obtain the following expressions for the amplitudes of reflected and transmitted waves of diffracting polarization:

$$E_r^+ = i\delta \sin k^- L \left\{ (4k^- \tilde{\lambda}^2/\tau) \cos k^- L + i[\tilde{\lambda}^2 + (2k^- \tilde{\lambda}/\tau)^2 - 1] \right.$$

$$\left. \sin k^- L \right\}^{-1} E_e^+$$

$$E_t^+ = 4e^{iqL}(k^- \tilde{\lambda}^2/\tau) \left\{ (\tilde{\lambda}/\tau) \cos k^- L + (i/4)[\tilde{\lambda}^2 + (2k^- \tilde{\lambda}/\tau)^2 - 1] \right.$$

$$\left. \sin k^- L \right\}^{-1} E_e^+ \qquad (4.32)$$

where the notation is as in Chapter 1. Under these conditions, the reflection coefficient R^+ and transmission coefficient T^+ for a wave with diffracting polarization are:

$$R^+ = (\delta^2 \sin^2 k^- L)[16(k^- \tilde{\lambda}^2 \tau)^2 + \delta^2 \sin^2 k^- L]^{-1}$$

$$T^+ = 1 - R^+ = (4k^- \tilde{\lambda}^2/\tau)^2 [16(k^- \tilde{\lambda}^2/\tau)^2 + \delta^2 \sin k^- L]^{-1}. \qquad (4.33)$$

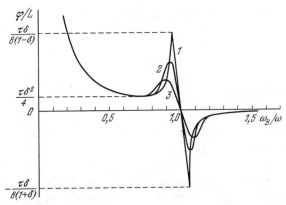

FIGURE 4.8. Dependences of the angle of the polarization plane rotation on frequency for sample with different thickness: $1 - L\delta/p \gg 1$, $2 - L\delta/p = 1$, $3 - L\delta/p = 0.5$. The curves are plotted for $\delta = 0.1$.

Thus, the wave with non-diffracting polarization has no reflection, $R^- = 0$ and $T^- = 1$, within this approximation.

In Eq. (4.33), in more detail, within the frequency range $\omega_B(1+\delta)^{-1/2} < \omega < \omega_3(1-\delta)^{-1/2}$ (i.e., when $|q^2 - \tau^2/4| < q^2\delta$) the value k^- becomes imaginary and $\sin k^- L$ in (4.33) becomes $sh|k^-|L$. When the crystal is sufficiently thick $(L > p/\delta)$ $sh|k^-|L \gg 1$ and $R^+ \approx 1$, i.e., light with right-handed circular polarization is totally selectively reflected (Fig. 1.8). The frequency width of the reflection band is $\Delta\omega = \omega_B[(1-\delta)^{-1/2} - (1+\delta)^{-1/2}] \approx \delta\omega_B$, i.e., it is defined by the anisotropy δ of dielectric permittivity. When the frequency moves away from the region of strong reflection outside the selective reflection band, the reflection does not decrease smoothly—it oscillates. The oscillations are due to light diffraction within a limited volume and unassociated with boundary reflection. These oscillations are not observed, as a rule, in experiments, because samples are not sufficiently perfect.

In thin crystals $(L \ll p/\delta)$

$$R^+ \simeq \delta^2 \frac{\sin^2[(q - \tau/2)L]}{(2q/\tau - 1)^2} \qquad (4.34)$$

i.e., reflected intensity is proportional to δ^2, while there is prominent broadening of the reflection curve (see Fig. 1.8); this result is also obtained from (2.16) within the kinematic approximation. The intermediate case where $L \sim p/\delta$ is also shown in Fig. 1.8.

Experimentalists often measure the difference between transmissions for opposite circular polarizations, i.e., the so-called circular dichroism $D = (T^- - T^+)/(T^- + T^+)$. This value for the non-absorbing cholesteric crystals can be expressed in terms of T^+ and R^+ as

$$D = \frac{1 - T^+}{1 + T^+} = \frac{R^+}{2 - R^+}. \tag{4.35}$$

4.1.6   Optical Rotation

General Features. It is well known [4.17] that optical rotation in a medium is due to the difference in phase velocities of waves with right- and left-handed circular polarizations [17]. In plane cholesteric texture, the phase difference for certain wavelengths may be large since the wave with circular polarization is subject to strong diffraction reflection while the other polarization interacts weakly with the crystal. Optical rotation in the selective reflection band of CLCs is stronger than intrinsic optical rotation (associated with molecular optical activity) due to diffraction effects and has some pecularities. One of these peculiarities is that the wave with diffracting circular polarization is strongly attenuated in the CLC due to diffraction reflection. When light is transmitted through a crystal its initial linear polarization changes to elliptical polarization which depends on crystal thickness. Thus, we consider the optical rotation of, say, the major axis of the polarization ellipse. In the case of light propagation along the optical axis, if incident polarization is linear, the ratio of ellipse axes is given by the following expression:

$$b = \frac{1 - \sqrt{T^+}}{1 + \sqrt{T^+}} \tag{4.36}$$

where T^+ is as in (4.33). The other pecularity is the strong optical rotation dependence on frequency both in and outside the absorption bands. Optical rotations for frequencies at different sides of Bragg's frequency ω_B are of opposing directions whereas those at Bragg's frequency vanish. This behavior of optical rotation seems rather strange but it has a simple explanation: the phase velocity of one of two circular diffracting eigenwaves presented in expansion (1.19) is smaller than that of the non-diffracting wave, the other greater. The exited diffracting wave is predominantly slow or fast depending on the sign of the difference between the frequency and ω_B. This causes frequency dependence of the optical rotation and its sign. Finally, there is one unusual property which distinguishes CLCs from other gyrotropic media—the angle of optical rotation in a CLC nonlinearly depends on sample thickness.

From solution of the boundary problem (4.33), we find that when the incident wave is plane-polarized, the amplitude of a wave leaving the crystal is

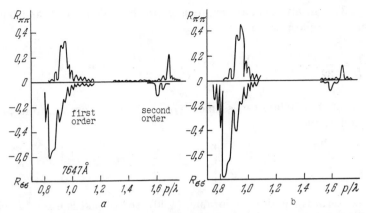

FIGURE 4.9. Measured (a) and calculated (b) frequency and polarization dependence in first and second order reflection from a mixture of cholesterics [13]. $R_{\pi\pi}(R_{\sigma\sigma})$ is the reflection coefficient of the π- or σ-polarization into π- or σ-polarization; incidence angle is 45°; sample thickness is $L = 11.47\,\mu$m; $L/p = 15$.

$$\mathbf{E}^+ = \frac{E_e}{\sqrt{2}}\left\{\mathbf{n}_- e^{i(k^+ - \tau/2)L}\right.$$
$$\left. + \mathbf{n}_+ \frac{k^- \tau e^{i\tau L/2}}{k^- \tau \cos k^- L + i[(k^-)^2 + \tau^2/4 - q^2]\sin k^- L}\right\}. \quad (4.37)$$

The difference between two terms in (4.37) gives the rotation angle

$$\varphi = \frac{1}{2}\left[(k^+ - \tau)L + arctg\left\{\frac{[(k^-)^2 + \tau^2/4 - q^2]tgk^- L}{k^-\tau}\right\}\right]. \quad (4.38)$$

This formula is more simple for thick crystals

$$\frac{\varphi}{L} = Re\frac{k^+ - k^- - \tau}{2}. \quad (4.39)$$

The optical rotation frequency dependence given by (4.38) and (4.39) is presented in Fig. 4.9. Far from the selective refection band, expressions (4.38) and (4.39) reduce to the deVries formula [18].

$$\frac{\varphi}{L} = \tau\delta^2[16\tilde{\lambda}^2(1 - \tilde{\lambda}^2)]^{-1}. \quad (4.40)$$

This formula has the disadvantage that in contrast to the exact formula (4.38) it tends to $\pm\infty$ in the selective reflection band.

Optical Rotation in Thin Layers. When reflection of a CLC is not too strong (samples are used routinely to study optical rotation), we obtain, from (4.38), an approximate expression which holds both in and outside the selective reflection band

$$\frac{\varphi}{L} = \frac{\delta^2 \tau}{32\tilde{\lambda}^3} \left[\left(\frac{\tau}{2\tilde{\lambda}k^-} \right)^2 (1 - \tilde{\lambda}) \left(1 - \frac{\sin 2k^- L}{2k^- L} \right) + \frac{1}{1 + \tilde{\lambda}} \right]. \quad (4.41)$$

For a rough estimate we assume $k^-/q = \tilde{\lambda} - 1$ and obtain, from (4.41), a simple formula

$$\frac{\varphi}{L} = \frac{\delta^2 \tau}{16\tilde{\lambda}(1 - \tilde{\lambda}^2)} \left[1 - \frac{(1 + \tilde{\lambda}) \sin 2q(1 - \tilde{\lambda})L}{2(1 - \tilde{\lambda})\tau L} \right]. \quad (4.42)$$

The first term is the deVries formula (4.40), the second is essential in the selective reflection band at $\tilde{\lambda} \to 1$. This formula yields the correct optical rotation frequency dependence at any frequency. If the sample is thin ($L < p/\delta$), formula (4.41) and (4.42) differ slightly from the exact (4.38) and apply if rotation angle $\varphi < 1$. We note again that in the selective reflection band formulae (4.38), (4.41) and (4.42) yield the nonlinear rotation angle dependence on thickness L. Specifically, at small L values, $\varphi \sim L^2$ near the rotation maxima (see Fig. 4.8) and $\varphi \sim L^3$ near the frequency at which the rotation changes sign.

The Limit $\tilde{\lambda} = \tau/2q << 1$. In this case waves 1 and 2 (or 3 and 4 depending on the direction of the incident wave) are excited in a crystal.

From (1.21) we obtain the following equation for transmitted wave amplitude (see (1.18))

$$E_t^+ = \left[E_e^+ \left(\cos \frac{(k^- - k^+)L}{2} + it \sin \frac{(k^- - k^+)L}{2} \right) \right.$$
$$\left. - iE_e^- \frac{\delta t}{2\tilde{\lambda}} \sin \frac{(k^- - k^+)L}{2} \right] e^{i(k^+ + k^- + \tau)L/2}$$

$$E_t^- = \left[E_e^- \left(\cos \frac{(k^- - k^+)L}{2} - it \sin \frac{(k^- - k^+)L}{2} \right) \right.$$
$$\left. - iE_e^+ \frac{\delta t}{2\tilde{\lambda}} \sin \frac{(k^- - k^+)L}{2} \right] e^{i(k^+ - k^- - \tau)L/2} \quad (4.43)$$

where $t = (1 + \frac{\delta^2}{\tilde{\lambda}^{-2}})^{1/2}$. Rotation angle φ nonlinearly depends on crystal thickness and generally, on the direction of initial polarization.

Under the assumed limitations for q, two cases are possible. If $\tau/1q = \tilde{\lambda} >> \delta$, (4.43) results in (4.40) and rotation angle ceases to depend on initial polarization direction.

When the wavelength is small, Mauguin's limit $\tilde{\lambda} << \delta$, the transmitted wave polarization, depends on molecular orientation at the CLC surface. As follows from (4.43), linearly polarized incident light becomes elliptically

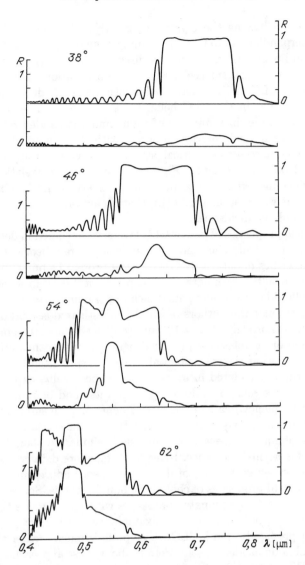

FIGURE 4.10. Experimental spectra of circularly polarized light reflection from a planar cholesteric layer of thickness 6 μm for oblique incidence. The upper and lower curves correspond respectively to left- and right-handed polarization for every incidence angle [23].

polarized upon leaving the crystal. If linearly polarized incident light is parallel (perpendicular) to CLC molecules at the front surface, transmitted light is also linearly polarized in the direction parallel (perpendicular) to molecules at the back surface. Hence, the polarization plane follows the orientation of CLC molecules and polarization angle is determined by the number of pitches in the CLC helix, i.e. $\varphi = 2\pi L/p$. The case $\tilde{\lambda} \ll \delta$ occurs, for example, in a mixture of right- and left-handed rotating CLC [19,20]. This analytical description agrees with the numerical calculations and experimental results for optical parameters of CLCs in the cases of both normal and oblique incidence. Figure 4.9 shows the correlation between the numerical and experimental data on cholesteric-nematic mixtures [13]. Specifically, it was first shown in [13] that higher-order reflections exist in the case of oblique incidence.

The optical properties of planar CLC layers were studied in detail in work done in Japan [21–24]. They showed that the dynamic theory of diffraction not only describes qualitatively CLC optical properties but also yields the quantitative results which agree with experimental data in a wide class of cholesteric liquid crystals and nematic-cholesteric mixtures (see Figs. 4.10–4.12). In particular the authors of [21–24] investigated second-order selective reflection thoroughly (Fig. 4.12) and confirmed the correspondence between second-order reflection polarized parameters and theoretical results above. We note that the experimental scattering spectra in Figs. 4.10–4.13 reproduce those calculated in every respect. Specifically, experiment confirms the features related to finite layer thickness and boundary reflection effects (the frequency beats of reflectivity depending on sample thickness). We omit the results which include reflection at dielectric boundaries for the case of oblique incidence; the analysis is similar to that described in Chapter 1 for normal incidence. Refer to [4] for more details.

Thus, we presented the use of diffraction theory approximations to describe the optical properties of cholesteric liquid crystals. Some papers [25–30] develop several approximate approaches which are outside the scope of our considered approximations. For example, the authors of [25,29,30] examine light propagation in a CLC whose pitch is much greater than wavelength for oblique incidence; some details about these approaches are found in [27,28]. We omit the details of these complex approximate methods since it is not clear that they are useful in the presence of exact numerical methods. Nevertheless, we mention an interesting effect revealed in [29,30] of the possible forward diffraction of light by chiral crystals with long pitches. They showed that, if diffraction conditions are satisfied, the reciprocal lattice vector can compensate for the birefringence-induced difference between two eigenwaves propagating in the same direction. Under these conditions it is convenient to trace the transformation between orthogonal eigenpolarizations which are almost linear [29]. We also note that the theory of linear wave transformation [80] effectively describes wave propagation in non-ideal chiral structures whose parameters vary continuously in space.

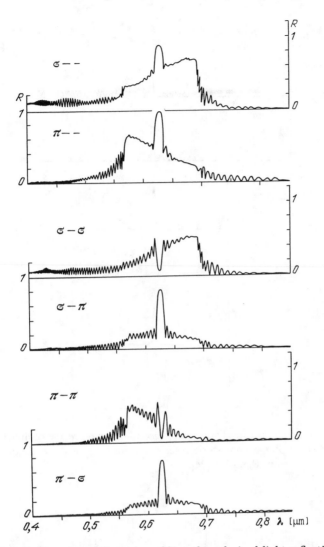

FIGURE 4.11. Experimental spectra of linearly polarized light reflection from a planar cholesteric layer of 18 μm thickness and incidence angle 45°. The polarizations of incident and reflected beams (the left and right symbols) are shown. Dash instead of the right symbol corresponds to the measurements without detection of the scattered beam polarization [23].

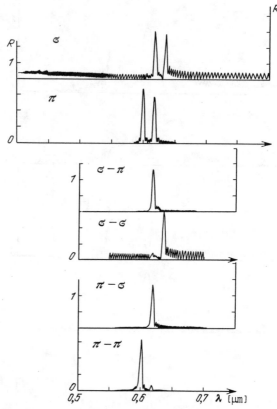

FIGURE 4.12. The measured spectra of second-order reflection of linearly polarized light from a planar cholesteric layer of 38 μm thickness and incidence angle 45°. Notation is as in Fig. 4.11 [24].

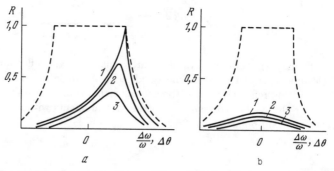

FIGURE 4.13. The qualitative form of the reflection spectrum of thick cholesteric samples for different beam polarizations at oblique incidence: (a) polarization subjected to absorption suppression effect; (b) polarization not subjected to suppression effect. Curves 1–3 correspond to order parameter values $S = 1$, 0.9, and 0.7 respectively; dashed curve corresponds to absence of absorption [7].

4.2 Absorbing Cholesteric Liquid Crystals

We assumed, up to this point, that cholesteric crystals do not absorb light. Actually quite often there is considerable absorption near the absorption bands of the crystal molecules or in a special admixture (e.g., a dye). Light absorption is sensitive to CLC helical structure such that it may vanish or there may be qualitative changes in its polarization properties. We consider here the main optical properties of absorbing CLC related to diffraction in the presence of anisotropic absorption by individual molecules and refer to [4] for an exhaustive discussion of this topic including circular dichroism in the absorption bands.

4.2.1 The Dielectric Tensor of Absorbing CLCs

Light absorption is described by the imaginary part of dielectric permittivity tensor. Hence, we use the same expression as above except that the principal values ϵ_1, ϵ_2, ϵ_3 of the tensor $\hat{\epsilon}$ are now complex. For example, if the absorption spectrum of a CLC includes several absorption bands which are linearly polarized at angles ψ_j to the molecular long axis, the imaginary parts of $\bar{\epsilon} = (\epsilon_1 + \epsilon_2)/2$ and $\epsilon_a = \epsilon_1 - \epsilon_2$ are [31]

$$Im\bar{\epsilon} = \sum_j \eta_j [1 + 1/2(1 - \frac{3}{2}\sin^2\psi_j)S],$$

$$Im\epsilon_a = 3S\sum_j \eta_j[1 - 3/2\sin^2\psi_j] \qquad (4.44)$$

where the value of η_j is proportional to the corresponding oscillator strength and S is the order parameter.

4.2.2 Suppression of Absorption

Consider the effect of absorption on the optical parameters in the selective reflection band [32–35]. Absorption usually reduces the reflection from a CLC. However, since the reflection from a CLC is connected with diffraction, these reflections decrease and light absorption may be diminished considerably in the selective reflection band. The effect of absorption suppression of diffraction nature, the Borrmann effect, is known for X-rays, Mössbauer gamma quanta and neutrons [36]. This effect in CLCs is considered in [34] for normal incidence and in [35] for an arbitrary incidence angle. It is shown therein that absorption may be suppressed completely if absorbing oscillators are directed along the molecular long axis (i.e., at $Im\epsilon_1 \neq 0$, $Im\epsilon_2 = Im\epsilon_3 = 0$) and the molecules are ideally ordered within the crystal ($S = 1$). In the case of normal incidence, complete suppression is also possible if oscillators are oriented across the long axis. In general,

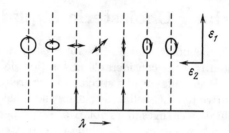

FIGURE 4.14. Orientation of total field polarization in a CLC for the diffracting eigenwave as a function of wave length [34].

complete suppression is governed by the following relationship

$$Im\epsilon_a = \pm 2Im\bar{\epsilon}. \tag{4.45}$$

Suppression of absorption means that, for a certain polarization and incidence angle, the intensity of light reflected from a thick crystal equals the incident intensity (Fig. 4.13). For example, in the case where absorbing oscillators lie along or across the long axis and normal incidence, reflection becomes unity at frequency $\omega = \omega_B \sqrt{2\epsilon_2/(Re\epsilon_1 + \epsilon_2)}$ for diffracting circular polarization; this follows directly from (1.16).

The physical cause of absorption suppression is that the electric field vector of wave superposition within the crystal (1.20) is, at every point of the CLC, perpendicular to the molecular long (or short) axes (see Fig. 4.14) and, hence, to the absorbing oscillators; however, there is no absorption of these waves. For real crystals there is partial absorption suppression; the peak of the reflection curve never reaches unity (Fig. 4.14) and its height strongly depends on order parameter S.

Absorption suppression also manifests itself in circular dichroism. Using (4.37) and (4.38), the value of circular dichroism for thick absorbing CLCs is expressed as

$$D = th[L\ Im(k^- - k^+)] \tag{4.46}$$

where $k^\pm$ are defined by (1.15). At $Im\bar{\epsilon} >> |\delta|$, this expression becomes

$$D = th\left[\frac{q^2L}{8\tau}\frac{Im\bar{\epsilon}(Re^2\epsilon_a - Im^2\epsilon_a) - \gamma Im\epsilon_a Re\epsilon_a}{\gamma^2 + 4Im^2\bar{\epsilon}}\right] \tag{4.47}$$

where $\gamma = 2[1 - (\omega_B/\omega)^2]Re\bar{\epsilon}$.

It follows from (4.47), that the value of D changes sign with frequency. This means that, in the diffractive reflection band the absorption not only is suppressed but also increases to more than its average value.

Absorption suppression and the frequency dependence of dichroism were observed experimentally in [37,38] when studying light intensity changes as a function of frequency while the light is transmitted through a CLC in the selective reflection band [37,38].

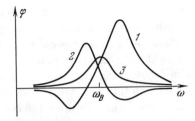

FIGURE 4.15. The qualitative dependence of the angle of polarization plane rotation on frequency for different values of absorption in a CLC: $1 - Re\,\epsilon_a > Im\,\epsilon_a$, $2 - Re\,\epsilon_a < Im\,\epsilon_a$, $3 - Re\,\epsilon_a = Im\,\epsilon_a$.

4.2.3   Optical Rotation

Anisotropic absorption strongly affects the frequency dependence of the angle of rotation of the polarization plane. A non-absorbing crystal in the selective reflection region exhibits both right- and left-handed rotation (see Fig. 4.8), but an absorbing crystal rotates predominantly in one direction [34,34]. This is because the absorptions of slow and fast waves in expansion (1.18) are very different; consequently, one of the waves is suppressed in CLCs and the curve of optical rotation becomes asymmetric or the sign of rotation remains unchanged.

The angle of optical rotation is determined, as for non-absorbing CLCs, by the phase difference between the addends in (4.37), except that $\hat{\epsilon}$ is now complex.

Under the assumptions of (4.47), we find the angle of polarization rotation

$$\varphi = \frac{q^2 L}{8\tau} \frac{\gamma(Re^2\epsilon_a - Im^2\epsilon_a) + 4Im\bar{\epsilon} \cdot Im\epsilon_a Re\epsilon_a}{\gamma^2 + 4Im^2\bar{\epsilon}}. \tag{4.48}$$

This formula produces the curves plotted in Fig. 4.15.

4.2.4   Borrmann Effect for Oblique Incidence

We briefly consider the Borrmann effect for oblique incidence light. In this case, solutions of the dynamic system (4.3) show that absorption may be suppressed completely [4,37], i.e., the imaginary part of the wave vector may vanish according to (4.7).

As in the case of normal incidence, complete absorption occurs, i.e., reflection R is unity, when a certain relationship exists between the absorption anisotropy and its average value; this relationship is (4.45) with a positive sign. In contrast to the case of normal incidence, complete absorption suppression occurs when absorbing oscillators are oriented along the molecular long axis, i.e., $\psi = 0$ in (4.44). If $\psi = \pi/2$, in particular, if oscillators are oriented across the long axis, it is impossible to have complete absorption suppression.

In addition, the degree of suppression strongly depends on the order parameter S, i.e., the suppression is complete only at $S = 1$ and decreases with the decrease of S. Figure 4.13 shows the calculated values of reflection for oblique incidence of light with different polarizations when absorbing oscillators lie along the molecular long axis. In Fig. 4.13a the polarization of the incident wave coincides with that of the direct wave in the eigensolution for which complete suppression occurs. Figure 4.13b corresponds to incident polarization of the second eigensolution; this figure shows that even though the order parameter S deviates slightly from unity, the reflection value decreases sharply at its maximum. That is, if S decreases from 1 to 0.9, the reflection decreases from 1 to 0.7.

It can be shown that reflection from an absorbing CLC may reach unity for a single value of frequency or angle. Reflection is complete when incident polarization (see (4.12)) is given by

$$tg\alpha e^{i\beta} = -i/\sin\theta. \qquad (4.49)$$

It was mentioned that this polarization coincides with that of the direct wave which is excited in a CLC. At the frequency value or angle in question, the total electric field of the corresponding eigensolution is, at any point in the CLC, perpendicular to the molecular long axis, i.e., the wave does not experience absorption. We note that the exact solution for non-absorbing CLC with this field configuration was first exposed by Dreher and Meier [9].

The fact that absorption cannot be completely suppressed when absorbing oscillators are perpendicular to the molecular long axis has a simple physical interpretation; it is impossible to produce a configuration such that the electric field is everywhere perpendicular to absorbing oscillators.

The Borrman effect for oblique incidence was experimentally observed in [39] but it was clearly present only when the direction of light propagation made a small angle with the optical axis ($\pi/2 - \theta < 19°$). No effect was observed at $\frac{\pi}{2} - \theta > 19°$, perhaps because the authors made no provision for the polarization dependence of the effect.

4.3   Chiral Smectic Liquid Crystals

Chiral smectic liquid crystals are another type of liquid crystal whose structure and optical properties are close to those of CLCs. Chiral smectic liquid crystals, C^* [1], differ from CLCs in their layered structure. The C^* molecules produce mono- and bi-molecular layers such that the molecular interactions within them is stronger than those between molecules of different layers. The mean orientation of all molecules within a layer is the same in a C^* but this orientation (direction) changes from layer to layer, as in CLCs. The difference between them is that the C^* director is not orthogonal to the helical axis. Hence, C^* structure is represented as in Fig.

1.3 if all molecular axes, keeping the azimuthal angle constant, are turned so that their axes make the same indirect angle with the helical axis. Since, C^* optical properties are described in the same way as those of CLCs we superficially mention the specific features of C^*; the details are found in [4].

4.3.1   Dielectric Permittivity Tensor of a C^*

Since the structures of C^*s and CLCs differ, their dielectric permittivity tensors differ in their dependence on coordinates along the helical axis Z. Specifically, when going along Z in a CLC, only two of the CLC tensor principal axes change orientation, but in a C^* all principal axes of $\hat{\epsilon}$ change orientation. In accordance with structure, the period of C^* dielectric properties along Z coincides with the helical pitch, while in a CLC this period is $p/2$.

The optical properties of a C^* are determined by its tensor $\hat{\epsilon}(z)$ [41]

$$\hat{\epsilon}(\mathbf{r}) = \begin{pmatrix} \epsilon_{11}^0 + \epsilon_a \cos 2\varphi & \epsilon_a \sin 2\varphi & \epsilon_a' \cos \varphi \\ \epsilon_a \sin 2\varphi & \epsilon_{22}^0 - \epsilon_a \cos 2\varphi & \epsilon_a' \sin \varphi \\ \epsilon_a' \cos \varphi & \epsilon_a' \sin \varphi & \epsilon_{33}^0 \end{pmatrix} \tag{4.50}$$

where

$$\epsilon_{11}^0 = \epsilon_{22}^0 = (\epsilon_1 + \epsilon_2 \cos^2 \Theta + \epsilon_3 \sin^2 \Theta)/2$$
$$\epsilon_{33}^0 = \epsilon_2 \sin^2 \Theta + \epsilon_3 \cos^2 \Theta,$$
$$\epsilon_a = (\epsilon_1 - \epsilon_2 \cos^2 \Theta - \epsilon_3 \sin^2 \Theta)/2$$
$$E_a' = (\epsilon_3 - \epsilon_2) \sin \Theta \cos \Theta \tag{4.51}$$

In these formulae Θ and φ are respectively the polar and azimuthal angles of one of the principal axes, namely, axis "3" of the dielectric tensor of an individual smectic layer with respect to the fixed spatial axes X, Y, Z, the X, Y axes lying in the plane of the layer and Z perpendicular to the layer. The direction of axis "3" is determined by the long axis of molecules in a layer, while axis "1" lies in the plane of the layer: ϵ_1, ϵ_2 and ϵ_3 are the principal values of the $\hat{\epsilon}$ tensor corresponding to the "1", "2" and "3" axes. The angle Θ is constant over C^* volume and φ changes with layer in agreement with helical structure

$$\varphi = 2\pi z/p \tag{4.52}$$

where p is helical pitch. Since φ depends on z, tensor $\hat{\epsilon}$ is a periodic function of z and is expanded into a Fourier series

$$\hat{\epsilon}(z) = \sum_{s=-\infty}^{\infty} \hat{\epsilon}_s \exp[is\tau z] \tag{4.53}$$

where $\tau = \frac{2\pi}{p}$. It follows from (4.50) and (4.52) that in a C^*, $\hat{\epsilon}_0$, $\hat{\epsilon}_{\pm 1}$ and $\hat{\epsilon}_{\pm 2}$ in (4.53) are nonzero and determined by the formulae

$$\hat{\epsilon}_1 = \hat{\epsilon}_{-1}^* = 1/4(\epsilon_2 - \epsilon_3)\sin 2\Theta \begin{pmatrix} 0 & 0 & \pm i \\ 0 & 0 & 1 \\ \pm i & 1 & 0 \end{pmatrix}$$

$$\hat{\epsilon}_{-2} = \hat{\epsilon}_2^* = 1/4(\epsilon_1 - \epsilon_2\cos^2\Theta - \epsilon_3\sin^2\Theta) \begin{pmatrix} 1 & \pm i & 0 \\ \mp i & -1 & 0 \\ 0 & 0 & 0 \end{pmatrix} \qquad (4.54)$$

$$\hat{\epsilon}_s \equiv 0, at \mid s \mid \geq 3.$$

The upper and lower signs in (4.54) correspond to the right- and left-handed helices, respectively. In the $\hat{\epsilon}_0$ tensor, only diagonal components are nonzero (see (4.50) and (4.51)):

$$(\hat{\epsilon}_0)_{11} = (\hat{\epsilon}_0)_{22} = \epsilon_{11}^0, \quad (\hat{\epsilon}_0)_{33} = \epsilon_{33}^0. \qquad (4.55)$$

Note that because of different periodicities in CLCs and C^*s we must indicate that $\tau = 2\pi/p$ in (4.53) rather than $4\pi/p$ as for CLCs. Note also that the Fourier expansion (4.54) does not include C^* periodicity related to layered structure because the corresponding period is small and appears only in X-ray scattering.

4.3.2   Second-Order Diffraction Reflection

As in (4.54), the second harmonic in the Fourier expansion of $\hat{\epsilon}(z)$ for a C^* coincides with, to within a factor, the fundamental harmonic of expansion (2.15) for a CLC. This means that second-order diffraction scattering in a C^* is qualitatively similar to that of first-order in a CLC. This case can be analyzed quantitatively using the equations of Section 4.1 with the Fourier components $\hat{\epsilon}_0$ and $\hat{\epsilon}_{\pm 2}$ from (4.54) and (4.55).

If light propagates along the C^* optical axis, according to the exact solution for this case, only the second-order diffraction reflection occurs. Then the optical properties of a C^* are similar to those of a CLC and are described by the expressions given in Section 1.2 in which $\bar{\epsilon}$ is substituted by

$$\bar{\epsilon}' = \frac{1}{2}(\epsilon_1 + \epsilon_2\cos^2\Theta + \epsilon_3\sin^2\Theta) - \frac{1}{8\epsilon_3}(\epsilon_2 - \epsilon_3)^2\sin^2\Theta$$

and δ is substituted by

$$\delta' = \frac{1}{2\bar{\epsilon}'}(\epsilon_1 - \epsilon_2\cos^2\Theta - \epsilon_3\sin^2\Theta) + \frac{1}{8\epsilon_3\bar{\epsilon}'}(\epsilon_2 - \epsilon_3)^2\sin^2\Theta. \qquad (4.56)$$

4.3.3   First-Order Reflection

Since the optical properties of a C^* differ from those of CLCs only for light propagation at an angle to the optical axis in first-order reflection, we restrict ourselves to first-order diffraction reflection in a C^*. If we substitute $\hat{\epsilon}$ in the form of (4.54) and (4.55) into (2.26) the equation system degrades into two independent systems:

$$\begin{cases} (\epsilon_\sigma - k_0^2/q_0^2)E_0^\sigma + fE_1^\pi = 0 \\ fE_0^\sigma + (\epsilon_\pi - k_1^2/q_0^2)E_1^\pi = 0 \end{cases}$$

$$\begin{cases} (\epsilon_\pi - k_0^2/q_0^2)E_0^\pi + fE_1^\sigma = 0 \\ fE_0^\pi + (\epsilon_\sigma - k_1^2/q_0^2)E_1^\sigma = 0 \end{cases} \tag{4.57}$$

where E^σ and E^π are amplitudes of the σ- and π-polarized waves, $\epsilon_\sigma = (\hat{\epsilon}_0)_{11}, \epsilon_\pi = (\hat{\epsilon}_0)_{11}\sin^2\theta + (\hat{\epsilon}_0)_{33}\cos^2\theta$, $f = (\epsilon_2 - \epsilon_3)\sin 2\theta\cos\theta$, $q_0^2 = (\omega/c)^2$ and $\frac{\pi}{2} - \theta$ is the angle between the optical axis and direction of light propagation. It follows from (4.57) that a σ-polarized yields a π-polarized wave after diffraction and vice versa.

The solution for (4.57) and the corresponding boundary problem is more simple than in the case of a CLC and similar to the analogous solution for X-ray diffraction. Hence, we give the final results for a plane-parallel sample of C^*.

4.3.4 Boundary Problem

When the helical axis is perpendicular to a C^* surface (Bragg's case), the angular (frequency) reflections bands for σ- and π-polarizations given by (4.57) coincide. The reflection is then polarization and angle (frequency) dependent as

$$R = |e_1^* \hat{\epsilon}_1 e_0|^2 \frac{\sin^2\left[\frac{q_0 L(\Delta_B^2 - f^2)^{1/2}}{2\bar{\epsilon}^{1/2}\sin\theta}\right]}{\Delta_B^2 - f^2\cos^2\left[\frac{q_0 L(\Delta_B^2 - f^2)^{1/2}}{2\bar{\epsilon}^{1/2}\sin\theta}\right]} \tag{4.58}$$

where e_0 and e_1 are the polarization vectors of the incident and diffracted waves, $\bar{\epsilon} = (\epsilon_\sigma + \epsilon_\pi)/2$, L is crystal thickness and parameter Δ_B characterizes the deviation from the Bragg condition as above:

$$\Delta_B = (\tau/2q_0^2)\left(2q_0\sqrt{\bar{\epsilon} - \bar{\epsilon}\cos^2\theta} - \tau\right).$$

Note that the result of summation of (4.58) over final polarization e_1 does not depend on e_0, i.e., only scattered wave polarization, rather than intensity, depends on initial wave polarization. The reflection band is limited by values $\Delta_B = \pm f$ and its center lies at $\Delta_B = 0$.

Polarization parameters are easily revealed by analyzing factor $|e_1^* \hat{\epsilon} e_0|^2$ in (4.58). That is, in first-order diffraction reflection from a C^*, the right-handed circulary polarized wave becomes left-handed polarized and vice versa; (at normal incidence in a CLC however, one of the circularly polarization waves is not reflected, the other is reflected and retains its polarization). Linearly polarized light remains linearly polarized, and angles

φ_0 and φ_1 between the scattering plane and incident planes and scattered light polarization, respectively are related as $\varphi_1 = \pi/2 - \varphi_0$.

When the helical axis is parallel to the sample surface (Laue case), the reflection bands for σ- and π-polarizations generally do not coincide although they may partially overlap. When $L \gg q_0 p^2 \cos\theta$, the reflection and its frequency (angle) dependence for the σ- and π-polarizations are presented as

$$R = f^2 \left[(\Delta_L \pm m)^2 + f^2 \right]^{-1} \sin^2 \left[\frac{q\sqrt{(\Delta_L \pm m)^2 + f^2}}{2\sqrt{\epsilon}\cos\theta} \right] \quad (4.59)$$

where $m = (\epsilon_\sigma - \epsilon_\pi)/2 = \cos^2\theta[\epsilon_1 + \epsilon_2(1 - 3\sin^2\Theta) + \epsilon_3(1 - 3\cos^2\Theta)]/4$ $\Delta_L = (\tau/2q_0^2)(\tau - 2q_0\sin\theta)$ and the "plus" and "minus' signs correspond to σ- and π-polarizations respectively. Note that at $m = 0$ reflection bands coincide and polarization properties are similar to those of the Bragg case. It follows from (4.59) that reflection coefficient R in the Laue case is nonmonotonically dependent on thickness L; these are the pendulum beats known in the theory of X-ray diffraction. These beats occur particularly in smectic liquid crystals with long period for diffraction in the forward direction [40].

These analytical results (exact and from dynamic diffraction theory) agree with the numerical calculations of smectic crystal optical properties [41,29]. Note that, for simplicity, we used a physically justified constraint for the $\hat{\epsilon}(z)$ tensor of a C^* viz. one of its principal axes lies in the plane layer. If that were not the case, no new effect and no principal difficulties in solving the problem would appear but the corresponding expressions would be cumbersome, thus we omit them.

In conclusion, we note that C^*s may be characterized by ferroelectric properties [42] which make their structure and optical properties sensitive to electric fields. Thus, C^*s with ferroelectric properties show promise for applications in electrooptical imaging devices. Hence, the electrooptics of C^*s is studied extensively [2,4,44] but this topic is out of the scope of this book.

4.4   Blue Phase of Liquid Crystals

The blue phase of CLCs exhibit the most sophisticated optical properties of all liquid crystals [4,5,45]. However, the nature and structure of the blue phases are not determined with confidence at the present time; but, it is clear that optical methods are the most useful in the study of blue phases in general and spatial structures of their dielectric properties in particular. Hence we do not assume a concrete form for the dielectric tensor of blue phases but place the emphasis at specific optical features which may help to reveal blue phase structure in the future experiments.

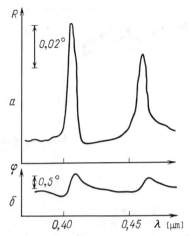

FIGURE 4.16. Dependences of selective reflection (upper curve) and optical rotation (lower curve) on wavelength in cholesteryl pelargonate at 91.35°C [78] corresponding to the transition from BPI to BPII. Sample thickness is 25 and 12 μm respectively.

4.4.1   Observed Properties of Blue Phases

The blue phase was observed for some CLCs in a narrow temperature range of less than or about one degree in the transition between isotropic liquid and usual cholesteric phase. The studies used several physical methods and showed that the blue phase is a thermodynamically stable state. However, in that narrow temperature range, some compounds exhibit up to three different blue phases labelled BPI, BPII and BPIII in the order they appear as temperature increases. Phase BPIII is the least investigated, hence we restrict ourselves to the two former cases.

Without going into details of blue phase properties (see [4,5,45]), we discuss the most unusual optical properties. First, the blue phase selectively reflects visible light—sometimes it appears as blue (Figs. 4.16, 17). In contrast to CLCs where only first-order selective reflection is strong, the blue phase yields several reflexes of comparable intensity. Definite circular polarization, as in CLCs, is selectively reflected backward from the blue phase. Second, the blue phase exhibits strong optical gyrotropy, the direction of optical rotation that inverts sign at the selective reflection frequency. Third, the blue phase is optically isotropic—it has no linear birefringence.

In recent experimental studies of selective reflection, the consensus was that the blue phase had a cubic symmetry with either a primitive or body-centered unit cell [5]. There are many proposed structures that agree with the experimental data or, at least, the main experimental facts about the blue phase. Specifically, these are the three-dimensionally periodic lattice of the locally biaxial dielectric permittivity tensor which follows from Lan-

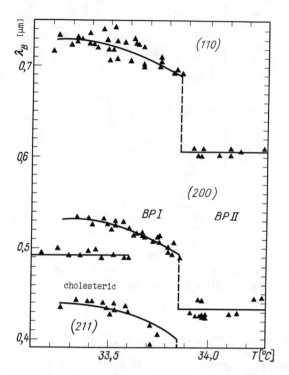

FIGURE 4.17. Selective reflection dependence on the temperature of a CLC and blue phases BPI and BPII for the 50–50% mixture of chiral and non-chiral biphenyls (BPI is supercooled below 33.6°C) [55].

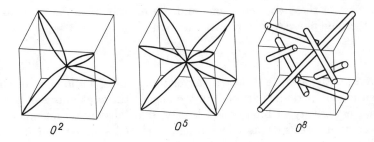

0^2 0^5 0^8

FIGURE 4.18. The structure of declinations in cubic cells of various space groups [79].

dau's theory [46,47,48] and the three-dimensional lattice of linear declinations in the director field (Fig. 4.18 [49]. Determination of blue phase structure requires analysis of all available experimental data and the preferential assignment of one of the model structures. If this is not possible from the available data we must recommend additional measurements which are necessary for the structure determination. The following sections of this chapter describe a way of achieving this goal.

4.4.2 Symmetry Restrictions for the Dielectric Permittivity Tensor

Since the dielectric permittivity tensor $\hat{\epsilon}(\mathbf{r})$ contains all the information about the optical properties of a blue phase, it is important to find its explicit form. However, blue phase structure is not known, therefore, we do not assume a specific form of $\hat{\epsilon}(\mathbf{r})$ but account for its properties based on known experimental facts. This is done by expanding $\hat{\epsilon}(\mathbf{r})$ in a Fourier series

$$\hat{\epsilon}(\mathbf{r}) = \sum_{\boldsymbol{\tau}} \hat{\epsilon}_{\boldsymbol{\tau}} e^{i\boldsymbol{\tau}\mathbf{r}} \qquad (4.60)$$

where $\boldsymbol{\tau}$ are reciprocal lattice vectors of the blue phase.

We set constraints for coefficients of (4.60) from known optical properties. First, the presence of selective scattering with several reflexes means that, in contrast to expansion (2.15) for a CLC, expansion (4.60) contains more than three non-zero terms. Also, the known polarization properties of selective scattering for these reflexes determine certain constraints for the corresponding Fourier-components $\hat{\epsilon}_{\boldsymbol{\tau}}$ of the tensor. Second, the optical activity and the sign inversion at the selective reflection frequency support the conclusion that blue phase optical activity is due to structural properties, Hence, the molecular gyrotropy in $\hat{\epsilon}$ may be neglected at least to a first approximation. Third, the absence of birefringence means that the zero harmonic $\hat{\epsilon}_0$ is proportional to the unit tensor.

In view of these considerations and the experimental data we assume that the blue phase has cubic structure with the tensor

$$\hat{\epsilon}(\mathbf{r}) = \hat{\epsilon}_0 + \hat{\epsilon}^a(x, y, z) \tag{4.61}$$

where $\hat{\epsilon}_0$ is the average dielectric permittivity and $\epsilon_{ik}^a(x, y, z)$ is the three-dimensionally periodic part of $\hat{\epsilon}(\mathbf{r})$. We neglect the weak effects of possible spatial periodicity of blue phase density, light absorption and molecular gyrotropy. The $\hat{\epsilon}^a(\mathbf{r})$ tensor is spurless, real and symmetric under these conditions. In conventional crystal optics, crystals are assumed homogeneous since the light wavelength is longer than the lattice period; here, only the homogeneous part of the dielectric permittivity tensor is essential and its symmetry is determined by the point group of the crystal and is well known from optics. For cubic structures (analyzed here), in accordance with the adopted approximations, the part of $\hat{\epsilon}_0$ that corresponds to the zero harmonic of (4.60) is proportional to the identity tensor.

The lattice period of a blue phase is comparable to light wavelength, hence we include the inhomogeneous part, $\hat{\epsilon}^a(\mathbf{r})$, whose local symmetry is not necessarily identical with cubic symmetry, which varies over the unit cell and is limited by the crystal space group. We use these symmetry restrictions to establish the most general form of $\hat{\epsilon}^a(\mathbf{r})$ for specific space groups describing blue phase structure. Since the blue phase is chiral, we restrict ourselves to space groups without inversion centers, i.e., enantiomorphic groups.

The symmetry properties permits us to determine the most general form of the blue phase tensor $\hat{\epsilon}(\mathbf{r})$ [50], i.e., the form which is invariant with respect to all transformations of the space group describing blue phase symmetry properties. According to experiment [5] the observed blue phase symmetry is described by cubic space groups.

In the study of blue phase $\hat{\epsilon}(\mathbf{r})$ symmetry properties we account for that fact that all cubic space groups have a third-order symmetry axis directed along the cube diagonal. As a result, all components of the $\hat{\epsilon}(\mathbf{r})$ tensor depend on two arbitrary periodic functions of coordinates $f_1(\mathbf{r})$ and $f_2(\mathbf{r})$:

$$\hat{\epsilon}(x, y, z) = \begin{pmatrix} f_1(x, y, z) & f_2(z, x, y) & f_2(y, z, x) \\ f_2(z, x, y) & f_1(y, z, x) & f_2(x, y, z) \\ f_2(y, z, x) & f_2(x, y, z) & f_1(z, x, y) \end{pmatrix}. \tag{4.62}$$

We find directly that the tensor of (4.62) is invariant under rotations of $120°$ and $240°$ around the third-order axis; this is equivalent to the cyclic permutation of coordinates X, Y and Z with a corresponding cyclic permutation of $\hat{\epsilon}(\mathbf{r})$ tensor components, e.g., $x \to y \to z \to x$, $\epsilon_{xx} \to \epsilon_{yy} \to \epsilon_{zz} \to \epsilon_{xx}$, etc.

In addition to third-order axes, a cubic group necessarily includes the other elements of symmetry, namely, the second- and fourth-order axes, screw axes and translations. These elements yield additional restrictions for functions $f_i(\mathbf{r})$ [50,51]; specifically, groups $T^1 - P23$, $T^2 - F23$, $T^3 - I23$, yield the following relationships for $f_i(\mathbf{r})$ because they have second order rotation axes:

$$f_1(x,y,z) = f_1(x,\overline{y},\overline{z}) = f_1(\overline{x},\overline{y},z) = f_1(\overline{x},y,\overline{z})$$
$$f_2(x,y,z) = f_2(x,\overline{y},\overline{z}) = -f_2(\overline{x},\overline{y},z) = -f_2(\overline{x},y,\overline{z}) \qquad (4.63)$$

where $\overline{x}_i = -x_i$. Since groups $T^4 - P2_13$ and $T^5 - I2_13$, have the screw 2_1 axes they yield

$$f_1(x,y,z) = f_1\left(\frac{1}{2}+x, \frac{1}{2}-y, \overline{z}\right) = f_1\left(\frac{1}{2}-x, \overline{y}, \frac{1}{2}+z\right)$$
$$= f_1\left(\overline{x}, \frac{1}{2}+y, \frac{1}{2}-z\right)$$
$$f_2(x,y,z) = f_2\left(\frac{1}{2}+x, \frac{1}{2}-y, \overline{z}\right) = -f_2\left(\frac{1}{2}-x, \overline{y}, \frac{1}{2}+z\right)$$
$$= -f_2\left(\overline{x}, \frac{1}{2}+y, \frac{1}{2}-z\right). \qquad (4.64)$$

All other cubic groups contain subgroups T^1 or T^4 hence relationships (4.63) and (4.64) are valid for them, plus an additional relationship. Namely, those relationships are: (4.63) and

$$f_i(x,y,z) = f_i(\overline{x},\overline{y},\overline{z}) \qquad (4.65)$$

for $O^1 - P432$, $O^3 - F432$ and $O^5 - I432$; (4.63) and

$$f_i(x,y,z) = f_i\left(\frac{1}{2}-x, \frac{1}{2}-z, \frac{1}{2}-y\right) \qquad (4.66)$$

for $O^2 - P4_243$; (4.64) and

$$f_i(x,y,z) = f_i\left(\frac{1}{4}-x, \frac{1}{4}-z, \frac{1}{4}-y\right) \qquad (4.67)$$

for $O^4 - F4_132$, $O^6 - P4_332$ and $O^8 - I4_732$; and (4.64) and

$$f_i(x,y,z) = f_i\left(\frac{3}{4}-x, \frac{3}{4}-z, \frac{3}{4}-y\right) \qquad (4.68)$$

for $O^7 - P4_132$. Beyond this, for body-centered groups ($I23$, $I2_13$, $I432$, $I4_132$) relationship $f_i(x,y,z) = f_i\left(\frac{1}{2}+x, \frac{1}{2}+y, \frac{1}{2}+z\right)$ holds and for face-centered groups ($F23$, $F432$, $F4_132$) the additional relationship $f_i(x,y,z) = f_i\left(\frac{1}{2}+x, \frac{1}{2}+y, z\right) = f_i\left(\frac{1}{2}+z, y, z+\frac{1}{2}\right) = f_i\left(x, \frac{1}{2}+y, \frac{1}{2}+z\right)$ applies. We note that relationships (4.65–4.68) are given for a coordinate system with axes directed along the edges of the unit cubic cell and origin at the point of maximum point symmetry [52]. No other relationship stems from the requirement that $\hat{\epsilon}(\mathbf{r})$ be invariant under transformations of enantiomorphic space groups. Relationships (4.53–4.68) and the relationships connected with the centering properties of unit cells are not sufficient to determine functions $f_i(\mathbf{r})$ although they impose stringent constraints on the $\hat{\epsilon}(\mathbf{r})$ tensor, its Fourier harmonics and phase relationships between different Fourier harmonics of (4.60).

4.4.3   Fourier Harmonics of $\hat{\epsilon}(\mathbf{r})$

Consider the vector of the reciprocal lattice in expansion (4.60) in form $\boldsymbol{\tau} = \left(\frac{2\pi}{d}\right)(h\mathbf{x} + k\mathbf{y} + \ell\mathbf{z})$ where $\mathbf{x}, \mathbf{y}, \mathbf{z}$ are the unit vectors of the coordinate axis, h, k and ℓ are arbitrary integers, so-called Miller indices and d is the dimension of the crystal unit cell. Then the Fourier harmonics are

$$\hat{\epsilon}_\tau = \frac{1}{d^3} \int \hat{\epsilon}(\mathbf{r}) \exp(-i\boldsymbol{\tau}\mathbf{r}) dr \tag{4.69}$$

and integration is over the volume of a unit cell. First of all formulae (4.62) and (4.69) yield that in cubic crystals

$$\hat{\epsilon}_{hk\ell} = \begin{pmatrix} f_1^{hk\ell} & f_2^{\ell hk} & f_2^{k\ell h} \\ f_2^{\ell hk} & f_1^{k\ell h} & f_2^{hk\ell} \\ f_2^{k\ell h} & f_2^{hk\ell} & f_1^{\ell hk} \end{pmatrix} \tag{4.70}$$

where

$$f_i^{hk\ell} = \frac{1}{d^3} \int f_i(x, y, z) \exp[2\pi i(hx + ky + \ell z)] dr. \tag{4.71}$$

Further, as follows from (4.70) and from the fact that the $\hat{\epsilon}^a(\mathbf{r})$ tensor has zero spur, the Fourier harmonics $\epsilon_{hk\ell}$, $\epsilon_{k\ell h}$ and $\epsilon_{\ell hk}$ are expressed by a single set of five numbers which generally are complex. Using the properties of $f_i(\mathbf{r})$ given by (4.63–4.68) we obtain restrictions on $f_i^{hk\ell}$ for reflexes of a special type presented in Table 4.1. These relationships simplify the study of blue phase optical properties and the phase transitions in it. R, I, and C are real, imaginary, and complex quantities, respectively; the letter c denotes chiral reflections while the asterisk denotes reflections that become allowed as a result of the local anisotropy of $\hat{\epsilon}$. The condition that reflections for body-centered lattices (T^3, T^5, O^5, O^8) is $h + k + \ell = 2n$ while, for face-centered lattices (T^2, O^3, O^4), the condition is that h, k and ℓ have the same parity.

The expressions for $\hat{\epsilon}(\mathbf{r})$ and its Fourier harmonics were obtained within the above specified single coordinate system. It is often convenient to consider each Fourier harmonic $\hat{\epsilon}_\tau$ in its own coordinate system in which one of the axes (say Z) is directed along $\boldsymbol{\tau}$ the two others perpendicular to $\boldsymbol{\tau}$. In such a system the ϵ_τ tensor is expanded with five basic tensors $\hat{\sigma}_m$ $(m = 0, \pm 1, \pm 2)$, each with the property that transformation of rotation

TABLE 4.1. Reflections in cubic groups and restrictions on the components $\hat{\epsilon}^\tau$.

Indicies of reflections	Types	ϵ^τ_{xx}	ϵ^τ_{yy}	ϵ^τ_{xy}	ϵ^τ_{xz}	ϵ^τ_{yz}	m
		T^1, T^2, T^3					
$h00$	c	R	R_1	0	0	I	$0, \pm 2$
$hk0$	c	R	R_1	R_2	I	I_1	$0, \pm 1_i, \pm 2$
hhh		0	0	C	C	C	0
hkl	c	C	C_1	C_2	C_3	C_4	$0, \pm 1, \pm 2$
		T^4, T^5					
$\qquad h = 2n+1$	$*$	0	0	I	R	0	± 1
$h00\ \ h = 2n$	c	R	R_1	0	0	I	$0, \pm 2$
$\qquad h = 2n+1$	c	I	I_1	I_2	R	R_1	$0, \pm 1, \pm 2$
$hk0\ \ h = 2n$	c	R	R_1	R_2	I	I_1	$0, \pm 1, \pm 2$
hhh		0	0	C	C	C	0
hkl	c	C	C_1	C_2	C_3	C_4	$0, \pm 1, \pm 2$
		O^6, O^8					
$\qquad h = 4n \pm 1$	$*$	0	0	$\mp iR$	R	0	1
$h00\ \ h = 4n+2$	$*,c$	0	R	0	0	I	± 2
$\qquad h = 4$		$-2R$	R	0	0	0	0
$hh0\ \ h = 2n+1$	c	I	I	I_1	R	$-R$	$0, \pm 2$
$\qquad h = 2n$	c	R	R	R_1	I	$-I$	$0, \pm 2$
$\qquad h = 4n \pm 1$		0	0	$(1 \mp i)R$	$(1 \mp i)R$	$(I \mp i)R$	0
$hhh\ \ h = 4n+2$		0	0	I	I	I	0
$\qquad h = 4n$		0	0	R	R	R	0
hkl	c	C	C_1	C_2	C_3	C_4	$0, \pm 1, \pm 2$
		O^1, O^3, O^5					
$h00$		$-2R$	R	0	0	0	0
$hh0$	c	R	R	R_1	I	$-I$	$0, \pm 2$
hhh		0	0	R	R	R	0
hkl	c	C	C_1	C_2	C_3	C_4	$0, \pm 1, \pm 2$
		O^2					
$\qquad h = 2n+1$	$*,c$	0	R	0	0	I	± 2
$h00\ \ h = 2n$		$-2R$	R	0	0	0	0
$hh0$	c	R	R	R_1	I	$-I$	$0, \pm 2$
$hk0$	c	R	R_1	R_2	I	I_1	$0, \pm 1, \pm 2$
$\qquad h = 2n+1$		0	0	I	I	I	0
$hhh\ \ h = 2n$		0	0	R	R	R	0
hkl	c	C	C_1	C_2	C_3	C_4	$0, \pm, \pm 2$

Table 4.1 (*continued*)

Indicies of reflections	Types	ϵ^τ_{xx}	ϵ^τ_{yy}	ϵ^τ_{xy}	ϵ^τ_{xz}	ϵ^τ_{yz}	m
				O^4, O^7			
	$h = 4n \pm 1$	*	0	0	$\pm iR$	R	0	-1
$h00$	$h = 4n + 2$	*,c	0	R	0	0	I	± 2
	$h = 4n$		$-2R$	R	0	0	0	0
	$h = 2n + 1$	c	I	I	I_1	R	$-R$	$0, \pm 2$
$hh00$	$h = 2n$	c	R	R	R_1	I	$-I$	$0, \pm 2$
	$h = 4n \pm 1$		0	0	$(1 \pm i)R$	$(1 \pm i)R$	$(1 \pm i)R$	0
hhh	$h = 4n + 2$		0	0	I	I	I	0
	$h = 4n$		0	0	R	R	R	0
	hkl	c	C	C_1	C_2	C_3	C_4	$0, \pm 1, \pm 2$

around τ by an angle φ reduces to multiplication by $\exp(im\varphi)$. That is, the $\hat{\epsilon}_\tau$ tensor is expanded over the irreducible representations of the group of rotations around τ, namely,

$$\hat{\epsilon}_\tau = \sum_{m=-2}^{2} \epsilon(\tau; m)\hat{\sigma}_m \tag{4.72}$$

where $\epsilon(\tau; m)$ are the expansion coefficients and

$$\hat{\sigma}_0 = \frac{1}{\sqrt{6}}\begin{pmatrix} -1 & 0 & 0 \\ 0 & -1 & 0 \\ 0 & 0 & 2 \end{pmatrix}, \quad \hat{\sigma}_{\pm 1} = \frac{1}{2}\begin{pmatrix} 0 & 0 & \pm i \\ 0 & 0 & 1 \\ \pm i & 1 & 0 \end{pmatrix},$$

$$\hat{\sigma}_{\pm 2} = \frac{1}{2}\begin{pmatrix} 1 & \mp i & 0 \\ \mp i & -1 & 0 \\ 0 & 0 & 0 \end{pmatrix}$$

(below we omit the index a at ϵ).

Expansion (4.72) is especially useful if τ is parallel to one of the crystal symmetry axes. In that case, there are selection rules [53] that characterize the space group for corresponding Fourier harmonics. These rules determine, for each group and every τ, the value of m such that $\epsilon(\tau; m \neq 0$ in (4.72). A specific example of this is one in which we suppose that vector τ is parallel to the second-order axis. Neither $\hat{\epsilon}_\tau$ tensor nor applicable Fourier component change with rotation by angle π about that axis. However, basis tensors change by a factor of $\exp(im\pi)$ after each rotation. For $m = \pm 1$ these factors are -1 and the terms with $m = \pm 1$ must be absent in (4.72). If τ is parallel to the third- or fourth-order axes, then (4.72) retains only the term with $m = 0$. The rules for the case when τ is parallel to the screw axis n_j are determined similarly. In this case, terms in (4.72) whose m's are such that $(m - jh)/n$ are not integers, vanish. (Here we assumed for simplicity that the screw axis is parallel to X and $\tau = 2\pi h\mathbf{x}/d$). The selection rules so determined were used to compile Table 4.1 (see last column).

4.4.4 The Explicit Form of $\hat{\epsilon}(\mathbf{r})$

Thus, symmetry restrictions are the most essential for Fourier harmonics of $\hat{\epsilon}_\tau$ with τ vectors parallel to symmetry axes. For cubic crystals these vectors are (h, o, o), (h, h, o) and (h, h, h). In addition, using (4.63–4.68), (4.70) and (4.71) we find the relationships between all equivalent harmonics and express them in terms of one of them. We call equivalent such Fourier harmonics that correspond to the τ vectors which transform into one other under the symmetry operations belonging to the crystal space group. Moreover, components (modes) of (4.72) with fixed m that correspond to equivalent Fourier harmonics are expressed via one of these components. We denote a set of such equivalent modes by $[hk\ell, m]$. However, symmetry restrictions reveal no relationship between non-equivalent Fourier harmonics, e.g., between Fourier harmonics with different $|\tau|$. Therefore, the number of parameters needed to determine the structure of the blue phase is generally infinite.

According to the experimental data, however, a few parameters are actually important. For example, BP (probably with group O^2) exhibits two reflections (100) and (110) (and equivalent to them) and BP (group O^8) exhibits three refections (110), (200) and (211) (equivalent). Polarization measurements [54–56] show that the dominant contribution to observed reflections originates from modes with $m = 2$. If we consider only these modes, we obtain, from (4.60) and (4.72), the following expressions for $f_i(r)$ which determine the spatial structure of the $\hat{\epsilon}(\mathbf{r})$ tensor for O^2 in accordance with (4.62):

$$f_1(x, y, z) = \epsilon(100; 2)(C_z - C_y) + \epsilon(110; 2)[2C_yC_z - C_x(C_y + C_z)]$$
$$f_2(x, y, z) = -\epsilon(100; 2)S_x + \epsilon(110; 2)\left[\sqrt{2}S_x(C_y - C_z) - S_yS_z\right], \text{ (4.73)}$$

for O^8

$$f_1(x, y, z) = \epsilon(110; 2)(2S_yC_z - S_xC_y - S_zC_x) + \epsilon(200; 2)(C_{2z} - C_{2y}) +$$
$$+ \frac{2}{3}\epsilon(211; 2)\left[2\sqrt{6}(C_{2z}S_xC_y - C_{2y}S_zC_x) - 2S_{2x}C_yS_z + S_{2y}C_zS_x\right.$$
$$\left. + S_{2x}C_xS_y\right],$$
$$f_2(x, y, z) = \epsilon(110; 2)\left[S_zC_y - \sqrt{2}(C_xC_y + S_zS_x)\right] - \epsilon(200; 2)S_{2x} +$$
$$+ \frac{2}{3}\epsilon(211; 2)\left[2C_{2y}S_zS_x - \sqrt{6}(S_{2z}S_xS_y + S_{2y}C_zC_x)\right.$$
$$\left. - 5S_{2x}S_yC_z - 2C_{2z}C_xC_y + 2C_{2y}S_zS_x\right] \quad \text{(4.74)}$$

where we denote $C_x = \cos(2\pi x/d)$, $S_x = \sin(2\pi x/d)$, etc. The coefficients $\epsilon(hk\ell; 2)$ in (4.73) and (4.74) are real for symmetry reasons, their absolute values and signs are determined either experimentally or theoretically. If the theory of phase transition is invoked, in addition to symmetry restrictions, the number of non-zero parameters in (4.60) and (4.72) may decrease thus refining blue phase structure (see [5]).

4.4.5   Optical Properties

The experimentally observed specific optical features of the blue phase and CLCs are connected with light diffraction at periodic structures. Since the blue phase has no birefringence, a theoretical description of its optical properties is more simple than that of a CLC and similar to X-ray diffraction theory in usual crystals except that polarization properties are different [50,53,57].

The experimentally observed reflections are identified using the Bragg condition (2.4), and possible structures of the blue phase particularly its possible space group are extremely limited [54,55,58,59]. This situation is similar to X-ray diffraction structural analysis in which an observed set of reflexes is used to determine the symmetry of a crystal and the orientation of crystallographic axes.

As shown below, more detailed information about blue phase structure is contained in the intensity and polarization properties of the reflections and their angular and frequency widths.

Kinematic Theory. Let us start our consideration of the optical properties of the blue phase from the simplest kinematic approximation.

From the standpoint of the simplest kinematic approximation (see (2.5) and (2.16)), the intensity $I_{\mathcal{T}}$ of a reflex and reflection coefficient R are proportional to the square of the corresponding Fourier harmonic

$$R(\mathbf{e}_0; \mathbf{e}_1) = \frac{I_{\mathcal{T}}}{I_0} |\mathbf{e}_1^* \hat{\epsilon}_\tau \mathbf{e}_0|^2 R_{\mathcal{T}} \qquad (4.75)$$

where I_0 is the incident intensity, $|\mathbf{R}_1^* \hat{\epsilon} \mathbf{e}_0|^2$ a polarization-structural factor (see below) and factor $R_{\mathcal{T}}$ depends on crystal dimensions and shape and determines the reflex angular and frequency width.

This formula reveals an essential property of light scattering by the blue phase—scattering shows a strong and complicated dependence on polarization that distinguishes light scattering in the blue phase from X-ray diffraction. The polarization dependence of R and the information on crystal structure are contained in the polarization-structural factor $|\mathbf{e}_1^* \hat{\epsilon}_\tau \mathbf{e}_0|^2$ which describes diffraction scattering of the $\mathbf{e}_0$-polarized wave into an $\mathbf{e}_1$-polarized wave. If we are not interested in the polarization of the diffracted wave, the factor $|\mathbf{e}_1^* \hat{\epsilon}_\tau \mathbf{e}_0|^2$ is summed over all $\mathbf{e}_1$ and the reflection coefficient becomes

$$R(\mathbf{e}_0) = (|\mathbf{A}_{\mathcal{T}}|^2 - q_{\mathcal{T}}^{-2}|\mathbf{q}_{\mathcal{T}}\mathbf{A}_{\mathcal{T}}|^2)R_{\mathcal{T}} \qquad (4.76)$$

where $\mathbf{A}_{\mathcal{T}} = \hat{\epsilon}_{\mathcal{T}}\mathbf{e}_0$, $\mathbf{q}_{\mathcal{T}} = \mathbf{q} + \boldsymbol{\tau}$ and $\mathbf{q}$ is the incident wave vector.

If the incident wave is polarized, the diffracted wave is polarized with polarization vector

$$\mathbf{e}_{\mathcal{T}} = \left[\mathbf{A}_{\mathcal{T}} - \mathbf{q}_{\mathcal{T}}(\mathbf{q}_{\mathcal{T}}\mathbf{A}_{\mathcal{T}})\mathbf{q}_{\mathcal{T}}^{-2}\right]\left|\mathbf{A}_{\mathcal{T}} - \mathbf{q}_{\mathcal{T}}(\mathbf{q}_{\mathcal{T}}\mathbf{A}_{\mathcal{T}})\mathbf{q}_{\mathcal{T}}^{-2}\right|^{-1}. \qquad (4.77)$$

Note that at $\mathbf{e}_1 = \mathbf{e}_{\mathcal{T}}$ the reflection in (4.75) is at a maximum.

Let us consider polarization properties of scattering in the blue phase in more detail [50]. If the Bragg condition is satisfied approximately for a certain reflex corresponding to the reciprocal lattice vector $\boldsymbol{\tau}$, the polarization properties of scattering are determined by the form of the Fourier harmonic $\hat{\epsilon}_{\mathcal{T}}$ in (4.75–4.77).

Note that, as in the case of Mössbauer diffraction with polarization-independent forward scattering amplitude (Section 3.4.5), polarization properties of diffraction scattering are conveniently described by introducing vectors of eigenpolarizations $\mathbf{n}_i$ and $\mathbf{n}'_i$ ($i = 1, 2$) (see (3.38)). For the blue phase, the operators $\hat{S}$ and $\hat{S}'$ in (3.38) are given in the form

$$S_{ik} = (\epsilon_{\tau ie})^* \epsilon_{\tau mk}(\delta_{m\ell} - q_\ell q_m/q^2)$$
$$S'_{ik} = (\epsilon_{\mathcal{T} i\ell})^* \epsilon_{\mathcal{T} mk}(\delta_{m\ell} - q_{\mathcal{T}\ell}q_{\mathcal{T}m}/q_{\mathcal{T}}^2). \qquad (4.78)$$

Expression (4.76) reaches its maximum and minimum values for the eigenpolarizations, and the diffractive scattering of each eigenpolarization $\mathbf{n}_i$ results the corresponding eigenpolarization of the diffracted wave $\mathbf{n}'_i$ because $(\mathbf{n}_2^{'*}\hat{\epsilon}_{\mathcal{T}}\mathbf{n}_1) = (\mathbf{n}_1^{'*}\hat{\epsilon}_{\mathcal{T}}\mathbf{n}_2) = 0$. The polarization-structural factor of an eigenpolarization is $|F_\sigma^\tau|^2$, where

$$F_\sigma^\tau = (\mathbf{n}_\sigma^{'*}\hat{\epsilon}_{\mathcal{T}}\mathbf{n}_\sigma). \qquad (4.79)$$

The kinematic approach permits the description and qualitative understanding of many specific features of blue phase optics. In the general case of an arbitrary relationship between components with different m values in (4.72), eigenpolarizations given by polarization vectors n_i and $\mathbf{n}'_i$ are elliptical and depend on the Bragg angle θ_B. We say that reflections with elliptical eigenpolarizations are chiral since they have different reflection coefficients for right- and left-handed polarized light; the chiral reflections are labelled "C" in Table 4.1. The eigenpolarizations of non-chiral reflections are linear.

When only one term with $m = 2$ or $m = -2$ is present in (4.72), maximum diffractive scattering is experienced by light whose eigenpolarization is the same as in the case of a CLC; within the kinematic approximation, we denote this polarization by $\hat{\mathbf{n}}_d$ (see (2.21)) and the polarization-structural factor in (4.76) is $|\epsilon(\tau_i \pm 2)|^2 (1 + \sin^2 \theta_B)|\mathbf{n}_d^*\mathbf{e}_i|^2/4$. For $\theta_B = \pi/2$ (backward scattering) in our case, only right-handed or only left-handed circular polarization is subject to diffraction. The experimentally observed polarization properties of scattering indicate that the amplitudes of modes with $m = 2$ or -2 are greater than those of all other modes.

Generally, if non-polarized light is diffractively scattered by the blue phase, the scattered light is partially polarized. Using (4.75–4.79) we find the degree of polarization of scattered light P_r as

$$P_r = \left| |F_1|^2 - |F_2|^2 \right| (|F_1|^2 + |F_2|^2)^{-1} \qquad (4.80)$$

where incident light is non-polarized. The scattered light polarization is described by vector $\mathbf{n}'_1$ if $|F_1| > |F_2|$, and $\mathbf{n}'_2$ if $|F_2| > |F_1|$. In a special case the scattered light may be completely polarized, even for unpolarized incident light was not polarized. That situation occurs when one of the eigenpolarizations is not subjected to diffraction scattering, i.e., one of the amplitudes F_1 or F_2 is zero (e.g., a single component with $m = 2$ or -2 is presented in (4.72)).

Circular and Linear Dichroism. Up till now we have considered selective reflection from the blue phase. When light reflection is at a maximum, the intensity of light transmitted through the sample is at a minimum; this fact was exploited in experimental studies of the blue phase [54,58,59]. Specifically, it is informative to measure the circular dichroism that depends on the difference between transmission coefficients T for the left-handed $(-)$ and right-handed $(+)$ polarized light as described in Section 4.15 (see (4.35) and Fig. 4.19).

When the sample does not absorb light, the values of D are easily expressed within the kinematic approximation in terms of the reflection coefficient $R_{\pm}$ for right- (left-)handed polarization:

$$D = \frac{1}{2}(R_- - R_+). \qquad (4.81)$$

Assuming that only the mode with $m = \pm 2$ is present in (4.72) equations (4.81) and (4.76) yield an expression for the circular dichroism averaged over wavelength for each individual reflex:

$$\int D d\lambda \sim |\epsilon(\boldsymbol{\tau}_j \pm 2)|^2 (1 + \sin^2 \theta_B)(\sin \theta_B)^{-1}. \qquad (4.82)$$

In addition to the circular dichroism, the blue phase exhibits linear dichroism D_e of diffraction origin. Linear dichroism is connected with the difference between reflection coefficients of two orthogonal linear polarizations, one of which, for example, is parallel to the plane of vectors $\mathbf{q}$ and τ, the other orthogonal to this polarization (4.11). It is essential that circular and linear dichroism have different dependence on components of the $\hat{\epsilon}_{\boldsymbol{\tau}}$ tensor and Bragg angle. Hence, if measured simultaneously these two effects provide additional information about $\hat{\epsilon}_{\boldsymbol{\tau}}$. If only the components with

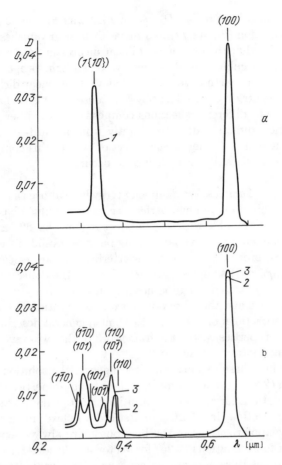

FIGURE 4.19. Circular dichroism spectra of BPII obtained from the mixture of cholesteryl nonanoate and cholesteryl chloride [58]: (a) normal incidence, (b), oblique incidence; curves 2 and 3 were obtained with different incidence directions.

$m = 2$ or -2 contribute to $\hat{\epsilon}_{\boldsymbol{\tau}}$, $D_e = \cos^2 \theta_B D/2 \sin \theta_B$, while in backward diffraction reflection ($\theta_B = \pi/2$) and linear dichroism vanishes.

We emphasize that the existence of linear dichroism does not contradict the assignment of cubic crystal symmetry. When light propagates exactly along the third- or fourth-order axis, it follows that linear dichroism must vanish for symmetry reasons. The physical cause for this is that the linear dichroisms due to different reflections compensate each other. However, if the propagation direction deviates slightly from a third- or fourth-order axis by an angle of the order of the reflection width, this compensation is violated and considerable linear dichroism occurs.

Optics of Perfect Samples. We assumed everywhere above that the scattering is small within an individual perfect region of the blue phase. However, now samples are available with large perfect regions [58–62], and for these regions the kinematic approximation may become invalid since it does not account for the processes of multiple scattering that are essential in thick ($L >> (\mathbf{qs})/(q^2|F_\sigma|)$) samples. Since the form of the dielectric permittivity tensor is complicated, there is neither an exact analytical solution of Maxwell equations and therefore nor an exact description of multiple scattering. The restrictions on $\hat{\epsilon}$ however, permit a general description of blue phase optical properties within the framework of the two-wave approximation of dynamic diffraction theory.

Since the blue phase exhibits no birefringence, the solutions of the dynamic system (2.26) have less complex polarization properties than the corresponding solutions for a CLC; in adddition they are analogous to those of higher-order reflections in a CLC and first-order reflexes in chiral smectics. Thus, the eigenpolarizations in the blue phase are always separated, i.e., the system of four equations (2.26) for vector amplitudes $\mathbf{E}_0$ and $\mathbf{E}_1$ split into two uncoupled systems of two equations for eigenpolarization wave amplitudes; these polarizations are not π- and σ-polarizations—they are determined by (3.38) where $\hat{S}$ and $\hat{S}'$ are given by (4.8). The systems are:

$$\begin{cases} (1 - k_0^2/q^2)E_0^\sigma + \tilde{F}_\sigma^{-\boldsymbol{\tau}} E_1^\sigma = 0 \\ \tilde{F}_\sigma^{\boldsymbol{\tau}} E_0^\sigma + (1 - k_1^2/q^2)E_1^\sigma = 0 \end{cases} \tag{4.83}$$

$$\tilde{F}_\sigma^{\boldsymbol{\tau}} = (\mathbf{n}_\sigma'^* \hat{\epsilon}_{\boldsymbol{\tau}} \mathbf{n}_\sigma)/\epsilon_0.$$

Note that the dynamic system (4.83) is completely equivalent to the corresponding system in X-ray diffraction theory. Thus, we use the results of X- and γ-ray theory to describe light diffraction in the blue phases as in first-order reflexes in chiral smectics and high-order reflexes in CLC. The only specifics of the diffraction in the blue phase is in its polarization properties.

The polarization properties of diffraction in the blue phase differ from those of a CLC. In a CLC, the eigenpolarizations of light propagating at an angle to the cholesteric axis vary in the angular (or frequency) selective

reflection band, while in the blue phase, polarization properties of eigenwaves remain essentially constant over the entire selective reflection band
since the reflections are relatively narrow.

Reflection and Transmission of Light. Using the analogous relationship between X-ray diffraction and blue phase optics, we give the final results for
the optical parameters of a plane-parallel layer of a blue phase illuminated
by a plane monochromatic wave (Fig. 2.5). We expand the amplitudes of
incident $\mathbf{E}^e$, diffracted $\mathbf{E}^r$ and transmitted $\mathbf{E}^t$ waves into eigenpolarizations

$$\mathbf{E}^e = \sum_{\sigma=1}^{2} E_\sigma^e \mathbf{n}_\sigma, \mathbf{E}^r = \sum_{\sigma=1}^{2} E_\sigma^r \mathbf{n}_\sigma^1, \mathbf{E}^t = \sum_{\sigma=1}^{2} E_\sigma^t \mathbf{n}_\sigma.$$

For Bragg geometry ($b < 0$, see Fig. 2.5)

$$E_\sigma^r = E_\sigma^e \tilde{F}_\sigma^\tau (\alpha + i\Delta ctg\ell)^{-1}$$
$$E_\sigma^t = E_\sigma^e \Delta(\Delta\cos\ell - i\alpha\sin\ell)^{-1} \tag{4.84}$$

and Laue geometry ($b > 0$, in Fig. 2.5)

$$E_\sigma^r = E_\sigma^e \tilde{F}_\sigma^\tau \Delta^{-1}\sin\ell$$
$$E_\sigma^t = E_\sigma^e (\cos\ell + i\alpha\Delta^{-1}\sin\ell) \tag{4.85}$$

where $\alpha = (\tau^2 + 2q_0\tau)/2q_0^2$; $\Delta = [\alpha^2 + \tilde{F}_\sigma^\tau \tilde{F}_\sigma^{-\tau}/b]^{1/2}$; $\ell = \Delta q_0^2 L/2(qs)$;
$b = (q_0 s)/(q_\tau s)$, and s is the inner normal to the sample surface; and
parameter α characterizes the deviation of wavelength (or incidence angle)
from that given by the Bragg condition (2.4). Assuming that deviations are
small, ($|\alpha| << 1$), we obtain that

$$\alpha = (\theta_B - \theta)\sin 2\theta_B \tag{4.86}$$

when incident angle θ changes and

$$\alpha = (\lambda - \lambda_B)\sin^2\theta\lambda_B^{-1} \tag{4.87}$$

when wavelength λ changes where $\lambda_B = 4\pi\sqrt{\epsilon_0}\sin\theta/|\tau|$. Quantities ℓ in
(4.74) is the dimensionless thickness of the crystal, while quantity Δ is
proportional to the diffraction correction to the wave vector. The relationships (4.84–4.85) completely solve the problem of light diffraction in the
blue phase within the two-wave approximation. For example, the reflection
coefficients for the waves with eigenpolarizations are

$$R_\sigma = |\tilde{F}_\sigma^\tau| b^{-1}\sin^2\ell(\alpha^2 + \Delta^2\cos^2\ell)^{-1} \tag{4.88}$$

for Bragg geometry and

$$R_\sigma = |\tilde{F}_\sigma^\tau|^2 b^{-1}\Delta^{-2}\sin^2\ell \tag{4.89}$$

for Laue geometry.

Polarization Parameters. If the incident wave has an arbitrary polarization vector e and polarization degree P, the reflection coefficient is

$$R(P, \mathbf{e}) = \frac{1}{2}(1 - P)(R_1 + R_2) + P(R_1|\mathbf{n}_1^* \mathbf{e}|^2 + R_2 \mid \mathbf{n}_2^* \mathbf{e} \mid^2). \qquad (4.90)$$

If $P = 0$ (non-polarized incident light), the degree of polarization of reflected and transmitted light becomes

$$P^r = |R_1 - R_2|(R_1 + R_2)^{-1}$$
$$P^t = |R_1 - R_2|(2 - R_1 - R_2)^{-1}. \qquad (4.91)$$

The reflected light polarization vector coincides with $\mathbf{n}_1'$, if $R_1 > R_2$, and with $\mathbf{n}_2'$, if $R_2 > R_1$. The transmitted light partial polarization is described by $\mathbf{n}_1$ at $R_2 > R_1$ and $\mathbf{n}_2$ at $R_1 > R_2$. If only one component with $m = 2$ or -2 is present in $\hat{\epsilon}_\tau$, polarization parameters are identical with those obtained within the kinematic approximation. Note that these expressions were obtained by neglecting light reflection at the dielectric boundary, i.e., when the blue phase and surrounding medium have the same average dielectric constant. When this assumption is not valid, the polarization parameters may be changed. There is no principal difficulty in solving the problem while accounting for dielectric boundaries (see also [4]).

Effects of Multiple Scattering. The formulae of dynamic theory (4.48–4.91) describe a series of effects related to multiple scattering which do not appear in the kinematic consideration. For example, in thick crystals, the frequency (or angular) width of the diffraction scattering region, i.e., the range of frequencies or angles in which the eigenpolarization waves experience selective reflection, is proportional to $|F_\sigma|$ (as seen in (4.84–4.85) and hence proportional to the corresponding harmonic in the Fourier expansion of $\hat{\epsilon}(\mathbf{r})$.

The dynamical approach also describes experimentally observed rotation of polarization plane in the blue phases and, particularly, the inversion of rotation sign (see Fig. 4.17). From (4.84) and (4.85) we find, for the case of circular polarizations $\mathbf{n}_i$ and $\mathbf{n}_i'$, an expression for the rotation angle of polarization plane φ_t for a wave transmitted through a layer of the blue phase

$$\varphi_t = \frac{1}{2} \left\{ \left[arctg \left(\frac{\alpha}{\Delta} tg\ell \right) \right]_{\sigma=1} - \left[arctg \left(\frac{\alpha}{\Delta} tg\ell \right) \right]_{\sigma=2} \right\}. \qquad (4.92)$$

The point at which the rotation changes sign corresponds to $\alpha = 0$. Note that in perfect samples of the blue phase, just as in CLCs, the rotation angle nonlinearly depends on sample thickness as follows from (4.22) (see [40]).

The other effect of multiple scattering is the pendulum beats which were considered in Chapter 3. These beats are a manifestation of light frequency dependence (a deviation of the angle from Bragg's value) of reflection and transmission coefficients at a fixed thickness, when all other parameters are fixed. The period of these dynamical beats is directly expressed by $|F_\sigma^T|$. These values are measurable directly without involving absolute reflection intensities.

Another result of the dynamic consideration is that the blue phase birefringence and linear dichroism due to multiple scattering does not contradict the fact that the blue phase structure belongs to the cubic space group. We are referring to diffraction birefringence with generally elliptical but circular eigenpolarizations for particular propagation directions (as in the case of light propagation along the CLC axis). The difference between eigenwave effective refraction indices Δn is proportional to the phase difference ψ between eigenpolarization waves transmitted through the crystal: $\Delta n = (q_0 s)\psi/q_0^2 L$, where ψ is given by (4.92) without factor $1/2$. The predominant presence of a linear component in the elliptical eigenpolarizations produces the linear birefringence. This birefringence exhibits strong frequency (angular) dependence and changes sign in the selective reflection band when the sign of α in (4.92) changes sign. In the special case when only components with $m = 2$ or -2 are present in ϵ_T, the linear dichroism and linear birefringence vanish at $\theta_B = \pi/2$ but increase when the Bragg angle decreases. Diffraction birefringence for non-chiral reflection is linear as described in [50]. Generally, diffractive birefringence is due to many reflexes, however, that case needs a special study [63,64].

Hence, we conclude that the experimentally observed linear birefringence [54,66] may have diffraction origin which does not deny blue phase cubic structure.

4.4.6   Structure Studies of the Blue Phase

In this section we consider possible ways of determining the structure of blue phases and the practical results obtained in this field. Structure studies of a blue phase (and any other crystal) include three stages: (i) determination of translational periodicity (or dimensions of a unit cell), (ii) determination of lattice symmetry properties, i.e., its space group, and (iii) exact determination of order parameter $\hat{\epsilon}(\mathbf{r})$ dependence on coordinates over the unit cell (or electron density if ordinary crystals are involved). These stages are listed in order of their complexity; and when (iii) is achieved, the structure is determined completely.

Optical structural analysis. Let us consider the most informative methods, namely, the optical methods. As shown above, every Fourier harmonic of the $\hat{\epsilon}(\mathbf{r})$ expansion in (4.60) results in reflective light scattering by the blue phase. Hence, the structure of the blue phase order parameter is determined completely if we experimentally delineate all Fourier harmonics and reconstruct $\hat{\epsilon}(\mathbf{r})$ from (4.60). Formally, expansion (4.60) may include an infinite number of harmonics. However, only few reflections have experimentally noticeable intensities (see [5]) and optical structure analysis reduces, in practice, to determination of a small number of tensor Fourier harmonics. Let us now discuss this issue in more detail. Specifically, in the most general case, every Fourier component is determined by twelve parameters because there are six independent components of the symmetric tensor $\hat{\epsilon}(\mathbf{r})$ which generally are complex quantities. We do not assume that $S_p(\hat{\epsilon}_{\boldsymbol{\tau}}) = 0$, i.e., we account for the spatial modulation of the isotropic part of the dielectric permittivity. The condition that an isotropic Fourier component exists for every space group is the same as for the component with $m = 0$ (for X-rays [67]). Note that the allowed reflections for cubic space groups given in Table 4.1 differ from the corresponding X-ray reflections [67]. Some reflections are allowed because the dielectric permittivity of the blue phase is anisotropic while that for X-rays is scalar. Note also that there may be density modulation of the isotropic part of $\hat{\epsilon}_{\boldsymbol{\tau}}$, however the practice is to neglect this effect.

If we use some additional factors or suggestions in studying the blue phase we reduce the number of parameters that require experimental determination. For example, if we use the symmetry properties of $\hat{\epsilon}_{\boldsymbol{\tau}}$ in cubic crystals, we reduce the number of independent parameters in $\hat{\epsilon}_{\boldsymbol{\tau}}$ for reflexes of a particular type (see Table 4.1). Also, if the results of phase transition theory are involved in the analysis of the blue phase, we reduce the number of free parameters. That is, within the framework of Landau's theory, for every $\hat{\epsilon}_{\boldsymbol{\tau}}$ in (4.72) there is one non-zero term with $m = 2$ or -2. Consequently, the number of free parameters in each harmonic reduces to two.

If we do not use additional non-optical information one parameter, $\hat{\epsilon}_{\boldsymbol{\tau}}$ phase, cannot be determined from light intensity and polarization measurements because they depend on quadratic combinations of $\hat{\epsilon}_{\boldsymbol{\tau}}$. In principle, it is possible to determine all other parameters. Measurements of scattering polarization parameters in a fixed geometry permit determination of seven parameters of $\hat{\epsilon}_{\boldsymbol{\tau}}$ [68]. This limitation originates from the fact that light waves are transverse and, e.g., the components with $m = 1$ cannot be measured in the case of backward scattering ($\theta_B = \pi/2$). Therefore, the scattered light intensity and polarization for two arbitrary directions are sufficient for determining $\hat{\epsilon}_{\boldsymbol{\tau}}$, apart from its phase.

The structural and symmetry properties of blue phases and their correlations with optical parameters discussed above result in the following conclusions: (i) the dimensions of a unit cell are defined by the frequency (angular) positions of selective reflection band (see (2.4)), and (ii) the space group of the structure is given by the set of $\boldsymbol{\tau}$ (Miller's indices of observed reflexes, i.e., by the set of selective reflection band. Note that in some cases one set of observed reflections may correspond to several space groups, as in X-ray structural analysis. The situation is more complicated in practice because few reflexes of the longest wavelengths are observed while, for example, the simple and body-centered groups are distinguishable from the seventh reflection only [65] in the order of increasing $|\tau|$. In that case, unambiguous space group determination requires additional information.

Polarization Measurement and the Phase Problem. Optical structure analysis of blue phases includes the possibility of obtaining more information than is possible from X-ray structure analysis. This is related to the fact that the routine polarization measurement for visible light is nearly impossible in X-rays. Measuring polarization properties of reflexes permits either complete reestablishment of space group or dramatic reduction in the number of competing options. Hence, it is informative to measure the chiral properties of certain reflections, i.e., the difference in scattering of opposite circular polarizations. For example, in groups O^5, O^2 and O^8 the second, third and sixth reflections are not chiral (see Table 4.1). In principle, all groups, except T^3 and T^5 [50,57], are distinguished by polarization measurements (the polarization properties of reflections in T^3 and T^5 coincide, only relative phases of certain reflections differ). The complete delineation of a structure requires determination of $\hat{\epsilon}_\tau$ for the observed reflections and accounting for relative phase. According to (4.60), this is equivalent to unambiguously establishing the coordinate dependence of order parameter $\hat{\epsilon}(r)$. We emphasize again that usual diffraction measurements including polarization in both X-ray and in visible regions, give no information about $\hat{\epsilon}_\tau$ phases which is necessary to determine $\hat{\epsilon}(\mathbf{r})$ in accordance with (4.60). There are sources of coherent light in the visible range, hence practical phase measurements in that range appear to be a less complex problem. Specifically, we can use the practical technique, often used in recording holograms, of detecting the interference between a subject beam (in our case, the diffracted beam) and a coherent reference beam of known phase. We note that for determining $\hat{\epsilon}(\mathbf{r})$, only relative phases of Fourier harmonics with different τ are needed. To find those phases one can use the multiwave diffraction, as in the case of X-rays [69,70]. When the Bragg condition (2.4) is satisfied for several reciprocal lattice vectors τ simultaneously, the intensity and polarization of diffracted waves carry information about relative phases corresponding to $\hat{\epsilon}_\tau$ (no coherent source needed in this case).

In blue phases, information is obtained from the selective scattering dependence on azimuthal angle, i.e., the dependence on the angle of sample rotation round the reciprocal lattice vector τ present in (2.4). Components $\epsilon(\tau; m)$ with $m = 0, \pm 1, \pm 2$ and the interference between them is responsible for the modulation of measured values with azimuthal period 2π, π, $2\pi/3$ and $\pi/2$. Hence, by measuring scattering parameter (e.g., the scattering intensity of two circular and two linear polarizations) dependence on the azimuth we can determine the amplitude and the relative phase of components $\epsilon(\boldsymbol{\tau}; m)$ in (4.72)

The Results of Structure Investigations. As mentioned above, the procedure for determination of blue phase structure may be executed on the basis of optical measurements alone without auxiliary information or assumptions. Experimental studies of pure substances and mixtures yield valuable information about blue phase structure, but they do not permit unambiguous structure determination. The consensus regarding the period of blue phases, is that blue phases occur in substances (mixtures with short pitches of the cholesteric phases ($p \lesssim 5000$ Å). In all substances studied, the ratio of maximum selective scattering wavelength λ_{Bp} (in medium) to cholesteric pitch is approximately the same, viz. $\lambda_{Bp_1} \approx 1.4p$, $\lambda_{Bp_2} \approx 1.2p$. The period of blue phases increases at lower temperatures. This effect is strong in BRI and is weaker [54] or completely vanishes [55] in $BPII$. According to theory (see [5]), period is almost independent of temperature and $\lambda_{BPI} \approx 1.3p$ and $\Lambda_{BPII} \approx 1..13p$. This slight discrepancy with experiment is probably related to the harmonics of $\hat{\epsilon}_\tau$ neglected in the theory.

Blue phase space groups are less clear and there are several space groups proposed as possible candidates for BPI and $BPII$. The only unanimous conclusion is that BPI nd $BPII$ have cubic structure. Many researchers (see [5]) think that BPI has a cubic body-centered lattice (group O^8) and $BPII$ has the primitive cubic cell (group O^2). The results of polarization measurements on oriented samples are the most convincing in this connection [58,59] because the reciprocal lattice vector orientation is established with respect to crystallographic axes. However, we cannot completely exclude the other possibilities because of the finite accuracy of measurements. Specifically, groups O^2 and O^8 have subgroups T^1 and T^5 respectively. Two sets of reflexes (100) and (110) observed in $BPII$ and attributed to the O^2 group may also be identified as (200) and (220) reflexes of any cubic group except O^1, O^2, O^3 and O^5 (see Table 4.1). A more reliable determination of group requires a larger number of reflection measurements.

Results which differ from those above were obtained when studying selective light scattering in cholesteric n-alcanoates ($3 \leq n \leq 18$) [65], one of the phases of which were assigned to the O^5 group. In [71], the $BPII$ lattice was assigned as body-centered since its third reflection is chiral. Discrepancies in space group designation for blue phases is apparently associated with the fact that a small number of reflections were considered, although it is possible that the symmetries of blue phases in different compounds or mixtures differ. Moreover, space group identification may be hampered by multiwave effects due to which reflection polarization properties differ from those predicated by theory, e.g., non-chiral reflexes may exhibit chirality. Some effects which make identification difficult may appear in mixtures with decomposed phases [72].

Until now, the optical methods did not reveal the complete structure of compounds, however some auxiliary concepts may change this. If we assume, according to theory, that the main structural contribution comes from modes $\epsilon(\tau; 2)$, we can calculate the ratio of the amplitude moduli of different harmonics from the data on the circular dichroism [58,59]. Taking the values of $\epsilon(\tau; 2)$ with the signs which follow from theory, we obtain $\hat{\epsilon}(\mathbf{r})$ from (4.73) and (4.74) to within a factor. We should note that for a series of mixtures of cholesteryl nanoate and cholesteryle chloride these ratios are, on average, $|\epsilon(110; 2)| : |\epsilon(200; 2)| : |\epsilon(211; 2)| \approx 1 : 1 : 0.33$ for BPI and $|\epsilon(100; 2)| : |\epsilon(110; 2)| \approx 2.5$ for $BPII$ [58,59,64], that is, they are close to values of Landau's theory. We can determine the signs of $\epsilon(\tau; 2)$, i.e., solve the phase problem experimentally using NMR, an effective and useful instrument for studying blue phase structure [73,74]. NMR spectra may contain information about the dielectric permittivity tensor $\hat{\epsilon}(\mathbf{r})$ and its spatial distribution in the unit cell. Using the values of $\hat{\epsilon}_\tau$ obtained by optical methods for calculation of NMR spectra and retaining $\hat{\epsilon}_\tau$ phases as free parameters, we can then fit the measured and calculated spectra, and thus find the phases and recover the distributions of $\hat{\epsilon}(\mathbf{r})$ in the structure involved.

To conclude this chapter we note that one of the main problems associated with optical studies of blue phases is the structure of $BPIII$ which has no long-range translational order [77]. One of the possible explanations for this is that $BPIII$ is the phase of strongly developed fluctuations in an isotropic liquid [46,57], since the estimates made [57] indicate that common fluctuations are not sufficiently large to cause the circular dichroism observed in this phase. It is also possible that $BPIII$ consists of an emulsion of BPI or $BPII$ in an isotropic liquid [76].

The optics of chiral liquid crystals, especially the optics of blue phases, remains the subject of intense investigation [90–115] because of the increasing number of possible new applications for this class of liquid crystals [92–95]; hence, the physics of this type of liquid crystal is considered interesting by many researchers [91–93]. The chiral liquid crystal, particularly the chiral smectics are regarded as prospective materials in LC display systems [94–95]. New applications of this material may be connected with surface guided optical waves [109–100] whose parameters are changed easily by weak external forces. The most significant point concerning the physics of chiral liquid crystals and blue phases is that primary methods of investigation are optical in nature.

There was recent progress in blue phase structure determination due to application of the Kossel diagram technique [99]. New BP phases induced by electric field were observed [98,100] by the electro-optics methods [96] which also enable us to observe [112] unusual electrostriction phenomena which along with the structure phase transitions in the electric field [98,111] found their theoretical explanation [97,102,112]. The previously unexplained reflections [114] which contradicted to the attributed to BP space groups found their explanation in the framework of many-wave optics [1,2,115]. The main attention in these investigations is paid to the $BPIII$.

Recently, it has been suggested that $BIII$ has a quasicrystal structure [85–87], but there was no experimental confirmation of this suggestion [96]. Very interesting possibilities for its structure investigations are opened by the method of fast freezing of blue phase which permits to transform the blue phase into solid state maintaining its liquid crystal structure properties [88]. However, the nature of $BPIII$ as well as its structure is not understood. It is why the fog phase ($BPIII$) remains to be a mystery which is more than a hundred years old and is an intriguing object of investigations.

References

1. P.G. de Gennes: *The Principles of Liquid Crystals* (Oxford, Charendon Press, 1974).

2. L.M. Blinov: *Electro- and Magnetooptics of Liquid Crystals* (Moscow, Nauka, 1978) (in Russian).

3. S. Chandrasekhar: *Liquid Crystals*, (Cambridge-New York-Melburn, Cambridge University Press, 1977).

4. V.A. Belyakov, A.S. Sonin: *Optics of Cholesteric Crystals* (Moscow, Nauka, 1982) (in Russian).

5. V.A. Belyakov, V.E. Dmitrienko: Uspakhi Fiz. Nauk **146**, 369 (1985) [Sov. Phys. - Usp. **28**, 535 (1985)].

6. M.A. Osipov: Fiz. Tv. Tela **27**, 1651 (1985).

7. V.A. Belyakov, V.E. Dmitrienko, V.P. Orlov: Uspekhi Fiz. Nauk **127**, 221 (1979) [Sov. Phys. - Usp. **22**, 63 (1979)].

8. S.M. Osadchii, V.A. Belyakov: Kristallografiya **28**, 123 (1983).

9. R. Dreher, G. Meier: Solid State Communication **13**, 607 (1973); Phys. Rev. **8A**, 1616 (1973).

10. D. Taupin: J. de Phys. Coll., C-4 **30**, 32 (1969).

11. E. Elachi, O. Yeh: J. Opt. Soc. Amer. **53**, 840 (1973).

12. V.A. Belyakov, V.E. Dmitrienko: Fiz. Tv. Tela **15**, 2724 and 3540 (1973) [Sov. Phys. - Solid State, **15**, 1811, 2364 (1974)].

13. D.W. Berreman, T.J. Sheffer: Phys. Rev. Lett. **25**, 577 (1970); Mol. Cryst. Liq. Cryst. **11**, 395 (1970); Phys. Rev. Ser. A., **5**, 1397 (1971).

14. R.W. James: *The Optical Principles of the Diffraction of X-rays* (London, Bell and Sons, 1967).

15. B.W. Batterman, H. Cole: Rev. Mod. Phys. **36**, 681 (1964).

16. Z.G. Pinsker: *Dynamical Scattering of X-rays in Crystals* (Berlin, Springer-Verlag, 1978).

17. M. Born, E. Wofl: *Principles of Optics* (Oxford, Pergamon Press, 1965).

18. H. DeVries: Acta Cryst. **4**, 219 (1951).

19. S. Chandrasekhar, G.S. Ranganath: Mol. Cryst. Liq. Cryst. **25**, 195 (1974).

20. G.S. Ranganath, S. Chandraskekhar, U.D. Kini, K.A. Suresh, S. Ramaseshan: Chem. Phys. Lett. **19**, 556 (1973).

21. H. Takezoe, Y. Ouchi, A. Sugita, M. Hara, A. Fukuda, E. Kuze, N. Goto: Japan J. Appl. Phys. **21**, L390 (1982).

22. H. Takezoe, Y. Ouchi, M. Hara, A. Fukuda, E. Kuze, N. Goto: Japan J. appl. Phys. **22**, L185 (1983).

23. H. Takezoe, Y. Ouchi, M. Hara, A. Fukuda, E. Kuze: Japan J. Appl. Phys. **22**, 1080 (1983).

24. H. Takezoe, K. Hashimoto, Y. Ouchi, M.Hara, A. Fukuda, E. Kuze: Mol. Cryst. Liq. Cryst. **101**, 329 (1983).

25. R.S. Akopyan, B.Ya. Zeldovich, N.V. Tabiryan: ZhETF **83**, 1770 (1982).

26. M.A. Paterson: Phys. Rev. **A27**, 520 (1983).

27. C. Oldano: Phys. Rev. **A31**, 1014 (1985).

28. C. Oldano, P. Allia, L. Trossi: J. Phys. **46**, 573 (1985).

29. S. Garoff, R.S. Meyer, R. Barakat: J. Opt. Soc. Am. **68**, 1217 (1978).

30. D.G. Khoshtariya, S.M. Osadchii, G.S. Chilaya: Kristallografiya **30**, 755 (1985).

31. M.J. Stephen, J.P. Straley: Rev. Mod. Phys. **46**, 617 (1974).

32. S. Chandrasekhar: Rep. Progr. Phys. **39**, 613 (1976).

33. G. Holzwarth, N.A.W. Holzwarth: J. Opt. Soc. Amer. **63**, 324 (1973).

34. R. Nitiananda, U.D. Kini, S. Chandrasekhar, K.A. Suresh: In: Proc. of Int. Liquid Cryst. Conference, Banglore-Pramana, Suppl., 1975 v. 1, p. 325.

35. V.A. Belyakov, V.E. Dmitrienko: FIz. Tv. Tela **18**, 2880 (1976) [Sov. Phys. - Solid State, **18**, 1681 (1976)].

36. Yu. Kagan, A.M. Afanas'ev: ZheTF **48**, 327 (1965).

37. K.A. Suresh: Mol. Crys. Liquid. Crys. **35**, 267 (1976).

38. S.N. Andronishidze, V.E. Dmitrienko, D.G. Khoshtariya, G.S. Chilaya: Pis'ma ZhETF **32**, 19 (1980).

39. S. Endo, T. Kuribara, T. Akahane: Japan J. Appl. Phys. **22**, 499 (1983).

40. S.M. Osadchii: Kristallografiya **28**, 758 (1983).

41. B.W. Berreman: Mol. Cryst. Liq. Cryst. **22**, 175 (1973).

42. S.A. Plkin: *Structural Transformations in Liquid Crystals*, (Moscow, Nauka, 1981) (in Russian).

43. V.A. Belyakov, V. E. Dmitrienko: In: *Light Scattering in Solids*, ed. J.L. Birman, H.Z. Cummins, K.K. Rebane (New York, London, Plenum Press, 1979) p.377.

44. V.E. Dmitrienko, V.A. Belyakov: ZhETF **78**, 1568 (1980) [Sov. Phys. -JETP **51** , 787 (1980).

45. H. Stegemeyer, K. Bergman: In: *Liquid Crystals of One and Two-Dimensional Order*, Ed. W. Helfrich, G. Heppke, (Berlin, Heidelberg, Springer-Verlag, 1980) p. 1611; H. Stegemeyer: Liq. Cryst. **1**, 3 (1986).

46. S.A. Brazovskii, S.G. Dmitriev: ZhETF **69**, 979 (1975).

47. S.A. Brazovskii, V.M. Filev: ZhETF **75**, 1140 (1978) [Sov. Phys. - JETP **48**, 573 (1978)].

48. H. Grebel, R.M. Hornreich, S. Shtrikman: Phys. Rev. A **28**, 1114 (1983).

49. S. Meiboom, J.P. Sethna, P.S. Andersen, W.F. Brinkman: Phys. Rev. Lett. **46**, 1216 (1981).

50. V.A. Belyakov, V.E. Dmitrienko, S.M. Osadchii: ZhETF **83**, 585 (1982) [Sov. Phys. - JETP **56**, 322 (1982)].

51. V.E. Dmitrienko: Acts. Cryst. **A40**, 89 (1984).

52. *Atlas of Space Groups of the Cubic System* (Moscow, Nauka, 1980) (in Russian).

53. R.M. Honreich, S. Shtrikman: Phys. Lett. **82a**, 345 (1981).

54. S. Meiboom, M. Sammon: Phys. Rev. Lett. **44**, 882 (1980); Phys. Rev. **24A**, 468 (1981).

55. D.L. Johnson, J.H. Flack, P.P. Crooker: Phys. Rev. Lett. **45**, 641 (1980); Phys. Lett. **82A**, 247 (1981).

56. J.H. Flack, P.P. Crooker, H.C. Svoboda: Phys. Rev. A. **26**, 723 (1982).

57. R.M. Hornreich, S. Shtrikman: Phys. Rev. A **28**, 1791 (1983).

58. V.A. Kizel, V.V. Prokhorov: Pis'ma ZhETF **38**, 283 (1983).

59. V.A. Kizel, V.V. Prokhorov: ZhETF **87**, 450 (1984) [Sov. Phys. - JETP **60**, 257 (1984)]; V.A. Kizel, V.V. Prokhorov, D.G. Khoshtariya, G.S. Chilaya: Kristallografiya **31** 130 (1986).

60. M. Marcus: Phys. Rev. A **25**, 2276 (1982).

61. R. Barbett-Massin, P. Cladis, P. Pieranski: Phys. Rev. A **31**, 3912 (1985).

62. H. Stegemeyer, F. Prosch: Phys. Rev. A **30**, 3369 (1984).

63. V.A. Belyakov, V.E. Dmitrienko: Kristallografiya **27**, 14 (1982).

64. V.A. Belyakov, E.I. Demiknov, V.E. Dmitrienko, V.K. Dolganov: ZhETF **89**, 2035 (1985) [Sov. Phys. - JETP **62**, 1173 (1985)].

65. A.J. Nicastro, P.H. Keyes: PHys. Rev. A **27**, 431 (1983).

66. P.H. Keyes, A.J. Nicastro, E.M. McKinnon: Nol. Crys. Liq. Crys. **67**, 715 (1981).

67. *International Tables for X-ray Crystallography*, vol. 1, (Birmingham, Kynoch Press, 1952) p. 306.

68. R.M.A. Azzam, N.M. Bashara: *Ellipsometry and Polarized Light* (Amsterdam, North-Holland, 1977).

69. B. Post: Acta Cryst. **A35**, 17 (1979).

70. L.D. Chapman, D.R. Yoder, R. Colella: Phys. Rev. Lett. **46**, 1578 (1981).

71. K. Tanimoto, P.P. Crooker: Phys. Rev. A **29**, 1596 (1984).

72. M. Marcus: Molec. Cryst. Liq. Cryst. **82**, 33 (1982).

73. E.T. Samulski, A. Luz: J. Chem. Phys. **73**(1), 142 (1980).

74. Z. Yanis, G. Chidichimo, J.W. Doane: Phys. Rev. A. **28**, 3012 (1983).

75. P.L. Finn, P.E. Cladis: Mol. Crys. Liq. Cryst. **84**, 159 (1982).

76. R.M. Hornreich, M. Kugler, S. Shtrikman: Phys. Rev. Lett. **48**, 1404 (1982).

77. E.I. Demikhov, V.K. Dolganovv: Pis'ma ZhETF **42**, 15 (1985).

78. K. Bergman, R. Pollmann, G. Scherer, H. Stegemeyer: Z. Naturefor., **34a**, 253 (1979).

79. S. Meiboom, M. Sammon, D.W. Berreman: Phys. Rev. A **28**, 3553 (1984).

80. V.V. Zheleznyakov, BV.V. Kocharovskii, V.V. Kocharovskii: Uspekhi Fiz. Nauk. **141**, 257 (1983).

81. R. Barbet-Massin, P.E. Cladis, Pieranksi: Phys. Rev. A **30**, 1161 (1984).

82. P. Pieranski, P.E. Cladis, R. Barbet-Massin: J. Physique Lett. **46**, L-973 (1985).

83. W. Kuczynski: Phys. Stat. Sol. (a) **96**, K127 (1986).

84. R. Barbet-Massin, P. Pieranski: J. Physique Coll. C3, suppl. to No. 3, **46**, C3-61 (1985).

85. V.M. Filev: Pis'ma ZhETF **43**, 523 (1986).

86. R.M. Hornreich, S. Shtrikman: Phys. Rev. Lett. **56**, 1723 (1986).

87. D.S. Rokhsar, J.P. Sethna: Phys. Rev. Lett. **56**, 1727 (1986).

88. J.A. Zasadzinski, S. Meiboom, D.W Berreman: Phys. Rev. Lett. **57**, 364 (1986).

89. V.A. Belyakov, S.M. Osadchii, V.A. Korotkov: Kristallografiya, **31**, 522 (1986).

90. V.A. Belyakov and V.E. Dmitrienko: *Optics of Chiral Liquid Crystals*, ed. by I.M. Khalatnikov, Soviet Scientific Reviews/Section A (London, Harwood Academic Publishers GmbH, 1989).

91. D.C. Wright, N.D. Mermin: Rev. Mod. Phys. **61**, 385 (1989).

92. G.S. Chilaya, V.G. Cigrinov: Kristallografiya **33**, 260 (1988).

93. G.S. Chilaya, L.N. Lisetski: Mol. Cryst. Liq. Cryst. **140**, 243 (1980).

94. L.A. Bersenev, V.G. Chigrinov, D.I. Dergachov et al.: Liq. Cryst. **5**, 1171 (1989).

95. A.V. Parfenov, V.G. Chigrinov: LIq. Cryst. **7**, 131 (1990).

96. P.P. Crooker: Liq. Cryst. **5**, 751 (1989).

97. R.M. Hornreich, S. Shtrikman: Liq. Crust. **5**, 777 (1989).

98. F. Porsch, H. Stegemeyer: Liq. Cryst. **5**, 791 (1989).

99. B. Jerome, P. Pieranski: Liq. Cryst. **5**, 799 (1989).

100. G. Heppke, B. Jerome, H.-S. Kitzerow, P. Pieranski: Liq. Cryst. **5**, 813 (1989).

101. P. Pieranski, P.E. Cladis, R. Barbet-Massin: Liq. Cryst. **5**, 829 (1989).

102. V.A. Belyakov, V.E. Dmitrienko: Liq. Cryst. **5**, 839 (1989).

103. V.E. Dmitrienko: Liq. Cryst. **5**, 847 (1989).

104. G. Voets, H. Martin, W. van Dael: Liq. Cryst. **5**, 871 (1989).

105. H. Zink, W. van Daeil: Liq. Cryst. **5**, 899 (1989).

106. H.F. Glesson, H. Cole: Liq. Cryst. **5**, 917 (1989).

107. G. Heppke, H.-S. Ktzerow, D. Ltzsh, H. Papenfuss: Liq. Cryst. **8**, 407 (1990).

108. H.-S. Kitzerow, P.P. Crooker, S.L. Kwok: Phys. Rev. A **42**, 3442 (1990).

109. V.A. Belyakov, V.P. Orlov: Poverkhnost **5**, 13 (1990).

110. V.A. Belyakov, V.P. Orlov: Mol. Cryst. Liq. Cryst. Lett. **8**, 1 (1991).

111. P. Pieranski, P.E. Cladis: Phys. Rev. A **35**, 3552 (1987).

112. H.-S. Kitzerow. P.P. Crooker, S.L. Kwok, G. Heppke: J. Phys. France **51**, 1303 (1990).

113. D. Lubin, R.M. Hornreich: Phys. Rev. A **36**, 849 (1987).

114. W. Kuczinski: Phys. Lett. A **110**, 405 (1987).

115. J.L. Birman, R.M. Hornreich, S. Shtrikman: Phys. Rev. A **39**, 1591 (1989).

5

Radiation of Fast Charged Particles in Regular Media

The physical properties of coherent radiation produced by a charged particle moving through a regular medium are similar to those of diffraction scattering processes. This type of radiation has been studied vigorously [1–6], however investigations in this field have been extended markedly due to the recent progress of accelerators, high-energy particle detection and the application of nuclear methods to solid-state physics. In addition to hard radiation of bremsstrahlung origin which appears when relativistic particles pass through a regular medium, coherent radiation of longer wavelengths is now scrutinized; in particular progress has been achieved in studying transition radiation in the X-ray range.

The theory of X-ray transition radiation was developed in the works by Garibyan and Yang [6] and by Baryshevskii et. al. [7]. However, the theory of radiation due to the periodicity of a medium is not comprehensive enough to explain all the cases of interest for fundamental physics and its applications [54–58]. In this chapter, we consider the main issues of coherent radiation theory of charged particles in regular media. The coherent radiation of fast charged particles that appears due to the spatial periodicity of a crystal is referred to as structure Cherenkov radiation in subsequent sections. There are other terms for this in the literature, e.g., parametric, quasi-Cherenkov, transition, or diffraction radiation in crystals. We we use the term "structure Cherenkov radiation" to emphasize the physical closeness of this radiation to Cherenkov radiation and to distinguish it from transition radiation which originates at the boundary between two media with different dielectric constants [8].

Structure radiation, as Cherenkov radiation, is induced by the electric field of a moving particle. In contrast to Cherenkov radiation which requires that $v > c_p$ [9,10] (v is particle velocity, c_p, phase velocity of photons in the medium) however, this inequality is not necessary for structure radiation. It is known [1] that the direction $\mathbf{k}$ and the frequency ω of an emitted photon in a periodic medium are related as

$$\frac{d}{c_p}\omega(\cos\theta - c_p/v) = 2\pi s \qquad (5.1)$$

where s is an integer and the other notation is given in Fig. 5.1 which

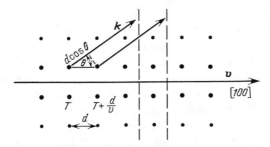

FIGURE 5.1. Schematics for determining the frequencies and directions of coherent emission in a crystal (a unit layer is shown with a dashed line).

shows particle motion in a crystal schematically. The left side of (5.1) is the difference between phases of photons emitted by two consecutive sites of the lattice separated by the distance of lattice constant d. At $s = 0$, Eq. (5.1) corresponds to the zero phase difference between interfering waves and is the condition for Cherenkov radiation, whereas $s \neq 0$ corresponds to structure radiation provided the medium is spatially periodic.

X-ray diffraction scattering in crystals and structure Cherenkov radiation are physically similar because the electromagnetic field of a fast particle may be described in terms of pseudophotons [11] and the coherent radiation of a particle passing through a crystal is a result of pseudophoton diffraction in the crystal lattice. The results of experimental and theoretical studies [1–7,12,13] are mainly related to coherent radiation of fast particles in media with simple types of periodicity, e.g., media with macroscopic periodicity (stacks of plates) or media (including crystals) with periodic modulation of scattering properties such as a coordinate-dependent dielectric constant. In a medium consisting of anisotropic scatterers, structure radiation exhibits a series of peculiarities [13] mainly related to radiation polarization properties.

At the beginning we consider structure Cherenkov radiation in media of complicated crystal structure within the framework of the kinematic approximation. The phenomena in question are the radiation of charged particles in cholesteric liquid crystals, coherent Colomb excitation of Mössbauer transition in a single crystal and the effect of spatial localization of particle trajectory on the parameters of structure radiation. The kinematic approximation describes radiation generation in a periodic medium but neglects regular medium effects on emitted radiation propagation. In some cases, it is not correct to neglect periodicity and introduce the averaged dielectric permittivity of the medium in the description of radiation propagation. In addition, medium periodicity profoundly influences photon propagation of some wavelengths and emission directions. Moreover, the kinematic approximation fails to describe particle induced radiation in thick perfect crystals—that is, the intensity angular distribution and polarization of radiation leaving a perfect crystal contradict kinematic theory prediction. Physically this failure is due to diffraction scattering of radiation in periodic structures. As in cases of X-ray, γ-ray, and light diffraction in liquid crystals, the Borrmann effect can reveal itself which results in an enhanced intensity of emitted radiation for some specific frequencies and directions. This makes interesting the cases when the radiation of a fast particle experiences the diffraction scattering.

Until recently, medium periodicity effects on radiation propagation were considered theoretically for ultrarelativistic particles and, mainly for transition radiation in the X-ray range (see [6,7] and references therein) by Garibyan and Yang and Baryshevskii et. al. Diffractive scattering of transition radiation in a crystal is treated. They make the point that this radiation exhibits enhanced intensity at some specific frequencies.

We describe the effect of diffraction scattering on radiation passing through a crystal and consider mainly media with complicated structure. Specifically, we focus on structure Cherenkov radiation, emission of gamma-quanta in coherent Coulomb excitation of the Mössbauer transitions in a single crystal, radiation of charged particles in CLCs, and the spatial localization effect of particle trajectory on structure radiation parameters. Note that we do not discuss channelled radiation—as it has been in several recent monographs and surveys [7, 15–17,59,60].

5.1   Structure Cherenkov Radiation (Kinematic Approximation)

5.1.1   Formulation of the Problem

The theory of coherent radiation produced from a fast charged particle in a periodic medium is discussed extensively in the literature (see [5–7]). This, radiation is described theoretically by solving the Maxwell equations for a medium with dielectric permittivity that varies periodically and current determined by particle motion. The explicit form of the dielectric permittivity and its spatial variation depends on the radiation-frequency and for any particular case can be determined using the microscopic approach. However, it is clear that spatial variation of dielectric permittivity $\hat{\epsilon}(\mathbf{r})$ must be in accord with medium structure symmetry as long as wavelengths are close to or less than medium period. Hence, coordinate dependence of $\hat{\epsilon}(\mathbf{r})$ is complicated in the general case and, as a rule, there is no exact analytic solution to the Maxwell equations. The exact solutions are known for a few particular cases [1,18,19]. Thus, the problem becomes the approximate description of particle-induced radiation in periodic media. Since the variable part of dielectric permittivity is practically always small, $[\hat{\epsilon}(\mathbf{r}) - \bar{\epsilon}]/\bar{\epsilon} << 1$, particle-induced coherent radiation with wavelengths comparable to medium period are adequately described using diffraction theory similar to that used above when the coherent scattering in periodic media was considered. Below, we apply this theory to structure radiation in media with complicated structures.

Let a charged particle move with constant velocity through a perfect crystal whose dielectric properties are given by $\hat{\epsilon}(\mathbf{r})$ where coordinate dependence is determined by translational properties and crystal symmetry, and crystal magnetic permeability is assumed by unity here and everywhere below. Let us determine the coherent radiation from the particle due to $\hat{\epsilon}(\mathbf{r})$ periodicity without touching upon transition radiation related to crystal boundaries.

Under these conditions, the Maxwell equations yield the following equation for the field $\mathbf{E}$:

$$- \operatorname{rot} \operatorname{rot} \mathbf{E} = \frac{1}{C^2}\hat{\epsilon}(\mathbf{r})\frac{\partial^2 \mathbf{E}}{\partial t^2} + \frac{4\pi}{c^2}\frac{\partial}{\partial t}\mathbf{j}_p \tag{5.2}$$

where c is the speed of light, $\mathbf{j}_p = e v \delta(vt - z)$ is particle current, e and v are particle charge and velocity, and z its coordinate at time t with Z-axis directed along the velocity. We represent field $\mathbf{E}$ in the form $\mathbf{E} = \mathbf{E}_0 + \mathbf{E}_1$, where $\mathbf{E}_0$ satisfies (5.2) with $\hat{\epsilon}(\mathbf{r}) \equiv \hat{\epsilon}_0$ and $\hat{\epsilon}_0$ is the coordinate-independent component of dielectric permittivity, i.e., $\mathbf{E}_0$ is the particle-induced field in a medium with coordinate-independent dielectric permittivity that coincides with the average value of $\hat{\epsilon}$ over the crystal.

For $\mathbf{E}_1$, (5.2) yields

$$- \operatorname{rot} \operatorname{rot} \mathbf{E}_1 = \frac{1}{c^2}(\hat{\epsilon}_0 + \hat{\epsilon}_1)\frac{\partial^2 \mathbf{E}_1}{\partial t^2} + \frac{1}{c^2}\hat{\epsilon}_1 \frac{\partial^2 \mathbf{E}_0}{\partial t^2} \tag{5.3}$$

where $\hat{\epsilon}_1(\mathbf{r}) = \hat{\epsilon}(\mathbf{r}) - \epsilon_0$ is the variable component of crystal dielectric permittivity. Accounting for crystal periodicity, we seek the solution of (5.3) in the form of (2.24).

For $\mathbf{E}_\tau$, we find from (5.3)

$$(1 - k_\tau^2/q^2)\mathbf{E}_\tau + \hat{\epsilon}_0^{-1} \sum_{\tau'}{}' \hat{\epsilon}_{\tau-\tau'}\mathbf{E}_{\tau'} + \frac{\mathbf{k}_\tau}{q^2}(\mathbf{E}_\tau \mathbf{k}_\tau)$$
$$= \hat{\epsilon}_0^{-1} \sum_{\tau'} \hat{\epsilon}_{\tau-\tau'}\mathbf{E}_0(\mathbf{k}_{\tau'}) \tag{5.4}$$

where $q^2 = \hat{\epsilon}_0 \omega^2/c^2$ is the square of the wave vector of wave with frequency ω in the medium with $\hat{\epsilon}_0$, $\mathbf{k}_\tau = \mathbf{k} + \tau$, $\mathbf{E}_0(k)$ is the Fourier component of $\mathbf{E}_0$, coefficients $\hat{\epsilon}_\tau$ are determined by the Fourier expansion of the dielectric tensor in form (2.22), and the primes at the sums mean that terms with $\hat{\epsilon}_0$ are excluded. Thus, if there are no restrictions on $\hat{\epsilon}(r)$, we must solve an infinite system of equations (5.4) with respect to $\mathbf{E}_\tau$ to find $\mathbf{E}_1$. For frequencies in the X-ray range, it is known that $\hat{\epsilon}_0$ is close to (but less than) unity and $\hat{\epsilon}_\tau \sim 10^{-6}$. Hence, there is a small parameter which allows us to retain a finite number of amplitudes $\mathbf{E}_\mathcal{T}$ in (2.24) and (5.4). We begin our consideration from the case where there is excitation of one strong wave $\mathbf{E_k}$. This case is referred to as kinematic. Then we consider the dynamic case in which the field of a particle excites more than one wave. To the kinematic approximation, (5.4) yields for $\mathbf{E_k}$:

$$\mathbf{E_k} = [\epsilon_0(1 - k^2/q^2)]^{-1} \sum_{\tau'} \hat{\epsilon}_{k-k_{\tau'}} \mathbf{E}_0(\mathbf{k}_{\tau'}). \tag{5.5}$$

This expression reaches its maximum for directions $\mathbf{k}$ which satisfy the conditions of coherent radiation (5.1) when $|\mathbf{k}|$ coincides with the modulus free wave vector with frequency ω in the crystal. This maximum is

$$\max|\mathbf{E_k}| = |\frac{q}{Im\,q} \sum_{\mathcal{T}'} \hat{\epsilon}_{k-k_{\tau'}} \mathbf{E}_0(\mathbf{k}_{\mathcal{T}'}, \omega)|. \tag{5.6}$$

5.1.2   Radiation of a Particle in a Crystal

The most favorable conditions for structure radiation occur when a charged particle is channelled in a crystal (see [7,15–17,20,21,59–61]. Hence, we consider this case and emphasize that in contrast to channelling radiation [15–17,59,60], structure radiation exists under channelling conditions and for any other direction of particle motion in a crystal. In the case of channelling, particle energy losses are minimal and its motion through the crystal is considered as straight and uniform regardless of path length (see Fig. 5.1). Structure radiation, like Cherenkov radiation, is a consequence of medium polarization induced by the electric field of a moving particle, i.e., its nature differs from that of a channelled particle for which radiation is due to the transitions between transverse motion levels in the channel [15–17]. A cursory kinematic consideration accounting for the coherence of photon emission by the lattice sites results in the relationships for coherent radiation in periodic media (5.1). Note that (5.1) is similar to the condition of constructive interference of waves emitted by the sites lying along the particle trajectory and excited by an electromagnetic wave of frequency ω and wave vector $\mathbf{k}_p = \omega\mathbf{v}/v^2$, i.e., with the phase velocity v. Relationship (5.1) yields the restriction (known from [11]) on admissible frequencies of emitted radiation at $s \neq 0$

$$\left|\frac{2\pi s c_p}{d(1 + c_p/v)}\right| \le \omega \le \left|\frac{2\pi s c_p}{d(s_p/v_{-1})}\right| \tag{5.7}$$

that defines the set of allowed discrete intervals of frequency. For fast particles these intervals overlap.

5.1.3   Photon Emission Cross Section

Let us find the expressions that determine structure radiation intensity and its connection with crystal structure parameters. This is done by solving

(5.2) for a particle moving through a single crystal; however, for the sake of a simple physical interpretation, at the beginning we use the microscopic approach at the beginning [22,23].

Let $f(\rho, \omega, \mathbf{k}, \mathbf{k})$ be the coherent photon emission amplitude with frequency ω, wave vector $\mathbf{k}$, and transfer of momentum $\mathbf{q}$ to the crystal via an atom with impact parameter ρ relative to the particle trajectory. The total differential cross section of photon emission in the crystal is presented as

$$\frac{d\sigma}{d\Omega_{\mathbf{k}}} = \sum_{\mathbf{q}} \left| \sum_{\rho} f(\rho, \omega, \mathbf{k}, \mathbf{q}) \sum_{\mathbf{r}_\rho} e^{i(\mathbf{k}_p - \mathbf{k})\mathbf{r}_\rho} \right|^2 \tag{5.8}$$

where $\mathbf{k}_p = \omega \mathbf{v}/v^2$, vector $\mathbf{r}_\rho$ determines instant atomic position with impact parameter ρ, and summation is over all crystal atoms and all possible values of momentum $\mathbf{q}$. Accounting for atomic thermal motion and summing in two steps; first, over atoms within a unit layer (see below) then over unit layers, we find

$$\frac{d\sigma}{d\Omega_{\mathbf{k}}} = \sum_{\mathbf{q}} \left| \sum_{\rho \mathbf{R}_\rho} f(\rho, \omega, \mathbf{k}, \mathbf{q}) e^{i(\mathbf{k}_p - \mathbf{k})\mathbf{R}_\rho} \right|^2 \left| \sum_m e^{i(\mathbf{k}_p - \mathbf{k})\mathbf{r}_m} \right|^2$$

$$e^{-2W} = \sum_{\mathbf{q}} |F(\omega, \mathbf{k}, \mathbf{q})|^2 \frac{\sin^2\left[\left(\frac{\omega}{v} - \frac{\omega}{c_p}\cos\theta\right)\frac{Nd}{2}\right]}{\sin^2\left[\left(\frac{\omega}{v} - \frac{\omega}{c_p}\cos\theta\right)\frac{d}{2}\right]}. \tag{5.9}$$

Here $\mathbf{R}_\rho$ determines equilibrium atomic position with impact parameter ρ within a unit layer, d is unit layer thickness, $\mathbf{r}_p$ determines p-th unit layer position, e^{-2W} is the Debye–Waller factor, and N is the number of unit layers along the particle trajectory.

5.1.4 The Structure Factor

Expression (5.9) yields the explicit cross section dependence on crystal structure and atomic thermal motion. The structure factor includes the sum over a crystal unit layer. A unit layer is one layer which includes all possible impact parameters on the given trajectory and has minimal possible thickness. For the trajectory shown in Fig. 5.1, the unit layer is as thick as a unit cell and is shown by a dashed line. In our example, a unit layer plays the same role as a unit cell in X-ray diffraction theory, and its thickness gives the value of d in (5.1). Note that a crystal may be built of its unit cell by all possible translations of this cell, while a crystal is comprised of unit layer if it is translated along the trajectory of the particle by distances that are multiples of d. The value of $d(\mathbf{k}_p)$ depends on particle velocity orientation with respect to crystallographic directions and is expressed in terms of the unit cell basis vectors a_i:

$$d(\mathbf{k}_p) = |m\mathbf{a}_1 + n\mathbf{a}_2 + p\mathbf{a}_3| \tag{5.10}$$

provided that direction $\mathbf{k}_p$ is given by indices (m, n, p) in this basis. If direction $\mathbf{k}_p$ is given by the indices (h, k, l) in the basis of the reciprocal lattice vector a'_i, $d(\mathbf{k}_p)$ is

$$d(k_p) = \frac{V}{S}|k\ell[\mathbf{a}'_2 \times \mathbf{a}'_3] + h\ell[\mathbf{a}'_3 \times \mathbf{a}'_1] + hk[\mathbf{a}'_1 \times \mathbf{a}'_2]| \tag{5.11}$$

where V is the volume of the unit cell and $S = 1$ if indices h, k, l have no common divisor and S coincides with the common divisor if it exists for any two indices. The value F is analogous to structure amplitude known in diffraction theory and has a simple physical meaning. For $\mathbf{k}$ vectors that satisfy the propagation condition of a free wave in a crystal, i.e., properly related to ω, F is the amplitude of a quantum emission from a unit layer of the crystal with frequency ω and wave vector $\mathbf{k}$.

This amplitude embodies structure Cherenkov radiation property dependence on crystal structure and particle velocity direction. In contrast to Cherenkov radiation whose intensity at fixed frequency is constant over the entire Cherenkov cone, structure radiation intensity at fixed frequency is given by (5.9) and depends on emission direction with respect to crystallographic directions. The second factor in (5.9) includes the sum over unit layers which is similar to the sum (2.5) over unit cells in the formulae of kinematic diffraction theory. This formula sets the condition (5.1) which limits possible frequencies and emission direction.

5.1.5   Temperature Dependence

Accounting for Eq. (5.1) and assuming that thermal motion is isotropic, we may present the third factor in (5.9) as

$$e^{-2W} = \exp\left(-\frac{\langle r^2 \rangle}{3}|\mathbf{k}_p - \mathbf{k}|^2\right) = \exp\left\{-\frac{\langle r^2 \rangle}{3}\left[\left(\frac{2\pi s}{d}\right)^2 + k^2\sin^2\theta\right]\right\} \tag{5.12}$$

where $\langle r^2 \rangle$ is the mean square radius of atomic thermal motion. We neglected the small difference between the thermal factor of structure radiation and the Debye–Waller factor (see [24]).

For radiation of wavelength $\lambda^2 \sim \langle r^2 \rangle$, the Debye–Waller factor affects the angular distribution of the radiation. Using (5.1) and (5.12) at $s \neq 0$ we may present it as

$$e^{-2W} = \exp\left\{-\frac{\langle r^2 \rangle}{3}\left(\frac{2\pi s}{d}\right)^2\left[1 + \frac{\sin^2\theta}{(\cos\theta - c_p/v)^2}\right]\right.. \tag{5.13}$$

It follows from (5.13) that the minimal value of the Debye–Waller factor is

$$\left(e^{-2W}\right)_{\min} = \exp\left\{-\frac{\langle r^2 \rangle}{3}\left(\frac{2\pi s}{d}\right)^2 \frac{1}{1-(v/c_p)^2}\right\} \qquad (5.14)$$

and it is reached at $\cos\theta = v/c_p$ and may differ from the maximal value $\exp\left[-\frac{\langle r^2 \rangle}{3}\left(\frac{2\pi s}{d}\right)^2\right]$ at $\theta = 0$. Equations (5.13) and (5.14) show that the thermal factor strongly affects the spatial distribution of emission, especially in the case of fast particles, and the corresponding effects are most prominent in the range of small angles. The radiation frequencies for fast particles are such that they permit the neglect of the difference between c_p and c as small compared with the difference between c and v. This omission yields

$$\left(e^{-2W}\right)_{\min} = \exp\left[-\frac{\langle r^2 \rangle}{3}\left(\frac{2\pi s}{d}\frac{E}{mc^2}\right)^2\right] \qquad (5.15)$$

at $\theta = \frac{mc^2}{E}$ where m and E are particle mass and energy.

It follows from (5.9) that the spatial distribution of structure radiation essentially depends on crystal structure, and the structure and Debye–Waller factors determine its explicit form. However, we can demonstrate that for forward emission (at small θ), the cross section dependence on θ is

$$\frac{d\sigma}{d\Omega_k} \sim \sin^2\theta \exp\left\{-\frac{\langle r^2 \rangle}{3}\left(\frac{2\pi s}{d}\right)^2\left[1 + \frac{\sin^2\theta}{(\cos\theta - c_p/v)^2}\right]\right\}. \qquad (5.16)$$

Hence, structure radiation intensity for the forward direction is zero and maximum intensity is at $\theta \neq 0$. For relativistic particles, the angle of a maximum intensity, $\theta_{\max}$ is found using (5.13,5.14):

$$\theta_{\max} \cong \left(\frac{mc^2}{E}\right)^2 \frac{d}{2\pi s\sqrt{<r^2>/3}}. \qquad (5.17)$$

Formulae (5.7–5.9) describe, at $s = 0$, Cherenkov radiation, known to occur in the optical range provided that $v > c_p$. At $s \neq 0$, radiation frequency given by (5.1) may, depending on s and particle velocity, lie in shorter wavelengths; condition $v > c_p$ does not hold in this case. In contrast to optical Cherenkov radiation whose Debye–Waller factor (5.12) is reduced to $\exp\left(-\frac{\langle r^2 \rangle}{3}k^2\sin^2\theta\right)$ and is always very close to unity because $k^2\langle r^2 \rangle << 1$, structure radiation intensity essentially depends on temperature at any wavelength. In this case, the exponent of (5.12) includes the temperature-dependent term $-(\langle r^2 \rangle/3)(2\pi s/d)^2$ which is generally not small compared with unity. The same term in the exponent suppresses the intensities of higher harmonics since it is proportional to s^2. The other case of higher harmonics suppression is related to the fact that the characteristic frequencies of medium polarization excited by the particle near its trajectory are

$v/[d(1 - \epsilon_{\beta^2})^{1/2}]$, where $\beta = v/c$. Therefore, if the harmonic frequency is above this value, the harmonic is suppressed because this frequency is absent in the crystal atom polarization.

5.1.6 Dimensions of the Emitting Region

Structure radiation intensity is mainly determined by the emission of atoms closest to the particle trajectory. Hence, when estimating intensity in the case of channelling, for instance, it is sufficient to account for emission of atoms located in the channel walls (in Fig. 5.1 those atoms are closed circles). That is, the sum in the structure factor in (5.9) is restricted to several terms with minimal values of impact parameter ρ. The Fourier component of the particle electric field E_ω related to frequency ω is important only when the distance from the trajectory is smaller than ρ_ω where

$$\rho_\omega = 2\pi v[\omega(1 - \epsilon_{\beta^2})^{1/2}]^{-1}. \tag{5.18}$$

Substituting values of the limiting frequencies (5.7) into (5.18), we find that, in the case of backward emission at minimal frequency, the distances at which the particle field includes the corresponding Fourier component $E_{\omega_{\min}}$ are restricted by the condition

$$\rho_\omega \leq \rho_{\omega_{\min}} = \frac{d}{s\sqrt{1 - \beta\sqrt{\epsilon}}}. \tag{5.19}$$

The corresponding restriction for the impact parameter at maximal frequency (forward emission) is

$$\rho \leq \rho_{\omega_{\max}} = \frac{d}{s\sqrt{1 + \beta\sqrt{\epsilon}}}. \tag{5.20}$$

Formulae (5.19) and (5.20) show that the dimension of the emitting region along the particle trajectory may be greater than interatomic distance d only for fast particles with $\beta \sim 1$ and close to the minimal frequency for the harmonic in question. If the particles are not very fast ($\beta < 1$), this dimension is always of the order of the interatomic distance. In the case of Cherenkov radiation ($s = 0$), the dimension off the effective emitting region is of the order of the wavelength.

5.1.7 Estimations of the Structure Radiation Intensity

The explicit form of the coherent amplitude $f(\mathbf{v}, \rho, \omega, \mathbf{k}, \mathbf{q})$ is different for different frequencies of emitted photons. Since the physical cause of radiation is the polarization of atoms by the particle electric field (Compton scattering of the particle field) and the polarization essentially depends on frequency, it is clear that for different ω, f is different. For example, when the radiation is in the optical range, the contribution to amplitude comes only from weakly bonded optical electrons, and hence, the amplitude f may be related to the dielectric permittivity at the optical frequencies. For a harder radiation, whose frequency is greater than all atomic frequencies, all atomic electrons produce similar contributions to f and the amplitude is proportional to the total electron density. In an intermediate case, the contributions from weakly and strongly bound electrons differ. In any case, we can distinguish in amplitude coordinate dependence, the factor depending only on the impact parameter ρ, and represent it as

$$f = \sum_e f_e n_e \tag{5.21}$$

where f_e is the amplitude of photon emission by the electron of the 1th type, and n_e is the density of electrons of the 1th type (optical electrons, conductivity electrons, K-electrons, etc.). Amplitude f_e, as a function of coordinates, depends only on the impact parameter ρ because its coordinate dependence is determined by the particle field.

Substituting (5.21) into (5.9) and changing to a continuous distribution of electron density in a crystal, we find the intensity of structure radiation per unit length of the trajectory:

$$\frac{d\sigma}{d\Omega_{\mathbf{k}}} = 2\pi e^{-2W} \sum_{s,q} \left| \int d\rho \sum_e f_e(\mathbf{v}, \rho, \omega, \mathbf{k}, \mathbf{q}) e^{-i\mathbf{k}_\perp \rho} n_{es\tau}(\rho) \right|^2$$
$$\times \delta \left(\frac{\omega}{v} - k\cos\theta - s\tau \right) \tag{5.22}$$

where $\mathbf{k}_\perp$ is the component of the photon wave vector perpendicular to the trajectory, $\tau = 2\pi \mathbf{v}/(vd)$,

$$n_{es\tau}(\rho) = \frac{1}{d} \int n_e(x, y, z) e^{-is\tau z} dz \tag{5.23}$$

is the spatial Fourier component of 1th type electron density in the crystal, and the Z-axis lies along the trajectory.

Let us estimate structure radiation intensity for different frequency ranges (different velocities of particles). In the case of optical frequency, structure radiation intensity is conveniently expressed via Cherenkov radiation intensity for the same crystal that corresponds to $s = 0$ in (5.1). It follows from (5.7) that, for the usual crystals, structure radiation frequency corresponds to the optical range only for slow particles with $sv/c_p \sim 10^{-3}$. For ions whose atomic mass is about 100, this corresponds to energies of about several tens of kiloelectron-volts. The range of allowed frequencies is about $(2\pi sv/d)(v/c_p) \sim opt^{-3}\omega_{pt}$ where ω_{opt} is the optical frequency. Using (5.22) and accounting for the dimension of the radiation-generating regions, we estimate the ratio of the optical structure radiation intensity to that of Cherenkov radiation at the same frequency as

$$\frac{I_s}{I_c} = \left(\frac{d}{s\lambda}\right)^2 \left(\frac{c}{v}\right)^2 \left(\frac{\epsilon_{s\tau}}{\epsilon - 1}\right)^2 \exp\left[-\frac{\langle r^2 \rangle}{3}\left(\frac{2\pi s}{d}\right)^2\right]$$

$$\sim \frac{1}{s^2}\left(\frac{\epsilon_{s\tau}}{\epsilon - 1}\right)^2 \exp\left[-\frac{\langle r^2 \rangle}{3}\left(\frac{2\pi s}{d}\right)^2\right] \tag{5.24}$$

where λ is the wavelength and $\epsilon_{s\tau}$ is the Fourier component of dielectric permittivity defined by (5.23), if $n(r)$ is substituted by $\epsilon(r)$ on the right side of the equation.

Integrating (5.24) over frequencies and accounting for (5.7), we obtain the following estimate for the ratio above:

$$\frac{\int I_s d\omega}{\int I_c d\omega} \simeq \frac{v}{c} s^{-2} \left[\frac{\epsilon_{s\tau}}{\epsilon - 1}\right]^2 \exp\left[-\frac{\langle r^2 \rangle}{3}\left(\frac{2\pi s}{d}\right)^2\right]. \tag{5.25}$$

Since $v/c \leq 10^{-3}$, structure radiation intensity is at least four orders less than Cherenkov radiation intensity [10]. The corresponding ratio for total intensity per particle (ratio of intensities integrated over particle trajectory length) is smaller because the free path length for particles emitting in the optical range at $s \neq 0$ is short, due to their low energy even under channelling conditions.

Let us now estimate the intensity for relativistic particles. If the frequency is greater than all atomic frequencies, the electrons in a crystal may be considered as free. Hence, using the method of pseudophotons, we express the radiation intensity in terms of the Compton scattering cross section. If the trajectory length (crystal thickness) is L, (5.8) yields the emission cross section in the form

$$\frac{d\sigma}{d\Omega_k} = |F|^2 \left(\frac{L}{d}\right)^2 e^{-2W} \tag{5.26}$$

where the frequency and emission direction are limited by (5.1). The structure of Eq. (5.26) reflects the coherent nature of emission; $|F|^2 e^{-2W}$ is the emission cross section of a photon by a unit layer. The emission cross section by the entire crystal is $(L/d)^2$ times greater than the cross section for a unit layer, whereL/d is the number of unit layers in the crystal along the particle trajectory.

Using the method of pseudophotons, we obtain the estimate for $|F|^2$:

$$\int |F|^2 d\omega_f \sim \frac{\Delta\omega_i}{\omega_i} \frac{e^2 c}{\hbar v} \ln\frac{\rho_{\max}}{\rho_{\min}} \cdot \frac{d\sigma(\omega_i \omega_f)}{d\omega_f} \cdot \Delta\omega_f |dn_{s\tau}\rho_{ef}^2|^2 \tag{5.27}$$

where $\omega_i, \omega_f, \Delta\omega_i, \Delta\omega_f$ are the frequency and frequency intervals essential for structure radiation of pseudophotons and emitted photons, respectively;

ρ_{max} and ρ_{min} are the maximum and minimum impact parameters for electrons at which the pseudophotons experience Compton scattering; ρ_{ef} is the effective diameter for the emitting region; and $\frac{d\sigma}{d\omega_f}$ is the differential cross section of Compton scattering.

From (5.26–5.27), we estimate that, at $1 - \beta \sim 10^{-3}$, one photon per particle is emitted if crystal thickness is

$$L \sim \frac{d10^6}{Z} \exp\left[-\frac{2\pi^2}{3} \frac{\langle r^2 \rangle s^2}{d^2} \right] \tag{5.28}$$

where Z is the crystal atoms' charge. Hence, $L \sim 10^{-2} - 10^{-1}$ cm, at $s = 1$, $Z \sim 50$.

5.1.8   On the Phenomenological Approach

It is natural that solutions to the phenomenological equations (Maxwell) in the corresponding (kinematic) approximation above yield the same results. In this case, the results should be expressed in terms of the dielectric permittivity tensor $\hat{\epsilon}$; the calculation for an explicit form of the $\hat{\epsilon}$ is a separate problem not covered here.

For example, if the right-hand side of (5.4) is transformed to include the structure amplitude of an elementary layer as defined above,

$$\frac{1}{\epsilon_0} \sum_{T'} \hat{\epsilon}_{T-T'} E_0(\mathbf{k}_T, \omega) = F(\mathbf{k}_p, \mathbf{k}_T) \sum_{\tau_z'} \delta(\mathbf{k}_p - \mathbf{k}_{\tau_z} - \boldsymbol{\tau}_z')$$

$$\mathbf{k}_p = \mathbf{v}(\omega/v^2) \tag{5.29}$$

where τ_z' represents different possible projections of the reciprocal lattice vectors on the particle velocity direction. When we use $\epsilon_1(\mathbf{r})$ and $\mathbf{E}_0(\mathbf{r}, \omega)$, $F(\mathbf{k}_p, \mathbf{k}_T)$ is expressed as

$$\mathbf{F}(\mathbf{k}_p, \mathbf{k}_T) = \left[(2\pi)^2 \epsilon_0 d(\mathbf{k}_p) \right]^{-1} \int \left[\int_0^d \epsilon_1(\mathbf{r}) e^{i(\mathbf{k}_p - \mathbf{k}_T)\mathbf{r}} \, dz \right]$$

$$\times \mathbf{E}_0(\mathbf{r}, \omega) \, dx dy. \tag{5.30}$$

This formula allows us to consider $|F(\mathbf{k}_p, \mathbf{k}_T|$ azimuthal dependence for various particle trajectories. This dependence is most prominent when the particle moves along a direction $\mathbf{k}_p$ that coincides with a direction of low crystallographic index. In that case, unit layer thickness $d(\mathbf{k}_p)$ is of the order of interatomic distance and the result of integration in (5.30), which corresponds to projection of $\epsilon_1(\mathbf{r}) e^{i\tau_z z}$ on the surface of the unit layer and is proportional to the projection of the electron density, is a clear function of X and Y. Hence, the amplitude $F(\mathbf{k}_p, \mathbf{k}_T)$ is a function of the orientation $\mathbf{k}_T$ (emission direction). If, on the contrary, $\mathbf{k}_p$ corresponds to a direction with high Muller indices, thickness $d(\mathbf{k}_p)$ is much greater than interatomic distance and the result of integration in (5.30) along dz, also proportional to the projection of electron density on the unit layer plane, is practically independent of X and Y, and $F(\mathbf{k}_p, \mathbf{k}_\tau)$ does not depend on X and Y. Thus, the radiation intensity is homogeneously distributed over a cone, as mentioned above. For example, within the framework of the kinematic approximation, intensity is proportional to $|F(\mathbf{k}_p, \mathbf{k}_\tau)|^2$. It follows from (5.30) that the value $F(\mathbf{k}_p, \mathbf{k}_\tau)$ not only depends on particle trajectory direction but also on the location of the particle trajectory within the crystal. This fact is essential for the description of structure radiation from a channelled particle. The specifics of radiation related to channelling exhibits itself in the angular distribution of radiation; the angular distribution as defined by (5.8) and (5.9) is directly associated with the localization of particle trajectory. A channelled particle cannot be described by a plane wave. Its wave function is localized within a crystal channel and particle motion is described adequately by classical formulae [20]. Hence, the module and the phase of amplitude f in the structure factor depend on distance ρ from the trajectory. All this is exhibited in the angular distribution; specifically, forward intensity is zero (see [5,16]). The wave function of a nonchannelled particle is a plane wave; however, similar treatment of coherent radiation from a particle [25] gives a different result for the angular distribution of radiation. In particular, forward intensity is nonzero. We note that structure radiation was recently observed [12] and studied experimentally [61–67].

5.2   Coherent Coulomb Excitation of a Mössbauer Transition in a Single Crystal

In this section, we consider quanta emitted by a particle in a crystal with wavelengths of the order of the lattice constant. However, here we do not assume that the particle wave function is localized in space (see [22,25,26]). Beyond this, tensor $\hat{\epsilon}(\mathbf{r})$ of (5.2) has another form because it is determined by the interaction between the field of a charged particle and the atomic nuclei in the crystal.

Many researchers have studied the problem of coherent X-ray emission by a fast charged particle moving through a crystal at a constant velocity and the effect of crystal structure on the angular and spectral distribution of radiation (see, e.g., [1,4,22,26]). The authors of [25] consider coherent radiation related to Compton scattering by the particle field at the crystal electrons for a particle represented by a plane wave. Here we consider coherent radiation due to Coulomb excitation of nuclei in a crystal by the field of a charged particle described by a plane wave [30]. In this case, the angular distribution of radiation differs from that of a particle moving in a classical trajectory. Coherent emission occurs only for some discrete directions, depending on crystal structure, energy of the excited nuclear level $E = \hbar\omega$, and particle velocity and determined by the following relation

$$\mathbf{k}_p - \mathbf{k} = \boldsymbol{\tau} \qquad (5.31)$$

where $\mathbf{k}$ is the emitted photon wave vector, $\mathbf{k}_p = \mathbf{v}\omega/v^2$, and $\boldsymbol{\tau}$ is the reciprocal lattice vector. We note here that the allowed directions of radiation from a particle moving in a classical path (straight trajectory) form a cone whose axis lies along the trajectory.

5.2.1 Coherent Coulomb Excitation in a "Frozen" Lattice

Let us consider coherent Coulomb excitation, neglecting thermal vibrations of the atoms in the crystal. The cross section of coherent emission of a gamma quantum is written as

$$\frac{d\sigma}{d\Omega_{\mathbf{k}}} = \int \left| \sum_{\ell f} M_{if}^{\ell}(\mathbf{p}, \mathbf{q}) f_{fi}^{\ell}(\mathbf{k}) e^{i(\mathbf{k}_p - \mathbf{k})\mathbf{r}_e} \right|^2 \delta(\epsilon_p - \epsilon_{p'} - \hbar\omega) \, d\mathbf{p} d\mathbf{q} \quad (5.32)$$

where i and f are quantum numbers describing initial and intermediate states of the excited nucleus, $\mathbf{p}$ and $\mathbf{p}'$ are initial and final particle momenta, ϵ is particle energy, $M_{if}(\mathbf{p}, \mathbf{q})$ is the matrix element of Coulomb excitation in which particle momentum is changed by $\mathbf{q}$ [27], $f_{f,i}(\mathbf{k})$ is the probability amplitude of a gamma quantum emission along the direction k which corresponds to transition from f into i, and $\mathbf{r}$ is the nuclear coordinate, index ℓ labels values related to the 1th site of the lattice, and summation is over all nuclei in the crystal.

If $F(\mathbf{p}, \mathbf{q})$ is the coherent amplitude of the process for an individual nucleus which corresponds to the change in particle momentum $\mathbf{p} - \mathbf{p}' = \mathbf{q}$, the coherent cross section of emission per crystal unit cell is represented in the form

$$\frac{d\sigma}{d\Omega_k} = \frac{(2\pi)^3}{V} \varphi_L(k) \int |F_0(\mathbf{pq})|^2 \, d^3q \sum_{\tau} \delta(\mathbf{k}_P - \mathbf{k} - \boldsymbol{\tau}) \qquad (5.33)$$

where V is unit cell volume and coherent amplitude $F_0(\mathbf{p}, \mathbf{q})$ takes the form

$$F_0(\mathbf{p}, \mathbf{q}) = \frac{1}{(1+\alpha)^{1/2}} \frac{2j_f + 1}{2j_i + 1} \chi(\pi, L) N^{1/2}(\mathbf{p}, \mathbf{q}). \qquad (5.34)$$

Here j_i and j_f are the spins of initial and excited nuclear states, respectively, α is the coefficient of gamma quanta internal conversion, $\chi(\pi, L)$ is the reduced nuclear matrix element [27], $N(\mathbf{p}, \mathbf{q})$ is the factor describing Coulomb excitation cross section dependence on momentum transferred from the particle to a nucleus, L is the multipolarity and π is the parity of the nuclear transition. The explicit form of $\varphi_L(\mathbf{k})$ which determines the angular distribution for quanta depends on the multipolarity of the nuclear transition. For example, in dipole transitions $\varphi_q(\mathbf{k}) = (1 + \cos^2 \theta)/2$ where θ is the angle between $\mathbf{k}$ and $\mathbf{p}$. Equation (5.33) shows that coherent emission occurs under the conditions of expression (5.31). This radiation is considered the result of diffraction by the particle field (of pseudophotons) at the crystal nuclei because the value $\mathbf{k}_p$ in (5.31) corresponds to the wave vector of the particle field harmonic of frequency ω.

Assuming condition (5.31) is fulfilled for a certain reciprocal lattice vector τ, we obtain the ratio of coherent cross section to total Coulomb excitation cross section in the form of $\sigma_c/\sigma = [(2j_f + 1)/(2j_i + 1)]\varphi_L(\theta_\tau)$ where θ_τ is the angle between $\mathbf{k}$ and $\mathbf{p}$ satisfying (5.31). Therefore, for an absolutely "frozen" lattice, the fraction of coherent excitation is $[(2j_f + 1)/(2j_i + 1)]\varphi_L(\theta_\tau)$ and may be of order of unity.

5.2.2   Influence of Thermal Motion of Atoms

The angular distribution of coherently emitted gamma quanta remains unchanged if the thermal motion of lattice atoms is taken into account; however, radiation intensity decreases and may be the order of the intensity for a "frozen" lattice for very low-energy nuclear transitions only. The amplitude of a wave of coherent Coulomb excitation taking into account the thermal motion is expressed by the quantity

$$F(\mathbf{p}, \mathbf{q}, \mathbf{k}) = F_0(\mathbf{p}, \mathbf{q})e^{iW(\mathbf{q})}e^{-W(\mathbf{k})}. \tag{5.35}$$

Here $F_0(\mathbf{p}, \mathbf{q})$ is given by (5.34), $e^{-W(\mathbf{q})}$ is the Debye–Waller factor, and $W(\mathbf{q}) = \frac{1}{2}q^2 \langle x^2 \rangle$ where $\langle x^2 \rangle$ is the mean-square amplitude of atomic thermal vibrations within the crystal.

Equation (5.35) gives the coherent term of the Coulomb excitation cross section in the form

$$\frac{\sigma_c}{\sigma} = \frac{2j_f + 1}{2j_i + 1}\varphi_L(\theta_\tau)R, R = e^{-2W(k)}\frac{\int N(\mathbf{p}, \mathbf{q})e^{-2W(\mathbf{q})}\, d^3q}{\int N(\mathbf{p}, \mathbf{q})\, d^3q}. \tag{5.36}$$

When calculating the factor R, it is convenient to take into account that the momentum transferred to the nucleus in Coulomb excitation is represented by $\mathbf{q} = \mathbf{q}_m + \mathbf{q}_\perp$ where $\mathbf{q}_m = \mathbf{k}_p = \mathbf{v}\omega/v^2$ is the minimal possible change of particle momentum and $\mathbf{q}_\perp$ is perpendicular to $\mathbf{v}$ (i.e., it is assumed that $\epsilon_p >> \hbar\omega = E$). Thus, we can write

$$R = e^{-2W(\mathbf{k})}e^{-2W(\mathbf{k}_p)}\frac{\int N(\mathbf{q}_m + \mathbf{q}_\perp)e^{-2W(\mathbf{q}_\perp)}\, d\mathbf{q}_\perp}{\int N(\mathbf{q}_m + \mathbf{q}_\perp)\, d\mathbf{q}_\perp}. \tag{5.37}$$

This formula shows that the value of the coherent cross section depends on the mean-square amplitude of atomic thermal vibrations $\langle x^2 \rangle$, particle velocity, and energy of the nuclear transition as $\exp\left[-2W\left(\frac{\omega}{v}\right) - 2W(k)\right]$. The cross section decreases with velocity decreases and increases in transition energy E and $\langle x^2 \rangle$ (i.e., crystal temperature).

5.2.3 Radiation Intensity

Another factor that strongly influences coherent cross section is the Coulomb excitation cross section dependence on transferred momentum. If the principal contribution to Coulomb excitation cross section comes from small values of transferred momenta $q \sim q_m$, that corresponds to dipole nuclear transitions, the last factor in (5.37) is about unity, and $R \sim \exp[-2W\left(\frac{\omega}{v}\right) - 2W(k)]$. Thus, in the case of low-energy nuclear levels and sufficiently fast particles, regardless of mass, the cross section of coherent Coulomb excitation for dipole transitions may be comparable to the total Coulomb excitation cross section. If, on the other hand, the momenta $q > q_m$ are essential in the Coulomb excitation, i.e., the scattering angles are large as in nonrelativistic case for nuclear transitions with the multipolarity higher than that of a dipole, the last factor in (5.37) decreases the coherent cross section. In this case, coherent cross section depends on the mass of the fast particle, that is, it rapidly decreases for greater masses, because the momentum transferred to the crystal increases and the Debye–Waller factor decreases. For example, if $N(\mathbf{p}, \mathbf{q})$ does not depend on the scattering angle (that approximately corresponds to quadrupole nuclear transitions [27]) the estimate for R is

$$R \geq \exp\left[-2W(k) - 2W\left(\frac{\omega}{v}\right)\right] \frac{\lambda_c^2}{\langle x^2 \rangle} \qquad (5.38)$$

where λ_c is the Compton length of the particle involved. The last factor in (5.37) increases with particle energy because the transversal momentum transferred to the crystal $q_\perp$ decreases. In the ultrarelativistic limit, R is equal to $\exp[-4W(k)]$ and depends neither on transition multipolarity nor on particle mass. This is so because in this limit the value $q_\perp$ becomes much smaller than q_m, since the scattering cross section is localized at angles smaller than $(mc^2/E)^2$ [11,29].

Since coherent radiation is strongly directed and its total intensity weakly depends on the direction of $\mathbf{k}$, satisfying condition (5.31), the phenomenon at hand may be used to produce highly monochromatic beams of gamma quanta with small angular divergence. We note that in addition to coherent emission of nuclear origin, there are quanta which are coherently emitted in directions defined by (5.31) due to scattering from particle field at the electrons of the crystal. The cross section for this process is described by formulae similar to those given above in this section, but the effect of thermal vibrations on the corresponding coherent amplitude differs slightly from (5.35); that is, the exponential $\exp[-W(\mathbf{q}) - W(\mathbf{k})]$ should now be substituted by $\exp[-2W(\mathbf{k}_p - \mathbf{k})]$. This form of the thermal factor, as in X-ray diffraction theory, results in a decrease of coherent cross section when scattering angle increases (the angle between $\mathbf{k}$ and $\mathbf{p}$). The differential intensity of this radiation at the frequency of nuclear transition is negligible compared with that of Coulomb excitation; the ratio is estimated as $(Z^2 r_e^2)(c/\omega)^{-2} \exp[-2W(\mathbf{k}_p - \mathbf{k}) + 4W(\mathbf{k})]$ where r_e is the classical electron radius and Z is the charge of crystal atoms. However, since the energy width of nuclear radiation is small and coincides with the width of a nuclear level Γ, the total (integrated over the frequency range $\sim \omega$) intensity of radiation induced due to scattering the particle field by electrons proves to be greater than the nuclear radiation intensity. Hence, to detect the nuclear radiation, one should use a detector with high energetic resolution, for example, a Mössbauer-type detector, and carry out the measurements in the range of nuclear diffraction maximum (see Chapt. 3).

Let us use the method of pseudophotons to evaluate the absolute intensity of Coulomb excitation of a Mössbauer transition in a crystal by relativistic electrons [30,31]. The number of quanta emitted by a single electron is

$$N \sim \frac{e^2}{\hbar c} \frac{\Gamma}{E} \left(\frac{\hbar c}{E a_0}\right)^2 \frac{L}{a_0} e^{-4W(E/\hbar c)} \frac{2j_f + 1}{(1+\alpha)(2j_i + 1)} \qquad (5.39)$$

where a_0 is the Bohr radius and L is the crystal thickness. The estimate from (5.39) shows that when relativistic electrons excite the Mössbauer 14.4 keV transition in Fe^{57}, one quantum is coherently emitted per 10^{12}–10^{13} electrons for a crystal as thick as $L = 10^{-4}$ cm.

Note that the thickness of $L = 10^{-4}$ cm was chosen because just this value is close to the absorption length of quanta of the corresponding energy. It may be sought that an additional increase of the crystal's thickness will not result in an increase of emission from the crystal. However, the dynamic effects of interaction of radiation with matter in the crystal can provide the emission of radiation from much thicker crystals.

5.3 Structure Cherenkov Radiation in a Cholesteric Liquid Crystal

Unlike the two previous sections concerned with X-ray structure radiation, the structure radiation of the optical wavelength range in cholesteric liquid crystals will be considered here.

Until now we discussed the radiation which is due to the spatial periodicity of electron (nuclear) density in a crystal. Here we shall demonstrate that fast particles moving through cholesteric liquid crystals induce radiation of the same nature, but this radiation is related not to the periodicity of crystal density but to the regular variation of the spatial orientation of the molecules in liquid crystals (see Chapt. 4). The radiation wavelength depends in this case on the period of the cholesteric helix and lies in the optical wavelength range.

Since the physical nature of radiation is connected to the anisotropy of polarizability of molecules in the crystal, the radiation intensity is determined by the value of this anisotropy. The polarization of structure radiation in CLC differs essentially from that of Cherenkov radiation. Namely, Cherenkov radiation is linearly polarized and this polarization does not depend on the emission angle, whereas in the case in question, the polarization is elliptical and the ratio of the polarization ellipse axes depends on the direction of emission. In particular, when a particle moves along the optical axis, the polarization varies from circular (for the emission along the optical axis) to linear (for the perpendicular emission).

5.3.1   Basic Equations

Let us consider the radiation induced by a charged particle moving with a constant velocity through a cholesteric liquid crystal. From elementary kinematic considerations [1,23], we obtain the relationship between the frequency ω, direction of emission, and the particle velocity v

$$\cos\theta = c_p/v \pm 4\pi/kp\cos\eta) \tag{5.40}$$

where c_p is the phase velocity of a photon with the wave vector $\mathbf{k}$, p is the pitch of a cholesteric helix, θ is the angle between $\mathbf{k}$ and $\mathbf{v}$, and η is the angle between the particle velocity and the optical axis.

When $v < c_p$, we must retain only the minus sign in (5.40). For the sake of simplicity, we restrict ourselves to considering particles moving along the optical axis Z. The dielectric permittivity of a liquid crystal may be written in the form of (1.5) where the parameter δ determines the anisotropy of dielectric permittivity. This parameter is small ($\delta \sim 10^{-2}$); hence, the problem may be solved using perturbation theory.

The particle field in CLC is different from that in a homogeneous medium. The correction to the field which is due to the anisotropy of CLC dielectric tensor (1.5) in linear approximation in δ is determined by:

$$\Delta\mathbf{D}_1 - \frac{1}{c_p^2}\frac{\partial^2\mathbf{D}_1}{\partial t^2} = -\operatorname{rot}\operatorname{rot}\hat{\epsilon}_1\mathbf{E}_0 \tag{5.41}$$

where $\mathbf{E}_0$ is the particle field in a homogeneous medium with the permittivity $\bar{\epsilon}$ (the formula for $\mathbf{E}_0$ is given, e.g., by [32]), and $\hat{\epsilon}_1$ is the linear in the δ part of the tensor (1.5).

5.3.2 Intensity and Polarization Properties of Radiation

At a long distance from the crystal, the solution of (5.41) is [19]:

$$\mathbf{D}_1(\omega, R) = \mathbf{k} \times \mathbf{k} \times \mathbf{G}(\omega) \tag{5.42}$$

where R is the distance to the observation point, and

$$\mathbf{G}(\omega) = -\frac{e^{ikR}}{4\pi R} \int \hat{\epsilon}_1 \mathbf{E}_0(\omega) e^{i\mathbf{kr}} \, d\mathbf{r}. \tag{5.43}$$

Integration in (5.43) is accomplished throughout the crystal's entire volume. To be more precise, we consider the case of $|\mathbf{v}| \lesssim c_p$. Using the explicit form of $\mathbf{E}_0(\omega)$ [32] for a crystal plate with thickness L and with a surface perpendicular to the optical axis, we obtain the expression for the number of photons emitted into a unit solid angle and in unit frequency interval per particle

$$\frac{d^2 n}{d\omega d\Omega} = |F(\theta)|^2 \frac{\sin^2\left[\frac{L}{2p}\left(\frac{\omega p}{c_p}\cos\theta - \frac{\omega p}{v} + 4\pi\right)\right]}{1/4\left(\frac{\omega p}{c_p}\cos\theta - \frac{\omega p}{v} + 4\pi\right)^2}. \tag{5.44}$$

The first factor in (5.44) is analogous to the square of the modulus of the structure amplitude in the X-ray diffraction theory; it is

$$|F(\theta)|^2 = \frac{1}{4\pi^2}\frac{e^2}{\hbar c}\frac{\omega\bar{e}\delta p}{v^2}\left[1 - I_0\left(\frac{v\sin\theta}{c_p\sqrt{1-(v/c_p)^2}}\right)\right]^2 \frac{1+\cos^2\theta}{\sin^2\theta}. \tag{5.45}$$

The second factor in (5.44) becomes, at $L \to \infty$, $\delta\left(\frac{\omega p}{c_p}\cos\theta - \frac{\omega p}{v} + 4\pi\right)$, from which the formula (5.40) ensues. The formulae (5.44) and (5.45) show that there is no emission in the exact forward and backward directions, since $|F(\theta)|^2 \sim \sin^2\theta$ at $\theta \ll 1$ and $\pi - \theta \ll 1$; this follows from the corresponding asymptotic of the Bessel functions. In addition, it follows from (1.5), (5.42), and (5.45) that the structure radiation from a cholesteric crystal is elliptically polarized. One of the axes of the polarization ellipse, a, lies in the plane of $\mathbf{k}$ and $\mathbf{v}$, and the other one, b, is perpendicular to this plane. The ratio of the axes is

$$\frac{a}{b} = \cos\theta. \tag{5.46}$$

After integrating (5.44) over frequencies and angles, we see that about 10^{-3} photons are emitted per particle when the crystal thickness is $L \sim 10^{-2}$ cm, the cholesteric pitch is $p \sim 10^3$ Å, the anisotropy parameter is $\delta \sim 10^{-2}$, and the particle's velocity is $v \sim 0.9c_p$. This estimate indicates that the structure radiation in CLC may be observed experimentally.

Note that in this section, in contrast to the earlier ones of this chapter, we solve the Maxwell equations using the phenomenological expression for $\hat{\epsilon}$. Naturally, the form of the formulae is similar in both cases. However, in the case of liquid crystal, the phenomenological form of $\hat{\epsilon}$ is more relevant to the problem since this value may be directly obtained from optical measurements, but the physics of structure radiation for a cholesteric crystal are completely similar to those for the more commonly encountered crystals considered above.

In this section we do not take into account the influence of the medium's periodicity on propagation of radiation emitted by the particle. It will be shown below that this influence may, in some cases, be very important. In such instances, a dynamical problem must be solved. It then turns out, in particular, that the results presented in this section cannot be used to describe the intensity of radiation emitted from a thick absorbing crystal.

5.4 Dynamic Theory of Coherent Radiation of Fast Charged Particles in Regular Media

In the above, we did not consider the effects of crystal periodicity on the propagation of emitted quanta; in other words, we used the kinematic approximation. It has already been mentioned that there are some cases in which periodicity strongly affects radiation properties, provided that for corresponding frequencies and directions, the diffraction scattering takes place in the crystal. In those cases, the kinematic approximation is not applicable even for a qualitative description and one must take into account the diffractive scattering within the crystal using dynamical diffraction theory described in previous chapters. For transition radiation in the X-ray range, diffractive scattering was examined in many earlier works (see [6,7]).

Below, we consider structure Cherenkov radiation with some attention to diffraction scattering. In the dynamic approximation of diffraction theory, radiation is considered for the frequencies corresponding to excitation of two plane waves in the crystal, with wave vectors $\mathbf{k}_1$ and $\mathbf{k}_2$ being connected by the Bragg condition $\mathbf{k}_2 = \mathbf{k}_1 + \boldsymbol{\tau}$ where $\boldsymbol{\tau}$ is the reciprocal lattice vector. The frequencies and emission directions are found for which the excitation of two waves in the crystal should be taken into account. Also, the corresponding polarizations were determined. It is thereby shown that in small vicinities of those frequencies and directions, the intensities and polarizations of radiation leaving the crystal may differ essentially from those of the radiation which does not meet the Bragg condition. The range of the corresponding angles (frequencies) is of the order of the dielectric permittivity anisotropy for the relevant frequency.

In particular, thick perfect crystals can exhibit the Borrmann effect (suppression of absorption) so that the intensities may prove to be greater than those without diffraction scattering. The optimal conditions for the Borrmann effect and the dependence of intensity on crystal thickness are analyzed. The requirements for the perfection of crystals needed to provide observable dynamic effects are discussed in brief.

Note also that in this section we do not impose restrictions on particle velocity. Because of this, the results obtained are applicable not only to ultrarelativistic particles, which are usually considered in theory and not generally affected by crystal regularity since channelling conditions are rather rigid, but also to particles with moderate energies which may meet the channelling conditions in wider ranges of directions and divergences of particle beams. The relationships given below can be directly applied to the radiation of channelling particles because the values directly involved include the spatial positions of particles in the crystal and depend on them.

Together with structure radiation, we consider below within the framework of dynamical theory the coherent Coulomb excitation of Mössbauer transition and Cherenkov radiation in cholesteric liquid crystal.

5.4.1   Dynamic Theory of the Structure Cherenkov Radiation

System of Dynamic Equation. The kinematic expression for $\mathbf{E_k}$ given in Sect. 5.2 is only valid when $|\mathbf{k}| \neq |\mathbf{k} + \boldsymbol{\tau}|$ for any reciprocal lattice vector. When $|\mathbf{k}| = |\mathbf{k} + \boldsymbol{\tau}|$ for a $\boldsymbol{\tau}$, i.e., when the Bragg condition is satisfied, we cannot take all the amplitudes $\mathbf{E_\tau}$ in the expansion (2.24) except one to be zero. In this case, the amplitude $\mathbf{E_{k+\tau}}$ turns out to be of the order of $\mathbf{E_k}$, and to find this amplitude, we must solve Eq. (5.4). Below we restrict ourselves to consideration of the case of two waves in which we retain only two amplitudes in (2.24). We denote those amplitudes as $\mathbf{E}_1$ and $\mathbf{E}_2$ and the corresponding wave vectors interrelated by the Bragg condition as $\mathbf{k}_1$ and $\mathbf{k}_2$. Then (5.4) reduces to [23]

$$(1 - k_1^2/q^2)\mathbf{E}_1 + \epsilon_0^{-1}\epsilon_{-\boldsymbol{\tau}}\mathbf{E}_2 + \frac{k_1}{q^2}(\mathbf{k}_1, \mathbf{E}_1) = \mathbf{F}_{\boldsymbol{\tau}_1}(\mathbf{k}_p, \mathbf{k}_1)\delta(k_p - k_{1z} - \tau_{1z})$$

$$\epsilon_0^{-1}\epsilon_{\boldsymbol{\tau}}\mathbf{E}_1 + (1 - k_2^2/q^2)\mathbf{E}_2 + \frac{k_2}{q^2}(\mathbf{k}_2, \mathbf{E}_2) = \mathbf{F}_{\boldsymbol{\tau}_1}(\mathbf{k}_p, \mathbf{k}_2)\delta(k_p - k_{1z} - \tau_{1z}) \quad (5.47)$$

where $\mathbf{k}_2 = \mathbf{k}_1 + \boldsymbol{\tau}$ and the other notation is as above (see also Fig. 5.2).

Multiplying the equations of (5.47) by the tensor $\delta_{ik} - \frac{1}{q^2}k_{pi}k_{pk}$ with $p = 1$ and $p = 2$ for the first and second equations, respectively, we obtain

$$(1 - k_1^2/q^2)\mathbf{E}_1 + \epsilon_0^{-1}\epsilon_{-\boldsymbol{\tau}}\left[\mathbf{E}_2 - \frac{k_1}{q^2}(\mathbf{k}_1\mathbf{E}_2)\right] = \left\{\mathbf{F}_{\boldsymbol{\tau}_1}(\mathbf{k}_p, \mathbf{k}_1) - \frac{k_1}{q^2}\right.$$

$$\left. [\mathbf{k}_1\mathbf{F}_{\boldsymbol{\tau}_1}(\mathbf{k}_p, \mathbf{k}_1)]\delta(k_p - k_{1z} - \tau_{1x})\right\} \quad (5.48)$$

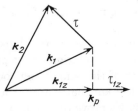

FIGURE 5.2. Schematics for determining the wave vectors in the dynamic expressions for structure radiation.

$$\epsilon_0^{-1}\epsilon_\tau \left[\mathbf{E}_1 - \frac{\mathbf{k}_2}{q^2}(\mathbf{k}_1\mathbf{E}_1) \right] + (1 - k_2^2/q^2)\mathbf{E}_2 = \left\{ \mathbf{F}_{\tau_1}(\mathbf{k}_p, \mathbf{k}_2) - \frac{\mathbf{k}_2}{q^2} \right.$$

$$\left. [\mathbf{k}_2\mathbf{F}_{\tau_1}(\mathbf{k}_p, \mathbf{k}_2)] \, \delta(k_p - k_{1z} - \tau_{1z}) \right\}.$$

The representation of the dynamic system in this form may be more convenient because the term $\mathbf{E}_i - \frac{\mathbf{k}_p}{q^2}(\mathbf{k}_p\mathbf{E}_i)$ at $k_p^2 \approx q^2$ differs very slightly from the component of $\mathbf{E}_i$ which is perpendicular to $\mathbf{k}_p$. An analogous transformation may be applied to Eq. (5.4). As is known (see Chapt. 2), the solution of a homogeneous system obtained from (5.48) by cancelling the right-hand sides describes the propagation of free waves in a crystal under the Bragg condition, the corresponding waves being transversal with an accuracy of about ϵ_τ/ϵ_0. In the solutions of the inhomogeneous system (5.48), the longitudinal and transversal components of the amplitudes $\mathbf{E}_i$ may, in general, be of the same order, but when $\mathbf{k}_1$ and $\mathbf{k}_2$ satisfy or almost satisfy the condition of propagation of free waves, the transversal components grow drastically, while the longitudinal ones remain small and do not vary strongly. In this range of $\mathbf{k}_1$ and $\mathbf{k}_2$, the solutions of both homogeneous and inhomogeneous systems prove to be transversal with an accuracy close to ϵ_τ/ϵ_0. Since we are considering the free waves induced by a particle, just this range of $\mathbf{k}_1$ and $\mathbf{k}_2$ is of interest. Hence, below we neglect the longitudinal components of the solution and from (5.48) obtain

$$\eta E_1^\mu + P_\mu \frac{\epsilon_\tau}{\epsilon_0} E_2^\mu = F_{\boldsymbol{\tau}_4}^\mu(\mathbf{k}_p, \mathbf{k}_1)$$

$$P_\mu \frac{\epsilon_\tau}{\epsilon_0} E_1^\mu + \eta' E_2^\mu = F_{\boldsymbol{\tau}_1}^\mu(\mathbf{k}_p, \mathbf{k}_2) \qquad (5.49)$$

where $\eta = 1 - k_1^2/q^2$, $\eta' = 1 - k_2^2/q^2$, and E_i^μ, F^μ are now the scalar amplitudes. $P_\sigma = 1$ for σ-polarization (orthogonal to the plane $\mathbf{k}_1\mathbf{k}_2$; $\mu = \sigma$) and $P\pi = \cos \mathbf{k}_1\hat{\mathbf{k}}_2$ for π-polarization (within the plane $\mathbf{k}_1, \mathbf{k}_2; \mu = \pi$).

Note that one more difference exists between the solution of the inhomogeneous system (5.48) and (5.49) and that corresponding to free fields, namely, in the inhomogeneous solutions, the projections of $\mathbf{k}_1$ and $\mathbf{k}_2$ on the direction of particle velocity are fixed for a fixed frequency.

Directions and Frequencies of Diffraction Scattering. The dynamic equations (5.48) and (5.49) yield the solution to the problem not only when the amplitudes $\mathbf{E}_1$ and $\mathbf{E}_2$ are of the same order, but also when one of them may be neglected, as in the kinematic approximation. In the latter case, the corresponding amplitudes prove to be small, as shown in the solutions. Hence, before arriving at a detailed solution of the dynamic equations, let us find the directions and frequencies of the waves $\mathbf{E}_1$ and $\mathbf{E}_2$, near which the amplitudes are, in fact, of the same order. This can be done by exploiting the fact that at a fixed frequency the wave vectors of free fields subjected to diffraction scattering differ from those not subjected to diffraction by a small value $|1 - k_\tau^2/q^2| \sim \epsilon_\tau/\epsilon_0 \sim 10^{-6}$. On the other hand, we neglect this difference and assume that $|k_1|^2 = |k_2|^2 = \epsilon_0 \frac{\omega^2}{c^2}$, we find that the frequencies and directions in question within the kinematic approximation are (see Fig. 5.2).

$$k_p - k_{1z} - \tau_{1z} = 0$$
$$|\mathbf{k}_1| = |\mathbf{k}_1 + \boldsymbol{\tau}| = \frac{\omega}{c}\sqrt{\epsilon_0}. \tag{5.50}$$

We denote the frequencies and angles $(\mathbf{k}_1, \hat{\mathbf{k}}_2)$ and $(\mathbf{k}_1, \hat{\mathbf{k}}_p)$ given by these formulas as ω_B, $2\theta_B$, and θ_1. For certain fixed $\boldsymbol{\tau}$ and τ_{1z}, the directions $\mathbf{k}_1$ and $\mathbf{k}_2$ form conical surfaces, the values θ_B, θ_1, and ω_B varying along those surfaces as

$$\omega_B = -a \pm \sqrt{a^2 - b^2}; \quad \sin\theta_B = \frac{c\tau}{2\omega_B}\cos\theta_1 = \frac{c}{\omega_B}\left(\frac{\omega_B}{v} - \tau_{1z}\right)$$

$$b = \left(\frac{v\tau}{2}\right)^2 \left[\left(\frac{\cos\hat{\mathbf{k}}_p\boldsymbol{\tau}}{\cos^2\phi + \cos^2\hat{\mathbf{k}}_p\boldsymbol{\tau}\cdot\sin^2\phi} + \frac{2\tau_{1z}}{\tau}\right)^2\right. \tag{5.51}$$

$$+ \left.\frac{\sin^2\hat{\mathbf{k}}_p\boldsymbol{\tau}}{(\cos^2\phi + \cos^2\hat{\mathbf{k}}_p\boldsymbol{\tau}\sin^2\phi)^2}\right] \cdot \left[1 - \left(\frac{v}{c}\right)^2 \frac{\sin^2\hat{\mathbf{k}}_p\boldsymbol{\tau}\cos^2\phi}{(\cos^2\phi + \cos^2\hat{\mathbf{k}}_p\boldsymbol{\tau}\cdot\sin^2\phi)^2}\right]^{-1}$$

$$a = \frac{v\tau}{2}\left[\frac{\cos\hat{\mathbf{k}}_p\boldsymbol{\tau}}{\cos^2\phi + \cos^2\hat{\mathbf{k}}_p\boldsymbol{\tau}\cdot\sin^2\phi} + \frac{2\tau_{1z}}{\tau}\right]$$

$$\cdot \left[1 - \left(\frac{v}{c}\right)^2 \frac{\sin^2\hat{\mathbf{k}}_p\boldsymbol{\tau}\cos^2\phi}{(\cos^2\phi + \cos^2\hat{\mathbf{k}}_p\boldsymbol{\tau}\cdot\sin^2\phi)^2}\right] \tag{5.52}$$

where ϕ is the difference of azimuthal angles of the vector $\mathbf{k}_1$ and $\boldsymbol{\tau}$ around the direction $\mathbf{k}_p$. Two stands for ω_B in (5.51), corresponding to two values of ϕ which differ by π. The spatial orientation of the vectors defined by (5.51) and (5.52) and the corresponding frequencies ω_B may be determined

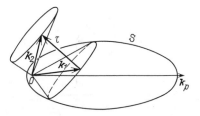

FIGURE 5.3. Geometrical schematics for determining the values $\mathbf{k}_1$ and $\mathbf{k}_2$, in whose vicinities structure radiation should be described dynamically.

from the geometrical scheme given in Fig. 5.3. The values and directions of $\mathbf{k}_1$ in Fig. 5.3 are determined by a vector that joins the origin of $\mathbf{k}_p$ with any point of the cross section of the surface S (determining wave vectors allowed for structure radiation) by the plane which is perpendicular to the vector $\boldsymbol{\tau}$ and shifted from the point O by $\tau/2$. Since $\mathbf{k}_2 = \mathbf{k}_1 + \boldsymbol{\tau}$, then the locus of the ends of the vector $\mathbf{k}_2$ is the first cross section shifted by $\boldsymbol{\tau}$. The surface of wave vectors for structural radiation is the surface given by Eq. (5.50). This surface S shown in Fig. 5.3 for a fixed τ_{1z} yields the frequencies and directions of the structure radiation within the kinematic approximation. The wave vector for any permitted direction of radiation is given by the section of the line along this direction from the point O (the origin of $\mathbf{k}_p$) to the surface S and the corresponding frequency is $c|\mathbf{k}_1|$. Thus, in the small vicinity of (on the order of $\epsilon_{\boldsymbol{\tau}}/\epsilon_0$) the direction and frequency determined by the formulas (5.51) and (5.52), the parameters of structure radiation are not governed by the kinematic approximation, and to find them, we should solve the dynamic system (5.47).

Solution of the Dynamic System. The solution of the system (5.48) determining the waves induced by the particle field in a crystal is

$$E_1^\mu = D_\mu^{-1}\left(F_1^\mu \eta' - P_\mu F_2^\mu \frac{\epsilon_{-\boldsymbol{\tau}}}{\epsilon_0}\right), \quad E_2^\mu = D_\mu^{-1}\left(F_2^\mu \eta - F_1^\mu P_\mu \frac{\epsilon_{\boldsymbol{\tau}}}{\epsilon_0}\right) \quad (5.53)$$

$$F_i^\mu = F^\mu(\mathbf{k}_p, \mathbf{k}_i) \quad (5.54)$$

$$D_\mu = \eta \eta' - P_\mu^2 \frac{\epsilon_{\boldsymbol{\tau}}\epsilon_{-\boldsymbol{\tau}}}{\epsilon_0^2}. \quad (5.55)$$

In order to discover the explicit form of the dependence of the solutions of (5.53) on frequency and $\mathbf{k}_1$, let us express η' from η by representing the frequency in the form $\omega = \omega_B(1 + \nu)$ where ν is a small parameter and ω_B is given by (5.51). Then we use the fact that $\mathbf{k}_2 = \mathbf{k}_1 + \boldsymbol{\tau}$ and the projections $\mathbf{k}_1$ and $\mathbf{k}_2$ on $\mathbf{k}_p$ are fixed and equal to $k_{1z} = k_p - \tau_{1z}$ and $k_{2z} = k_p - \tau_{1z} - \tau_z$. Hence, $\mathbf{k}_1$ and $\mathbf{k}_2$ may be represented in the form $\mathbf{k}_1 = \mathbf{k}_{10} + \xi k_{10}\mathbf{s}$ where $\mathbf{s}$ is the normal to $\mathbf{k}_p$ lying in the plane $\mathbf{k}_1, \mathbf{k}_p$, and $\mathbf{k}_{10}$ is an auxiliary vector (see Fig. 5.4) such that $|\mathbf{k}_{10}| = \omega/c$ and the

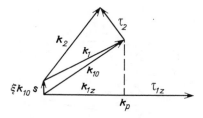

FIGURE 5.4. Definition of the auxiliary values $\mathbf{k}_{10}$ and $\mathbf{s}$ used to solve the dynamic system.

projection $\mathbf{k}_{10}$ on $\mathbf{k}_p$ coincides with the projection on the same direction of the vector $\mathbf{k}_{1B}$ defined by (5.51) and (5.52).

Using this relationship, we find

$$\eta' = \frac{\boldsymbol{\tau}(2\mathbf{k}_{10} + \boldsymbol{\tau})}{k_{10}^2} + \eta/b = 2g\nu + \eta/b = 2\alpha + \eta/b \qquad (5.56)$$

where

$$B = \frac{\cos \widehat{\mathbf{k}_1 \mathbf{s}}}{\cos \widehat{\mathbf{k}_2 \mathbf{s}}},$$

$$g = \frac{\sin \theta_B}{\beta^2}\left[\beta \cos \boldsymbol{\tau}\hat{\mathbf{k}}_p + \cos \phi \frac{\sin \boldsymbol{\tau}\hat{\mathbf{k}}_p}{\sin \theta_1}\left(\frac{c\tau_{1z}}{\omega_B} + \beta^2 - 1\right)\right]$$

and ϕ is the difference of azimuthal angles measured around v of the vectors $\boldsymbol{\tau}$ and $\mathbf{k}_{10}$, $\beta = \frac{v}{c}$.

Using (5.56), we represent D_μ in the form $D_\mu = -b^2(\eta - \eta^+)(\eta - \eta^-)$ where η^+, η^- are the roots of the equation

$$\eta(2\alpha + \eta/b) - P_\mu^2 \frac{\epsilon_{\boldsymbol{\tau}}\epsilon_{-\boldsymbol{\tau}}}{\epsilon_0^2} = 0. \qquad (5.57)$$

For $\eta^\pm$, we have

$$\eta^\pm = b\left(-\alpha \pm \sqrt{\alpha^2 + P_\mu^2 \frac{\epsilon_{\boldsymbol{\tau}}\epsilon_{-\boldsymbol{\tau}}}{\epsilon_0^2}b^{-1}}\right). \qquad (5.58)$$

We have expressed the roots of (5.57) through the phenomenological parameters ϵ_τ and ϵ_0 in order to simplify the expressions in question. For the applications below, it would be expedient to write down the roots of an equation similar to (5.57) for the case in which the phenomenological parameters are substituted by the expressions obtained from microscopic consideration. It is known (see, e.g., [33]) that in this case, we must use $\epsilon_0 = 1$ in the Maxwell equations and substitute η and η' by $\eta + F_{11}$ and $\eta' + F_{22}$, and $\epsilon_{\mathbf{k}_i - \mathbf{k}_q}$ by F_{iq} where $F_{iq} = \frac{4\pi}{Vk^2}F(\mathbf{k}_i, \mathbf{k}_q)$, $F(\mathbf{k}_i, \mathbf{k}_q)$ is the structure amplitude known in X-ray diffraction theory. The value $\eta^{\pm}$ in (5.57) must be taken in the form

$$
\eta^{\pm} = b\left[-\left(\alpha + \frac{F_{11}}{2b} + \frac{F_{22}}{2} \right) \right.
$$

$$
\left. \pm \sqrt{\left(\alpha + \frac{F_{11}}{2b} + \frac{F_{22}}{2} \right)^2 + \frac{1}{b}[2F_{11}\alpha - (F_{11}F_{22} - P_\mu^2 F_{12}F_{21})]} \right] \quad (5.59)
$$

As follows from (5.53), (5.55), and (5.58), the amplitudes E_i found from the solution of the dynamic system (5.49) are rapidly varying functions of frequency ω and $\mathbf{k}_{1\perp}$, and they differ essentially from those obtained within the kinematic approximation (5.5). The physical consequences of this difference will be discussed below.

Boundary Problem. Let us now consider the structure radiation which is emitted when a charged particle moves perpendicularly to the plane parallel crystal of thickness L. In this case, the radiation field induced by the particle in the crystal $\mathbf{E}(\mathbf{r})$ is a superposition of the solutions $\mathbf{E}^H$ of the inhomogeneous system ϵ^{fr} describing the free fields in a crystal

$$
\mathbf{E}(\mathbf{r}) = \mathbf{E}^H(\mathbf{r}) + \mathbf{E}^{fr}(\mathbf{r}) \tag{5.60}
$$

where $\mathbf{E}$ are the superpositions of fields corresponding to the wave vectors $\mathbf{k}_1$ and $\mathbf{k}_2$, i.e., they are the four-component vectors. The amplitudes of the free fields are to be found from the boundary conditions and expressed through the solutions of (5.49) obtained above. In this case, just as in X-ray diffraction, it is convenient to consider the Laue case (in which both waves $\mathbf{E}_1$ and $\mathbf{E}_2$ exit from the crystal through the same surface, Fig. 5.5) and the Bragg case (when those waves exit through different surfaces, Fig. 5.6). In the Laue case, the boundary conditions may be written as

$$
\int \left[\mathbf{E}_1^H(k_1) + \mathbf{E}_1^{fr}(k_1) \right] dk_{1z} = 0,
$$

$$
\int \left[\mathbf{E}_2^H(k_2) + \mathbf{E}_2^{fr}(k_2) \right] dk_{2z} = 0 \tag{5.61}
$$

where the index "fr" labels the solutions of the homogeneous system obtained by cancelling the right-hand sides of (5.49). In the Bragg case, the boundary conditions are:

$$
\int \left[\mathbf{E}_1^H(k_1) + \mathbf{E}_1^{fr}(k_1) \right] dk_{1z} = 0,
$$

$$
\int \left[\mathbf{E}_2^H(k_2) + \mathbf{E}_2^{fr}(k_2) \right] e^{ik_{2z}L} dk_{2z} = 0. \tag{5.62}
$$

The formulae (5.61) and (5.62) include the amplitudes with the explicit arguments since we want to emphasize dependence on $\mathbf{k}_i$. From (5.61) and

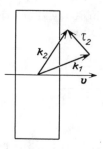

FIGURE 5.5. Laue case.

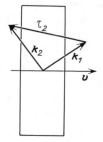

FIGURE 5.6. Bragg case.

(5.62), it follows that the wave vectors of the waves to be tailored must have coinciding components along the surface $\mathbf{k}_{i\perp}$. Hence, the free solution to (5.49) is taken such that both the eigensolutions of the system have the same values of $\mathbf{k}_{1\perp}(\mathbf{k}_{2\perp})$. Just those solutions are used in X-ray diffraction theory. It is known that in those solutions the corrections to the vacuum values of wave vectors are determined by the formulas (5.58) in which $\mathbf{s}$ in the expression for b must be substituted by the normal $\mathbf{n}$ to the crystal's surface and $\mathbf{k}_1 = \mathbf{k}_1^{vac} + |k_1^{vac}|\dfrac{\mathbf{n}\eta}{2\cos \mathbf{nk}_1}$. Thus, the conditions (5.61) relate the amplitude $\mathbf{E}_i$ to the amplitude $\mathbf{E}_{i\pm}^{fr}$ where the index $\pm$ denotes their relation to a corresponding root of the secular equation given by (5.58). Hence, the field in the crystal is a superposition of three fields $\mathbf{E}_i, \mathbf{E}_{i+}^{fr}, \mathbf{E}_{i-}^{fr}$. The values of $\mathbf{E}_{i\pm}^{fr}$ are:

$$
E_{1\pm}^{fr} = \frac{\pm E_1(\tilde{\eta}_\pm + F_{11}) + F_{12}E_2}{\tilde{\eta}_+ - \tilde{\eta}_-}, \quad E_{2\pm}^{fr} = E_{1\pm}^{fr}\frac{\tilde{\eta}_\pm + F_{11}}{F_{12}}. \tag{5.63}
$$

In the Bragg case,

$$
E_{1\pm}^{fr} = \mp \left\{ E_1(\tilde{\eta}_\pm + F_{11}) \exp\left(-\frac{i\omega L \tilde{\eta}_\pm}{2c} \right) \right.
$$
$$
\left. + E_2 F_{12} \exp L \left\{ i \left[\frac{\omega}{v} - \tau_{1z} - \frac{\omega}{v}\left(1 - \frac{k_1^2 c^2}{\omega^2} \right)^{1/2} \right] \right\} \right\}
$$

$$\cdot \left[(\tilde{\eta}_- + F_{11}) \exp\left(-\frac{i\omega L \tilde{\eta}_-}{2c} \right) - (\tilde{\eta}_+ + F_{11}) \exp\left(-\frac{i\omega L \tilde{\eta}_+}{2c} \right) \right]^{-1},$$

$$E_{2\pm}^{fr} = \pm \frac{\tilde{\eta}_\pm + F_{11}}{F_{12}} E_{1\pm}^{fr}. \tag{5.64}$$

The value of $\tilde{\eta}_\pm$ in (5.63) and (5.64) are determined by (5.58) if we take

$$\alpha = \frac{1}{2k_{10}^2} \left\{ \tau \left[2 \left(sk_{1\perp} + \mathbf{n}\sqrt{k_{10}^2 - k_{1\perp}^2} \right) + \tau \right] \right\},$$

$$B = \frac{\cos \hat{k_1 \mathbf{n}}}{\cos \hat{k_2 \mathbf{n}}},$$

where $\mathbf{n}$ is the normal to the crystal's surface, and E_i is given by (5.53) and (5.57). If the obtained fields are tailored at the crystal boundaries with the external fields using a relationship similar to (5.61) and (5.62), then the amplitudes of the waves leaving the crystal in the Laue case are

$$E_q^{ex} = E_q \exp \left\{ i \left[\frac{\omega}{v} - \tau_{1z} - \frac{\omega}{c} \left(1 - \frac{k_{1\perp}^2 c^2}{\omega^2} \right)^{1/2} \right] L \right\}$$

$$+ \sum_{p=\pm} E_{qp}^{fr} \exp \left[-\frac{i\omega L \tilde{\eta}_p}{2c} \right] \tag{5.65}$$

where $\mathbf{E}^{fr}$ is given by (5.63) and $q = 1, 2$. In the Bragg case,

$$E_q^{ex} = E_q \exp \left\{ i \left[\frac{\omega}{v} - \tau_{12} - \frac{\omega}{c} \left(1 - \frac{k_{1\perp}^2 c^2}{\omega^2} \right)^{1/2} \right] L \right\}$$

$$+ \sum_{p=\pm} E_{qp}^{fr} \exp \left[\frac{i\omega L \tilde{\eta}_p}{2c} \right] \tag{5.66}$$

where E_{qp}^{fr} is given by (5.64). Since the expressions (5.63–5.66) are valid both for σ- and π-polarizations, we omit the polarization index μ. However, it must be kept in mind that because the roots of (5.58) differ for σ- and π-polarizations, the emission and propagation of waves with those polarizations for thick perfect crystals may be qualitatively different (see below). As seen from (5.65) and (5.66), the intensity and polarization for the waves leaving a crystal at a fixed direction or frequency exhibit beats depending on crystal thickness. In contrast to the pendulum solution of the X-ray diffraction theory which gives one beating period for the σ- and π-polarizations, structure radiation has three beating periods in every polarization since three waves are interfering.

Emission from Thick Perfect Crystals. It was shown above that in the crystals with finite thickness L, the polarization and intensity of radiation in a fixed direction are rather complicated nonmonotone functions of L. Here we consider the case of thick crystals $\left(L\frac{\omega}{c}ImF_{11} \gg 1\right)$ which are completely nontransparent at the frequency under consideration out of the Bragg condition. In this case, only one or two terms in (5.65) and (5.66) are not zeros and analysis is greatly simplified. We will demonstrate below that the structure radiation intensity from a thick perfect crystal may be either greater or less than that for an imperfect crystal of the same size.

Let us determine the conditions which provide the extremal values of E_q^{ex}. Since the amplitudes of all waves in (5.65–5.66) are expressed through the solutions of an inhomogeneous problem, we first analyze the dependence of the solutions of the inhomogeneous system (5.49) on ω and $\mathbf{k}$.

The frequencies for which the extremal values of E_i occur correspond, as follows from (5.53), (5.57), and (5.58), to the extremal values of $|Im\eta_+|$ as functions of frequency and may be found from

$$\frac{d}{d\alpha}|Im\eta_\pm| = \frac{d}{d\alpha}\left|Im\left\{-\frac{1}{2}(F_{11} + btsF_{22})\pm\right.\right.$$

$$\left.\left. b\sqrt{\left(\alpha + \frac{F_{11}}{2b} + \frac{F_{22}}{2}\right)^2 + \frac{1}{b}[-2F_{11}\alpha - (F_{11}F_{12} - P_\mu^2 F_{12}F_{21})]}\right\}\right|. \quad (5.67)$$

The analysis of (5.67) shows that the maximal and minimal values of E_i occur for σ-polarization ($p_\sigma = 1$). For example, for $b = 1$ (i.e., when $\tau\mathbf{k}_p = 0, \pi$ and the reflecting plane is perpendicular to $\mathbf{k}_p$) the imaginary part of $\eta_\pm$ is at a minimum at $\nu = 0$ and under the assumption $F_{ik} = F_{ii}e^{-W}$

$$Im\eta_- \cong \frac{1}{2}[1 - e^{-2W}]ImF_{ii} \quad (5.68)$$

where e^{-2W} is the Debye–Waller factor for scattering from the direction $\mathbf{k}_1$ to the direction $\mathbf{k}_2$. For this frequency, the maximum of the ratio of E_{ik} to the corresponding kinematic approximation amplitude reaches the value

$$\max\left|\frac{E_i}{E_i^{kin}}\right| \sim (1 - e^{-2W})^{-1}. \quad (5.69)$$

When $b = -1$, i.e., when $\tau\mathbf{k}_p = \pi/2$ and the reflecting surface is parallel to $\mathbf{k}_p$, the maximum of E_i is at the frequency corresponding to

$$\nu = \left[\cos\phi\frac{\sin\theta_B}{\sin\theta_1\beta^2}\left(\frac{c\tau_{1z}}{\omega_B} + \beta^2 - 1\right)\right]^{-1}ReF_{11}(1 + e^{-W}).$$

For those frequencies

$$\max\left|\frac{E_i}{E_i^{kin}}\right| \cong \left[e^{-W}(1 - e^{-W})\right]^{-1}. \quad (5.70)$$

Also, at $b = -1$ there is a deep minimum of radiation intensity at the frequency corresponding to $\nu = \left[\cos\phi\frac{\sin\theta_B}{\beta^2\sin\theta_1}\cdot\left(\frac{c\tau_{1z}}{\omega_B} + \beta^2 - 1\right)\right]^{-1} ReF_{11}$ where

$$\min\left|\frac{E_i}{E_i^{kin}}\right| \cong \frac{ImF_{11}}{ReF_{11}} \tag{5.71}$$

These estimates illustrate how the radiation intensity from a thick perfect crystal is distributed over angles and frequencies. Let us now write down the integral parameters.

The total flux of radiation power passing through the plane perpendicular to the particle's trajectory in the direction k_i is

$$W_i = \frac{c}{(2\pi)^6}\int\left[\sum_{\mu=\pi,\sigma}|E_{i\mu}^{ex}|^2\right]dk_{i\perp}\cos k_1\hat{k}_p\,d\omega. \tag{5.72}$$

In the Laue case for thick crystals, only one term in (5.65) remains nonzero due to the absorption of free fields and $E_i^{ex} = E_i$. Using the latter relationship and expressing in (5.72) the integral over $k_{i\perp}$ through corresponding residues [see (5.53) and (5.57)], we find for a fixed frequency the ration of flux calculated using the kinematic and dynamic approximations. The maximum of this ratio is, as mentioned above, determined by σ-polarization. For $b = -1$ and the frequency corresponding to the maximum of E_i, we find

$$\max\left(\frac{W_i}{W_i^{kin}}\right) \cong \left(\frac{ImF_{11}}{ReF_{11}}\right)^{1/2}\left[e^{-w}(1 - e^{-w})\right]^{-3/2}. \tag{5.73}$$

The range of frequencies or angles in which this ratio is of the order of that in (5.73) is

$$\Delta\nu = \left[\frac{\sin\theta_B}{\beta^2\sin\theta_1}\left(\frac{c\tau_{1z}}{\omega_B} + \beta^2 - 1\right)\cos\phi\right]^{-1}. \tag{5.74}$$

In the Bragg case, only two terms in the expression for E^{ex} (5.65) are nonzeros and corresponding expressions for the flux may be represented in the form

$$w_1 = \frac{c}{(2\pi)^6}\int\sum_\mu\left|E_1 + E_2\frac{F_{12}}{\tilde{\eta}_+ + F_{11}}\right|^2\cos k_1\hat{k}_p\,dk_{1\perp}$$

$$w_2 = \frac{c}{(2\pi)^6}\int\sum_\mu\left|E_2 + E_1\frac{\tilde{\eta}_- + F_{11}}{F_{12}}\right|^2\cos k_2\hat{k}_p\,dk_{2\perp}. \tag{5.75}$$

The ratio of the maxima of these expressions to the kinematic flux for $f = 1$ is estimated as

$$\frac{w_i}{w_i^{kin}} \cong (1 - e^{-2W})^{-2}. \tag{5.76}$$

The estimate (5.76) is valid in the following frequency (or angular) range corresponding to the extreme frequency or angle:

$$\Delta\nu = \frac{\beta}{\sin\theta_B}(1 - e^{-2W})^{1/2} Re F_{11}. \tag{5.77}$$

These estimates hold for σ-polarization. To obtain those for π-polarization, we have only to substitute e^{-2W} with $P_\pi^2 e^{-2W}$; hence, extreme values are smaller for π-polarization.

Discussion. The relationships obtained in this section indicate that at some selected ranges for frequencies near ω_B, the results of dynamic theory differ quantitatively from those of a kinematic approach. The differences are related to angular distribution, polarization properties, and intensity of radiation from a crystal. As follows from (5.53), (5.65), and (5.66), the angular distribution at a fixed frequency may have, in contrast to the kinematic approximation, several maxima in the angular range $\Delta\theta \sim F_{12}$, i.e., it may have a fine structure. In crystals with finite thickness and at some fixed directions and frequencies, the intensity and polarization as functions of the thickness exhibit beats. In contrast to the corresponding beats in X-ray diffraction which have only one period for every eigenpolarization (π and σ), in this case three periods exist. In the quantities integrated over angles, there are two periods of beats for each polarization.

From a practical viewpoint, the most interesting are the results related to thick perfect crystals—namely, the increase of radiation intensity from a perfect crystal near the frequency ω_B. This effect is connected to the suppression of absorption within a crystal due to the Borrmann effect, rather than an increase of the radiation losses of a particle in a perfect crystal. When two coherently coupled waves are excited in a crystal, an anomalous transmission can occur (Borrmann effect) [34,35]. In this case, the thickness of the layer from which the radiation can exit is not $(kIm F_{11})^{-1}$ but is much greater $(kIm\eta_-)^{-1}$. We must keep in mind, nevertheless, that the rate of the Borrmann effect depends on the crystal structure and it may be completely realized in a crystal with a primitive unit cell and for some special reflections on a complicated crystal for which the phase factors in the structure amplitude are the same in every site of the unit cell.

As was shown in Chapt. 3, the suppression of absorption is stronger the smaller the difference of the amplitudes F_{11} and F_{12} is, or, in other words, the closer to unity the Debye–Waller factor of the corresponding reflection is. This means that at higher temperatures the suppression becomes weaker and its temperature dependence is determined by the Debye–Waller factor. Thus, the suppression of absorption and the increase of intensity from an ideal crystal are limited by the values of the Debye–Waller factor. For these effects to be realized in a nonideal (mosaic) crystal, disorientation of its crystallites should be small and satisfy the restrictions determined by (5.74) and (5.77).

Since the suppression effect is the most pronounced for σ-polarization, the yield of polarized radiation in the Borrmann effect is at a maximum for those directions $\mathbf{k}_1$ and $\mathbf{k}_2$ which are connected to the reciprocal lattice vector $\boldsymbol{\tau}$, making a considerable angle with the particle's velocity. In this case, the radiation leaving the crystal is almost completely polarized and corresponds to σ-polarization. As follows from the dependence of the Debye–Waller factor on $\mathbf{k}_2 - \mathbf{k}_1 = \boldsymbol{\tau}$, the emission from a crystal is maximum, if the other conditions are the same for the minimum value of the reciprocal lattice vector $\boldsymbol{\tau}$. The dependence of $F_{\boldsymbol{\tau}_1}(\mathbf{k}_p, \mathbf{k}_i)$ on the angles $\mathbf{k}_i \hat{,} \mathbf{k}_p$, $\mathbf{k}_p \hat{\boldsymbol{\tau}}_1$, and Debye–Waller factor $e^{-2W(\mathbf{k}_p - \mathbf{k}_i)}$ [23,24] results in the fact that the maximum coherent radiation losses at an X-ray frequency and, as a result, the maximum emission from the crystal are achieved for a particle moving along the main crystallographic direction, provided that at least one of the angles $\mathbf{k}_i \hat{} \mathbf{k}_p$ is not small. As follows from this discussion, the optimal conditions for the anomalously high intensity of a crystal may be fulfilled only for a small number of reciprocal lattice vectors $\boldsymbol{\tau}$ and $\boldsymbol{\tau}_1$ and we may expect an enhanced emission only for a few directions determined by Eq. (5.50). We should note also that the direction of enhanced radiation is, in general, at large angles relative to particle velocity. In particular, the enhancement of emission may occur in the back semisphere where incoherent radiation intensity for a fast particle is at a minimum.

5.4.2 Coherent Coulomb Excitation of a Mössbauer Transition

At the present time, a good deal of interest is paid to coherent X- and gamma radiation induced by charged particles moving through a stratified medium or crystal. In this connection, the experiments were initiated which were mainly related to possible applications of those radiations for detecting high-energy charged particles [5,6]. Usually, coherent radiation is considered which is connected to the polarization of medium electrons by the particle field. In addition to the above-mentioned mechanism of coherent radiation, another exists, considered in Sect. 5.2, which is based on the coherent Colulomb excitation of low-lying nuclear levels by the particle [30,31,36–38]. In contrast to the electron mechanism, the nuclear one yields the radiation at a fixed frequency, which is determined by the energy of the level excited, and with a narrow linewidth coinciding with the width of a nuclear level. Since the emission line of coherent nuclear radiation is very narrow and radiation is sharply directed, this mechanism can be used, in principle, to create a Mössbauer radiation source which does not decay in time and possesses high differential (spectral and angular) brightness [39]. However, the intensity integrated over angles remains very low; hence, we need to find a means to increase the total intensity of coherent nuclear radiation. One possible solution is to perform the coherent Coulomb excitation when the emitted gamma quanta are subjected to dynamic diffraction.

Hence, in this section we consider briefly the coherent radiation of a fast charged particle which is due to the Coulomb excitation of Mössbauer nuclei in a crystal when, in contrast to Sect. 5.2, the emitted quanta experience diffraction scattering in the crystal's periodic structure. It turns out that the diffraction of coherent radiation in a crystal essentially changes the angular distribution of this radiation along some individual directions and the absorption of radiation may be anomalously low (Borrmann effect); this results in a considerable increase of the radiation intensity as compared with that along the directions out of the range of diffractive scattering. Using the above results, we analyze here the conditions which provide the maximum yield of coherent gamma radiation from a crystal. After that, the results obtained are discussed in connection with the problem of increasing the differential brightness of gamma sources using the coherent Coulomb excitation of Mössbauer nuclei.

The System of Dynamic Equation. Let a charged particle move with constant velocity through a perfect crystal containing nuclei with a low-lying nuclear level. We will look for coherent radiation from the crystal, due to the Coulomb excitation of the nuclear (Mössbauer) level. We assume here that the emitted quanta experience diffractive scattering in the crystal. In contrast to [30,31] and Sect. 5.2 where diffraction scattering was not taken into account, the radiation field in this case is described by a superposition of two plane waves $\mathbf{E}_1, \mathbf{E}_2$ whose wave vectors $\mathbf{k}_1$ and $\mathbf{k}_2$ are related by the Bragg condition $\mathbf{k}_2 = \mathbf{k}_1 + \boldsymbol{\tau}$ where $\boldsymbol{\tau}$ is the reciprocal lattice vector.

By analogy with the above section (see also Chapt. 3) we obtain a system of equations (5.48, 5.49) for the amplitudes $\mathbf{E}_1$ and $\mathbf{E}_2$ from which the values $\mathbf{F}_\tau(\mathbf{k}_p, \mathbf{k}_i)$ must be determined with due regard to the resonance interaction of gamma quanta with Mössbauer nuclei (see Sect. 2.2 and Chapt. 3). The procedure of solving this system and the boundary problem is very similar to that of Sect. 5.4.1. Hence, we will not repeat here this solution and describe only the physical consequences of those solutions [38].

Note first of all that at fixed $\boldsymbol{\tau}$ and $\boldsymbol{\tau}_1$, the dynamic system (5.48, 5.49) should be used only in a small vicinity of two directions k_{1B} which are determined below, while structure radiation in other directions is described with the kinematic approximation. The directions sought are obtained with the formula:

$$\omega_M = \mathbf{k}_1\mathbf{v} + \boldsymbol{\tau}_1\mathbf{v}, \quad |\mathbf{k}_{1B}| = |\mathbf{k}_{1B} + \boldsymbol{\tau}| = \frac{\omega_M}{c} \qquad (5.78)$$

where $\omega_M = E/\hbar$ is the frequency corresponding to the energy of the Mössbauer transition E. The first formula of (5.78) determines the surface of a "kinematical" cone on which the direction $\mathbf{k}_{1B}$ must lie, the cone's angle being $2\theta_k$ where $\cos\theta_k = \frac{\omega_M - \boldsymbol{\tau}_1\mathbf{v}}{\omega_M v/c}$. Using the second formula of (5.78), we find that the difference of azimuthal angles of $\mathbf{k}_{1B}$ and $\boldsymbol{\tau}$ measured around

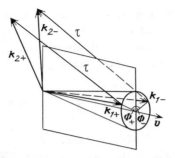

FIGURE 5.7. Dynamic directions for coherent Coulomb excitation of a Mössbauer transition; in the kinematic case, emission directions lie on the cone surface.

$\mathbf{v}$ is, as is seen in Fig. 5.7,

$$\phi_B = \pm \arccos \left(\frac{\cos \mathbf{k_1}\hat{\tau} - \cos \hat{\mathbf{v}}\tau \cos \theta_k}{\sin \theta_k \sin \hat{\tau}\mathbf{v}} \right). \tag{5.79}$$

In the vicinity of the directions $\mathbf{k}_{1B}$, the nuclear diffraction of gamma radiation in a crystal takes place.

The analysis of this problem in its general form shows that in a small vicinity of the directions $\mathbf{k}_{1B}, \mathbf{k}_{1B} + \tau$ given by (5.78, 5.79), the parameters of nuclear structure radiation differ essentially from those obtained with the kinematic approximation. The above-mentioned relates to the angular distribution, polarization, and intensity of radiation. The absorption of radiation in the crystal may sometimes be anomalously low or high and, correspondingly, the radiation yield is anomalously high or low. In the general case, no analytical solution can be written explicitly because when solving the problem, we encounter a secular fourth-order equation. Hence, we illustrate below these results using those obtained in the case of an unsplitted Mössbauer line, which allows the general relationships to be simplified [38]. Such analysis may be carried out by substituting the structural amplitudes which take into account the nuclear resonance scattering of gamma quanta in the formulas of the previous section. Then it turns out that the amplitude $\mathbf{E}_q^{ex}$ of the wave leaving the sample strongly changes within the range of diffraction scattering and its extreme values (divided by the value obtained using the kinematic approximation) are given by

$$\left| \frac{E_q^{ex}}{E_q^{kin}} \right| \sim \left| \frac{ImF_{11}}{Im\eta_\pm} \right|. \tag{5.80}$$

This means that the ratio (5.80) is at a maximum for σ-polarization.

In the case of σ-polarization, the imaginary part of $\eta_\pm$ becomes identically zero if $\left| \frac{F_{12}}{F_{11}} \right| = 1$ for a certain τ. This means that for σ-polarization, the effect of the suppression of nonelastic channels of gamma quanta interaction with nuclei [40] may manifest itself in full strength. In this case, the ratio (5.80) formally reaches infinity. However, we must keep in mind that gamma quanta interact not only with nuclei but also with electrons, the latter interaction always resulting in absorption. If this fact is taken into account, the maximum of the ratio (5.80) is

$$\max \left| \frac{E_q^{ex}}{E_q^{kin}} \right| \cong \frac{ImF_{11}}{F_{11}^R (1 - e^{-W})} \tag{5.81}$$

where e^{-2W} is the Debye–Waller factor and $\frac{V}{4\pi} k_{1B}^2 F_{11}^R$ is the electron component of the structure amplitude for scattering at a zero angle. The ratio $|E_q^{ex}/E_q^{kin}|$ remains of the same order as in (5.81), provided that the angles (or the values of α) lie within the range $(1 - e^{-2W})F_{11}^R$ near α_m. Since the suppression effect is most pronounced for σ-polarization, the emission from a thick crystal in the vicinity of the directions $\mathbf{k}_{1B}$ and $\mathbf{k}_{2B}$ given by (5.78, 5.79) is due mainly to σ-polarization, whereas the increase of intensity is given by (5.81).

Discussion. As follows from the above results under the conditions of diffraction scattering of nuclear structure radiation, i.e., in small vicinities of certain specific dynamical directions, the intensity of radiation leaving a perfect crystal turns out to be greatly increased due to the suppression (Borrmann) effect. The limitation of intensity growth proves to be related to electron absorption which, as known from [33] under the Borrmann effect, is less than that outside the Bragg condition and much less than nuclear resonance absorption. Hence, in the case of sufficiently perfect crystals, it seems rather realistic an enhancement of the intensity along the mentioned directions by two to three orders of magnitude as compared with the estimations of [30,31] (see also Sect. 5.2). Moreover, the radiation emitted along those dynamic directions is almost completely polarized.

In the case of split Mössbauer lines (magneto-ordered crystals or crystals with an inhomogeneous electric field at Mössbauer nuclei), the emission along dynamic directions is, in general, similar to that discussed above, but polarization and angular parameters are more complicated. For example, angular ranges can exist in which the suppression effect develops for any polarization, and in which this effect occurs for a polarization other than linear. Under those conditions, the polarization of emission along dynamic directions is shown to depend on the magnetic or electric structure of the crystals and is, in general, elliptical. Note also that in a crystal with hyperfine structure of a Mössbauer line, the nuclear radiation and radiation which is due to its origin to the electrons of crystal are spatially separated for the emission directions corresponding to magnetic and quadrupole reflections [33]. This fact may be very important for using nuclear structure radiation to generate Mössbauer radiation, because it allows us to obtain a collimated beam of Mössbauer quanta without an X-ray background at a close frequency.

5.4.3   Cherenkov Radiation in Chiral Liquid Crystals

Introductory Remarks. The peculiarities of coherent radiation induced by fast charged particles which were considered above using the examples of shortwave radiation are of great interest for scientific study and practical applications. However, these effects in short wave are very difficult for experimental observations (few papers about the experimental observations of structure radiation in crystals have been published until now [12]). In this connection, the coherent radiation of fast charged particles in chiral liquid crystals and the conditions of its diffraction become interesting in a broad sense. Structure radiation frequency in CLC, as follows from Sect. 5.3, lies, as a rule, in the optical range in which the experimental methods are well elaborated. Hence, the general feature of coherent radiation induced by charged particles under the conditions of diffraction are very convenient to study experimentally using chiral liquid crystals; moreover, the angular and relative frequency intervals where the corresponding effects must exhibit themselves are much wider in optics.

In Chapt. 4 we considered the great variety of CLCL optical properties which are sometimes absolutely unique [41,42]. It is natural that these unusual properties manifest themselves in Cherenkov radiation and in other kinds of coherent radiation of fast charged particles. Below we consider the coherent radiation of particles in cholesteric liquid crystals.

We have already mentioned in Chapt. 4 that both cholesteric crystals and smectic chiral crystals have the one-dimensional spatial periodicity of their properties with the period about optical wavelength. The periodicity of dielectric properties is due to the spatial variations of the orientation of the principal axis of the dielectric permittivity tensor, rather than in the most spatial modulation of dielectric permittivity, as in the usual periodic media. We have already seen that the spatial periodicity of the dielectric tensor results in diffraction of light at some specific frequencies and, as a consequence, in very interesting peculiarities of optical properties at these frequencies (wavelengths). For the same reason, Cherenkov radiation at those frequencies has certain peculiarities and unusual polarization properties.

In addition to the well-known Cherenkov radiation cone, whose axis is along the particle's velocity, in chiral liquid crystals there is a diffraction cone (Fig. 5.8) whose axis differs from the former one [43,44]. The spectrum of radiation emitted into the diffraction cone is essentially different from that of the Cherenkov cone, since it contains only frequencies which experience diffraction in the liquid crystal. Also, the polarization properties of those radiations are very unusual. In general, Cherenkov radiation from chiral liquid crystals is elliptically polarized, the ellipticity depending on the angle between the helical axis and the direction of emission.

The differential parameters of the radiation (relative to frequency and angle) [44,45] also differ strongly from those of Cherenkov radiation in

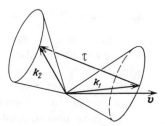

FIGURE 5.8. Diffraction cone for Cherenkov radiation.

homogeneous media. In some geometrical configurations, there are certain frequency (angular) intervals without radiation losses which are due to the fact that in periodic media the frequency intervals exist where the propagation of radiation is forbidden. At the boundaries of those intervals, the differential losses diverge as the square root of frequency [45–47].

It is interesting to note that the radiation losses integrated over frequency and angle coincide, practically speaking, with those in a homogeneous medium whose dielectric permittivity equals the mean value of that for the liquid crystal. The corresponding difference in integral losses is small since the dielectric anisotropy of liquid crystals is small, about $\delta = 0.01$–0.1. Thus, if considered with the zero approximation in δ, the diffraction effects do not change the total radiation losses of Cherenkov radiation in a liquid crystal, as compared with the isotropic liquid phase of the same substance, but these effects strongly influence the angular and frequency distributions of radiation.

Similar peculiarities are characteristic of structure Cherenkov radiation in chiral crystals, which was considered in Sect. 5.3. In contrast to Cherenkov radiation, as was mentioned, this radiation has no threshold at the particle velocity.

Emission from a Sample with Finite Dimensions. Since the optical properties of cholesteric and chiral smectic crystals are very similar (see Chapt. 4), we, to be definite, will consider below the coherent radiation induced by charged particles in cholesteric crystals. The results thereby obtained are in many cases applicable to chiral smectic crystals, but in some cases they require minor modifications.

Let us consider the Cherenkov radiation of a relativistic charged particle with charge e which moves with constant velocity through a cholesteric crystal, the particle velocity v being greater than the phase velocity of light c_p in this medium. The equations describing the coherent radiation arising in this case have the general form of (5.2).

The specific properties of Cherenkov radiation in cholesteric crystals are connected to the form of the dielectric permittivity tensor $\hat{\epsilon}(\mathbf{r})$ (1.5).

Generally speaking, in addition to the wave with the wave vector $\mathbf{k}_1$ whose propagation direction lies on the Cherenkov cone, there are more waves in a cholesteric crystal, these waves being related to the former wave by the Bragg condition. Since the dielectric anisotropy δ is small, we may, in the first approximation, take into account only one such wave. Hence, we will look for the radiation field (its temporal and spatial Fourier components) as a superposition of two plane waves $\mathbf{E}(\mathbf{k}_1, \mathbf{k}_2) = (\mathbf{E}_1 e^{i\mathbf{k}_1\mathbf{r}} + \mathbf{E}_2 e^{i\mathbf{k}_2\mathbf{r}})e^{i\omega t}$ where $\mathbf{k}_1$ lies in the Cherenkov cone and $\mathbf{k}_2 = \mathbf{k}_1 + \mathbf{r}$. Equation (5.2) with regard to (1.5) yields the system

$$(\hat{\epsilon}_0 - k_1^2/q^2)\mathbf{E}_1 + \hat{\epsilon}_{-1}\mathbf{E}_2 = \frac{ie}{2\pi^2\omega}[\mathbf{v} - \mathbf{k}_1(\mathbf{k}_1\mathbf{v})/q^2]\delta(\omega - \mathbf{k}_1\mathbf{v})$$

$$\hat{\epsilon}_1\mathbf{E}_1 + (\hat{\epsilon}_0 - k_2^2/q^2)\mathbf{E}_2 + 0 \qquad (5.82)$$

where $\hat{\epsilon}_i$ are the harmonics of the Fourier expansion of (2.15).

The radiation field for a finite sample is a superposition of the solutions of (5.82) and the solutions of a homogeneous system obtained from (5.82) by cancelling the right-hand sides. The coefficients of this superposition are to be determined from the boundary conditions, as was done above for the case of structure radiation in crystals.

The frequency ω_B and directions k_1, k_2, in the vicinities of which two waves are emitted, are determined by the conditions:

$$\omega_B - (\mathbf{k}_1\mathbf{v}) = 0, \quad |\mathbf{k}_1 + \boldsymbol{\tau}| = |\mathbf{k}_1| = \frac{\omega_B}{c_p} = \frac{\omega_B\sqrt{\hat{\epsilon}}}{c}. \qquad (5.83)$$

If we neglect the frequency dispersion of $\hat{\epsilon}$ in (5.83), then we obtain the following dependence of ω_B on the direction for $\mathbf{k}_1$ that lies on the Cherenkov cone surface:

$$\omega_B = \tau c_p/(2\cos\widehat{\mathbf{k}_1\boldsymbol{\tau}}). \qquad (5.84)$$

The frequencies ω_B given by (5.83) and the corresponding wave vectors $\mathbf{k}_1$ and $\mathbf{k}_2$ may also be found geometrically (see Fig. 5.8).

For the arbitrary direction of the particle velocity with respect to the cholesteric axis, the solutions of (5.82) and of the boundary problem cannot be represented explicitly in analytical form since the secular equation of the system (5.82) is, in general, of the fourth order. The analysis of the general solution indicates that the maxima of the E_i amplitudes as functions of $\mathbf{k}_1$ in the solution of the inhomogeneous system (5.82), corresponding to the intensity maxima, occur when the frequencies and emission directions are such that the determinant of (5.82) is at a minimum. The solution of the boundary problem shows that there are two characteristic cases of the emission pattern from a sample. In one case, beats appear in the intensities of $\mathbf{E}_1$ and $\mathbf{E}_2$ waves when the crystal thickness or particle velocity is varied. In the other case, the values of $\mathbf{E}_1$ and $\mathbf{E}_2$ strongly depend on frequency in the range $|\nu| \sim \delta$ where $\nu = \frac{\omega}{\omega_B} - 1$. In order to illustrate these issues, we analyze below the case allowing the general relationships to be simplified.

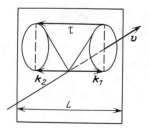

FIGURE 5.9. Geometry of Cherenkov radiation which provides backward diffractive scattering.

We consider the case when a particle moves at an angle to the surface of a sample, as thick as L, whose cholesteric axis is perpendicular to the surface and there is a direction on the Cherenkov cone which coincides with the cholesteric axis (Fig. 5.9). Using (5.84), we find for this direction that $\omega_B = \tau c_p/2$ and the vector $\mathbf{k}_2$ is also directed along the cholesteric axis but is opposite $\mathbf{k}_1$. Near these values of ω_B, $\mathbf{k}_1$, and $\mathbf{k}_2$, the system (5.82) splits into two independent systems (each of two equations), one of them describing the wave of the diffracting circular polarization and the other a wave of the opposite circular polarization which does not experience diffraction in CLCL [14,43]. The solution of (5.82) for diffracting polarization may be written explicitly with regard to the boundary conditions [43]. The amplitudes E_i oscillate and strongly depend on frequency so that their minima occur at $\nu = 0$ and maxima at $|\nu| = \delta/2$. The intensity of radiation leaving the crystal in the directions of the Cherenkov and diffraction cones, I_1 and I_2, is given by the following expressions (see also Fig. 5.10):

$$I_1 = I_c \frac{\nu^2 - (1/2)(\delta/2)^2[1 + \sin 2\kappa Lq/(2\kappa Lq)]}{\nu^2 - (\delta/2)^2 \cos^2 \kappa Lq}$$

$$I_2 = I_c \frac{(1/2)(\delta/2)^2 p[1 - \sin 2\kappa Lq/(2\kappa Lq)]}{\nu^2 - (\delta/2)^2 \cos^2 \kappa Lq} \qquad (5.85)$$

where $I_c = L \frac{e^2 \omega_B \sin^2 \hat{\mathbf{k}_1 \tau}}{2\pi^2 c_p^2 \cos \mathbf{k}_1 \tau}$ is the spectral density of radiation for a corresponding homogeneous sample (or in a cholesteric above the temperature point of transition into isotropic liquid), $\kappa = \omega_B/c_p = \frac{\omega_B \sqrt{\epsilon}}{c}$, $q = \sqrt{\nu^2 - (\delta/2)^2}$.

Thus, as follows from the above formulae, the intensities of radiation at the Cherenkov and diffraction cones as functions of frequency rapidly vary near ω_B and experience sharp beats for $|\nu| \sim \delta$ because of the diffractive scattering of light with the diffracting polarization. At $|\nu| >> \delta$, the radiation intensity tends to I_c in the Cherenkov cone and zero in the diffraction cone. The radiation at the diffraction cone is circularly polarized and corresponds to diffracting polarization, whereas that in the Cherenkov cone

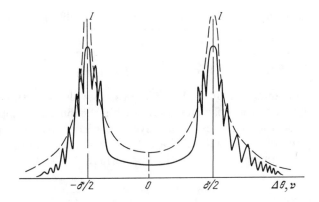

FIGURE 5.10. Angular (frequency) dependence of the intensities of waves emitted into the diffraction cone in an unlimited cholesteric crystal (dashed line) and in a sample with finite dimensions (solid line) at $b > 0$ [45].

is determined by the interference of both eigenpolarizations and changes with frequency, tending to the linear polarization at $|\nu| \gg \delta$. Note that the maximum intensity of radiation (the quantity which is differential in frequency and azimuthal angle) is reached, practically speaking, at the boundary of the selective reflection band at $\nu^2 - (\delta/2)^2 = (\pi/\kappa L)^2$ for both Cherenkov and diffraction cones, this maximal intensity being equal to $I_1 = I_2 = \frac{1}{2} I_c (\delta \kappa L / 2\pi)^2$. That is, it is $(\delta \kappa L / 2\pi)^2$ times grater than the corresponding value for a homogeneous medium. Surely, we must keep in mind that this result is only valid for a sample whose thickness does not lead to significant absorption.

Thus, the periodicity of cholesteric crystals changes qualitatively Cherenkov radiation near certain selected frequencies and directions. The angular and frequency ranges in which the above-mentioned changes take place are of the order of δ. Hence, we can say that not only integral parameters of radiation but also a differential one are accessible through experimental observations.

Until now we have considered the effect of first-order diffraction scattering on particle radiation. When the emission direction does not coincide with the cholesteric axis, high-order scattering may also take place. In this case, the wave vectors at the Cherenkov and diffraction cones are inter-related as $\mathbf{k}_2 = \mathbf{k}_1 + s\boldsymbol{\tau}$ where s is an integer other than ± 1, and the above-mentioned peculiarities of Cherenkov radiation exhibit themselves at the frequencies close to $s\omega_B$. The solution of the corresponding high-order reflection problem, according to the results of cholesteric optics in Chapt. 4, shows that the frequency (angular) intervals, where diffraction scattering is essential, are much narrower than in the first-order case and their widths decrease as $\delta^{|s|}$. Hence, the conditions under which first-order diffractive scattering is expected are the most favorable for experimental studies.

5.4.4   The Shift of Cherenkov Radiation Threshold

Along with Cherenkov radiation, Eq. (5.2) describes also structure Cherenkov radiation which was considered above within the kinematic approximation. In this case, the frequency ω and wave vector $\mathbf{k}_1$ of structure radiation are given by the relationships which are the generalization of the emission condition for Cherenkov radiation [the first formula in (5.83)]:

$$\omega - (\mathbf{k}\mathbf{v}) = s(\mathbf{v}\boldsymbol{\tau}). \tag{5.86}$$

The projection of the momentum of an emitted photon on particle velocity direction, unlike in (5.83), differs from the momentum of a pseudophoton ω/v by the projection of the reciprocal cholesteric lattice vector in the same direction.

Structure radiation, like Cherenkov radiation, at some specific frequencies and directions can meet the conditions of diffraction at the cholesteric periodic structure. In this case, it exhibits the same peculiarities as Cherenkov radiation—namely, two directions of emission, forbidden frequency ranges, specific features of the differential parameters of emission, etc.

If a particle moves through an infinite sample with the velocity $v < c_p$, then only structure radiation arises. If $v > c_p$, both structure and Cherenkov radiations are generated. We should emphasize that at frequencies such that Cherenkov radiation experiences diffraction in CLC, distinguishing the two types of radiation makes no sense since the frequencies and directions of emission at the diffraction cone coincide with those for structure radiation [see (5.83) and (5.86) with $\mathbf{k}_2 = \mathbf{k}_1 + \boldsymbol{\tau}$, $s = -1$]. These facts result in some specific features of Cherenkov radiation when the particle velocity v is close to the threshold value c_p.

Following [48], we assume now that the particle velocity only slightly differs from c_p, i.e., $|c_p - v| \sim \delta c_p$. Furthermore, for the sake of simplicity we assume that the particle moves along the cholesteric axis. Then the diffraction cone is directed backward, i.e., its axis is along the particle velocity and its apex angle coincides with that of a Cherenkov one. Since the velocity is close to the threshold, i.e., θ is small, Cherenkov radiation is directed almost directly forward in the Cherenkov cone and almost directly backward in the diffraction cone. Under these assumptions, the system (5.82) reduces to the following system of equations for the scalar amplitudes E_1 and E_2 which describe the diffraction of circular (diffracting) polarization:

$$(1 - k_1^2/q^2)E_1 + (\delta/2)E_2 = \frac{iev\theta}{2\pi^2\omega\overline{\epsilon}}\delta(\omega - \mathbf{k}_1\mathbf{v})$$
$$(\delta/2)E_1 + (1 - k_2^2/q^2)E_2 = 0. \tag{5.87}$$

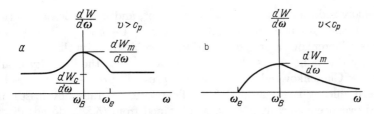

FIGURE 5.11. Differential radiation losses in CLC for particle velocity close to the threshold velocity. (a) Above the threshold velocity. (b) Below the threshold velocity.

Now instead of solving the boundary problem as was done above, we, as in [45,49], calculate the radiation energy losses of a particle per unit length of its trajectory.

Using the solution of (5.87) and calculating the work of slowing down which the field exerts on the particle, we find the spectral density of radiation losses at $v > c_p$ to be

$$\frac{dW}{d\omega} = \begin{cases} \dfrac{dw_c}{d\omega} = \dfrac{e^2\omega}{c^2}[1 - (c_p/v)^2], & \omega > \omega_e \\[3mm] \dfrac{e^2\omega\delta}{c^2} \cdot \dfrac{\nu + \sqrt{(\nu-\gamma)^2+1}}{(\nu-\gamma+\sqrt{(\nu-\gamma)^2+1})(\sqrt{(\nu-\gamma)^2+1}} & \omega < \omega_e \end{cases} \tag{5.88}$$

where $\nu = 2(\omega - \omega_B)/(\delta\omega_B)$, $\gamma = [1 - (c_p/v)^2]/\delta$

$$\omega_e = \omega_B \left[1 + \frac{4(v-c_p)^2 + \delta^2 c_p^2}{8c_p(v-c_p)} \right].$$

As follows from (5.88), the spectral density of radiation losses is described by the Tamm–Frank formula when particle velocity is greater than c_p and the frequency $\omega > \omega_c$ and lies outside the range $|\omega - \omega_B| \leq \delta\omega_B$. If, on the contrary, the frequency is within $|\omega - \omega_B| < \delta\omega_B$, the spectral density $dW_c/d\omega$ essentially exceeds the spectral density of radiation losses in a homogeneous medium $[dW/d\omega = (1/2)(1 + \sqrt{1+\gamma^{-2}})\,dw_c/d\omega]$, see Fig. 5.11a.

For $v < c_p$, the radiation losses are other than zero only at $\omega > \omega_e$ where ω_e^L is the lower frequency of the structure radiation [see (5.7)], which is described by the second line of (5.88). As follows from (5.88), if the velocity satisfies the condition $c_p - v \leq \delta c_p$, the spectral density of losses $dW/d\omega$ has a sharp maximum at $\nu = 0$ (see Fig. 5.11b) in the frequency range $|\omega - \omega_B| \sim \delta\omega_B$, the value for $dW/d\omega$ being of the same order as when the velocity exceeds $c_p(v/c_p - 1 \sim \delta)$. When the velocity falls much below the threshold values $c_p(v/c_p - 1 > \delta)$, the losses decrease rapidly and the spectral distribution corresponds to the values characteristic of structure

Cherenkov radiation, i.e., $dW/d\omega \sim e^2\omega\delta^2/c^2$, and it is at least δ^{-1} times less than that of Cherenkov radiation at $v/c_p - 1 \sim \delta$.

Thus, Cherenkov radiation in periodic media shifts its threshold in velocity by the value $\Delta c_p \approx \delta c_p$. However, at $v < c_p$, the intensity characteristic of Cherenkov radiation near the threshold $v/c_p - 1 \sim \delta$ takes hold in the frequency range around $\delta\omega_B$. Therefore, the radiation losses integrated over frequency at $v < c_p$, as obtained from (5.88), do not diverge even if the ϵ frequency dispersion is neglected, these losses being about $W = [e^2\omega_B\delta^2/(4c^2)]|\ln(1 - v/c_p)|$.

The shift of the velocity threshold of Cherenkov radiation by δc_p corresponds to the decrease of the energy threshold by

$$\Delta E = E_t[1 - \sqrt{(\epsilon - 1)/[\epsilon - (1 - \delta)^2]}],$$

where E_t is the threshold energy for emission in a homogeneous medium. This decrease in the threshold energy may be very essential in the case of ultrarelativistic threshold energies.

The physical cause for this shift is connected to the diffraction of electromagnetic waves in a periodic medium. At the frequencies which correspond to the strongest diffraction interaction with the medium, the dielectric permittivity proves to differ from $\bar{\epsilon}$ by $\sim \delta\bar{\epsilon}$ and the Cherenkov radiation threshold is shifted. Also, below the threshold for the homogeneous phase, which is determined by the mean dielectric permittivity $\bar{\epsilon}$, the radiation intensity distribution differs from that above the threshold and is concentrated close to ω_B in the frequency range $\sim \delta\omega_B$.

Note that the objects, which are convenient for experimental detection of the threshold shift, are the cholesterics with substantial dielectric anisotropy. Since the value of δ in a cholesteric may be as high as 0.1, the threshold shift in velocity may be about 10^{-1}. It is also important that for CLC a situation can be easily created (by a slight variation of temperature near the CLC phase transition temperature) for which the particle velocity is below the threshold in isotropic liquid and above it in the cholesteric phase.

Conclusion. Thus, Cherenkov radiation in cholesteric liquid crystals or, more generally, the coherent radiation of charged particles in chiral liquid crystals possesses a series of interesting peculiarities. Many of these peculiarities are common for all periodic media. Hence, a discussion of these effects is beyond the scope of the problem of Cherenkov radiation in liquid crystals only. Since the parameters of liquid crystals, such as their dielectric anisotropy δ and helical pitch p, can be easily varied and the effects in question are available for optical measurements, then chiral liquid crystals are a convenient model for studying the coherent radiation of charged particles in periodic media. For example, it is convenient to investigate the theoretically predicted effect of the increase of the X-ray intensity generated in crystals under certain conditions [6] by the use of chiral liquid crystals in optics. Also, it seems expedient to carry out experiments on the shift of the Cherenkov radiation threshold in periodic media [48] with the example of a chiral liquid crystal.

We should also note that some of the effects discussed above find analogies in other physical phenomena, although their discussion using Cherenkov radiation is simpler. For example, enhancing the intensity of the emission of radiation emitted by a particle at the frequency near the boundary of a forbidden zone is analogous to the growth of the efficiency of a nonlinear optical frequency transformation in periodic media [50,51] (see Chapt. 6).

Although there are many theoretical predications relating to the radiation of particles in chiral liquid crystals which also have analogies to radiations in other periodic media and possible applications [1,5–7,52–54], until now there has been no experimental study in this field which presents a challenge to the art of experimentalists.

Unlike structure radiation in liquid crystals, structure radiation in conventional crystals has attracted greater attention from investigators. After the first observation of structure radiation [12], several groups of researchers continued to study this phenomenon with different objects and for different energies of electron beams. Some interesting properties of this radiation, as predicted by the theory outlined, were observed [62–67]. In [63,64] the angular distribution of structure radiation in silicon and diamond was investigated and a fine structure in the intensity angular distributions discovered. High-energy structure radiation emitted from a silicon single crystal in an almost forward direction was observed in [64]. Polarization measurements were also performed with structure radiation from a silicon single crystal [65]. More fine effects due to the many-wave diffraction effect in the emission of structure radiation were observed in [66,67]. The progress of experimental investigations has stimulated further theoretical works [68–70]. The influence of a multiple scattering of emitting particles [68], and of many-wave diffraction conditions [69] on structure radiation were theoretically studied, as well as the possibility of using structure radiation for a solution of the phase problem (determination of the phase of X-ray structure amplitude) [70]. The above-mentioned results are only first achievements in this field and without a doubt Cherenkov structure radiation will be investigated further and find its place in physical investigations and various applications.

References

1. M.L. Ter-Mikaelyan: *Effect of the Medium on Electromagnetic Processes at High Energies* (Erevan, AN ArmSSR, 1969) (in Russian).

2. B.M. Bolotovskii, G.V. Voskresenskii: Uspekhi Fiz. Nauk **94**, 377 (1968).

3. F.G. Bass, V.M. Yakovenko: Uspekhi Fiz. Nauk **86**, 189 (1965).

4. G.M. Garibyan, Yang Shi: ZhETF **61**, 930 (1971).

5. V.L. Ginzburg, V.N. Tsitovich: *Transition Radiation and Transition Scattering* (Moscow, Nauka, 1984) (in Russian).

6. G.M. Garibyan, Yang Shi: *X-Ray Transition Radiation* (Erevan, AN ArmSSR, 1983) (in Russian).

7. V.G. Baryshevskii: *Channelling, Radiation and Reactions in Crystals at High Energies* (Minsk, BFU, 1982) (in Russian).

8. V.L. Ginzburg, I.M. Frank: ZhETF **16**, 15 (1946).

9. P.A. Cherenkov: Doklady AN SSSR **2**, 451 (1934); Uspekhi Fiz Nauk **93**, 385 (1967).

10. I.E. Tamm, I.M. Frank: Doklady AN SSSR **14**, 107 (1937); Uspekhi Fiz. Nauk **93**, 388 (1967).

11. V.B. Berestetskii, E.M. Lifshits, L.P. Pitaevskii: *Relativistic Quantum Theory, Part 1* (Moscow, Nauka, 1968).

12. S.A. Vorob'ev, B.N. Kalinin, S.L. Pak, A.P. Potylitsin: Pis'ma v ZhETF **41**, 3 (1985); Yu. N. Adishchev, S.A. Vorob'ev, V.G. Baryshevskii et al.: Pis'ma v ZhETF **41**, 295 (1985).

13. V.A. Belyakov, V.P. Orlov: Phys. Lett. A **42**, 3 (1972).

14. V.A. Belyakov, V.E. Dmitrienko, V.P. Orlov: Pis'ma v ZhTF **1**, 978 (1975).

15. N.P. Kalashnikov, V.S. Remizovich, M.I. Ryazanov: *Collisions of Fast Charged Particles in Single Crystals* (Moscow, Atomizdat, 1980) (in Russian); N.P. Kalashnikov: *Coherent Interaction of Charged Particles in Single Crystals* (Moscow, Atomizdat, 1981) (in Russian).

16. A.I. Akhiezer, N.F. Shul'ga: Uspekhi Fiz. Nauk **132**, 561 (1982).

17. V.A. Bazylev, N.K. Zhevago: Uspekhi Fiz. Nauk **137**, 605 (1982).

18. E.I. Katz: ZhETF **59**, 1854 (1970) [Sov. Phys. - JETP **32**, 1004 (1971)].

19. L.D. Landau, E.M. Lifshitz: *Electrodynamics of Continuous Media* (Moscow, Nauka, 1982).

20. I. Linhard: Mat. Fys. Madd. Dan. Vid. Selski. **34**, No. 14 (1965).

21. D.S. Gemmel: Rev. Mod. Phys. **46**, 129 (1974).

22. V.A. Belyakov: Pis'ma v ZhETF **13**, 254 (1971); Proceedings of All-Union Meeting on Physics of Interactions of Charged Particles with Single Crystals (Moscow, MGU, 1972, p. 214) (in Russian) [JETP Lett. **13**, 179 (1971)].

23. V.A. Belyakov, V.P. Orlov: Proceedings of the 3th All-Union Meeting on Physics of Interactions of Charged Particles with Single Crystals (Moscow, MGU, 1977 pp. 69,77) (in Russian).

24. A.A. Grigor'ev, V.A. Belyakov: FTT **17**, 1577 (1975).

25. V.G. Kudryavtsev, M.T. Ryasanov: Pis'ma ZhETF **11**, 507 (1970).

26. Yu. M. Kagan, F.I. Chukhovskii: Pis'ma ZhETF **5**, 166 (1967).

27. K. Adler et al.: Rev. Mod. Phys. **28**, 432 (1956).

28. M.A. Andreeva, R.N. Kuzmin: *Mössbauer Gamma-Optics* (Moscow, MGU, 1982) (in Russian).

29. V.I. Baier, V.M. Katkov, V.S. Fadin: *Radiation of Relativistic Electrons* (Moscow, Atomizdat, 1973) (in Russian).

30. V.A. Belyakov, V.P. Orlov: Phys. Lett. **44A**, 463 (1973).

31. V.A. Belyakov, V.P. Orlov: Proceedings of the 5th All-Union Meeting on Physics of Interaction of Charged Particles with Single Crystals (Moscow, MGU, 1974, p. 131) (in Russian).

32. B.M. Bolotovskii: Uspekhi Fiz. Nauk **62**, 201 (1957).

33. V.A. Belyakov: Uspekhi Fiz. Nauk **115**, 553 (1975) [Sov. Phys. - Usp. **18**, 267 (1975)].

34. R.W. James: *The Optical Principles of the Diffraction of X-Rays* (London, Bell and Sons, 1967).

35. Z.G. Pinsker: *Dynamical Scattering of X-Rays in Crystals* (Berlin, Springer Verlag, 1978).

36. *Mössbauer Spectroscopy II*, ed. by U. Gonser (Berlin, Springer-Verlag, 1981).

37. A.V. Kolpakov: Yadern. Fiz. **16**, 1003 (1972).

38. B.A. Belyakov, V.P. Orlov: *Methods and Apparatus for Precise Measurements of the Parameters of Ionizing Radiation* (Moscow, VNIIFTRI, 1976, p. 53) (in Russian).

39. A.I. Chechin et al.: Pis'ma v ZhETF **37**, 531 (1983).

40. A.M. Afanas'ev, Yu. M. Kagan: ZhETF **48**, 327 (1965).

41. V.A. Belyakov, A.S. Sonin: *Optics of Cholesteric Liquid Crystals* (Moscow, Nauka, 1982) (in Russian).

42. V.A. Belyakov, V.E. Dmitrienko: Uspekhi Fiz. Nauk **146**, 369 (1985) [Sov. Phys. - Usp. **28**, 535 (1985)].

43. N.V. Shipov: ZhETF **86**, 2075 (1984).

44. V.A. Belyakov, V.E. Dmitrienko, V.P. Orlov: Uspekhi Fiz. Nauk **127**, 221 (1979)[Sov. Phys. - Usp. **22**, 63 (1979)].

45. N.V. Shipov, V.A. Belyakov: ZhETF **75**, 1589 (1978) [Sov. Phys. - JETP **48**, 802 (1978)].

46. Yu. N. Adishechev, S.A. Vorobiev, B.N. Kalinin: ZhETF **90**, 829 (1986).

47. E.I. Katz: ZhETF **61**, 1686 (1971).

48. V.A. Belyakov, N.V. Shipov: ZhETF **88**, 1547 (1985) [Sov. Phys. - JETP **61**, 923 (1985)].

49. L.D. Landau, E.M. Lifshitz: *Electrodynamics of Continuous Media* (Moscow, Nauka, 1982) (in Russian).

50. V.A. Shipov, V.A. Belyakov: Phys. Lett. A **86A**, 94 (1981); ZhETF **82**, 1159 (1982).

51. S.V. Shiyanovskii: Ukrainskii Fiz. Zhurnal **27**, 361 (1982).

52. V.V. Fedorov, A.I. Smirnov: ZhETF **76**, 866 (1979).

53. I.M. Dykman, L.A. Prokhnitskii: Fiz. Tv. Tela **22**, 2420 (1980).

54. N.V. Shipov: Izvestiya Vuzov, Radiofizika **28**, 1043 (1985).

55. V.G. Baryshevsky, I.D. Feranchuk: J. Physique **44**, 913 (1983).

56. I.D. Feranchuk, A.V. Ivashin: J. Physique **46**, 1981 (1985).

57. V.A. Belyakov: Nucl. Instr. Meth. in Phys. Res. **A248**, 20 (1986).

58. A.M. Afanas'ev, M.A. Aginyan: ZhETF **74**, 570 (1978).

59. Y.-H. Ohtsuki: *Charged Beam Interaction with Solids* (London, New York, Taylor & Francis, 1983).

60. M.A. Kumakhov: *Radiation of Channelled Particles in Crystals* (Moscow, Energoatomizdat, 1986) (in Russian).

61. V.A. Bazylev, N.K. Zhevago: *Radiation of Fast Particles in the Matter and in External Fields* (Moscow, Nauka, 1987).

62. Yu. N. Adishev, R.D. Babadjanov, S.A. Vorob'ev et al.: ZhETF **93**, 1943 (1987).

63. A.N. Didenko, Yu. N. Adishev, S.A. Vorob'ev et al.: DAN SSSR **296**, 1360 (1987).

64. D.I. Ade'shvili, S.V. Mlazhevih, V.F. Boldyshev et al.: DAN SSSR **298** 844 (1988).

65. Yu. N. Adishev, V.A. Verzilov, S.A. Vorob'ev et al.: Pis'ma ZhETF **48**, 311 (1988).

66. V.P. Afanasenko, V.G. Baryshevski, R.F. Zuevski et al.: Pis'ma ZhETF **15**, 33 (1989).

67. V.P. Afanasenko, V.G. Baryshevski, S.V. Gatsikha et al.: Pis'ma ZhETF **51**, 213 (1990).

68. V.G. Baryshevski, A.O. Grubich, L.T'en Khai: ZhETF **48**, 311 (1988).

69. T.B. Ha, I.Ya. Dubovskaya: Phys. Stat. Sol. **155**, 685 (1989).

70. A.V. Ivashin, I.D. Feranchuk: Kristallografiya **34**, 39 (1989).

6

Nonlinear Optics of Periodic Media

This chapter is devoted to nonlinear optical phenomean in periodic media, mainly the case in which radiation wavelength is close to the spatial period of the nonlinear medium. Under this condition, effects can appear which are not observed in the usual crystals where the spatial period is much smaller than optical wavelength so that periodicity does not affect nonlinear optical phenomena [1,2].

To be more specific, we will consider as an example the generation of high optical harmonics. Here the periodicity of a medium leads not only to the well-known modification of the phase-matiching conditions in harmonic generation [3,4] (reciprocal lattice vector of the periodic medium can now be included in the phase-matching conditions), but also to the theoretically predicted enhancement of the efficiency of the nonlinear transformation of optical frequency when the harmonic generation in a nonlinear periodic medium undergoes diffraction [5]. For the sake of simplicity, we will consider the generation of high harmonics in periodic media within the approximation of a nondepleted pumping wave.

Before considering concrete examples of the generation of high harmonics, we note that the descriptions of charged particle radiation discussed in Chapt. 5 and of nonlinear harmonic generation (in the case of a given pumping wave) are somewhat analogous. In both cases, the problem is reduced to the solution of inhomogeneous Maxwell equations, the inhomogeneity being due either to particle current [see (5.2)] or nonlinear polarization induced by the pumping wave [see (6.2) below]. Therefore, the radiation of charged particles and the nonlinear generation of high harmonics in periodic media possess qualitatively similar features. One of the pecularities they have in common, namely, the enhancement of efficiency of nonlinear generation under diffraction conditions, was already mentioned. Analogous to this effect is the enhancement of the charged particle radiation intensity under diffraction conditions described in Chapt. 5.

6.1 Enhancement of Efficiency of the Nonlinear Transformation of Optical Frequencies in Periodic Medium

Considerable attention has recently been paid to the nonlinear optics of periodic media [3,6–8]. These media are of interest because they provide wider potentialities in phase-matched frequency transformation than uniform media. In this section it will be shown that a periodic medium has a greater number of phase matchings than a uniform medium and provides more efficient nonlinear frequency transformation. We will consider these phenomena taking second harmonic generation (SHG) in one-dimensional periodic media as an example. As we will see, maximum enhancement of the SHG for collinear geometry is reached when the frequency (2ω) of the wave generated under phase matching coincides with that of the frequency boundaries ω_e of the stop band (see Chapt. 1). If the absorption can be neglected, the maximum of SHG phase-matched intensity is proportional to the fourth power of the sample thickness L and its ratio to the SHG intensity in a homogeneous medium is proportional to $(\delta L \omega / c)^2$, where δ is the amplitude of the nonlinear medium dielectric constant spacial modulation.

As was mentioned above, an analogous increase in Cherenkov radiation intensity also occurs at the frequencies coinciding with the boundary frequencies of a forbidden band [9,10]. Since the specific features of the periodic medium are not important here, we, as was already mentioned, will choose for simplicity's sake a one-dimensional periodic nonlinear medium in which second harmonic generation occurs [5]. Let a pumping wave with frequency ω and wave vector $\mathbf{k}$ propagate in a plane-parallel sample of thickness L (Fig. 6.1), the dielectric properties of the medium outside the sample being described by the tensor $\hat{\epsilon}$, and those of the sample being harmonically modulated in space with small amplitude δ and described by the tensor

$$\hat{\epsilon}(\mathbf{r}) = \hat{\epsilon}(1 + 2\delta \cos \boldsymbol{\tau} \mathbf{r}) \tag{6.1}$$

where $\boldsymbol{\tau}$ is the reciprocal lattice vector of the periodic structure ($\tau = 2\pi/d$, d being the period). We will assume that the sample's properties allow for nonlinear generation of the second harmonic and the direction of $\mathbf{k}$ is close to the second harmonic Bragg's direction. Then we can apply the dynamic diffraction theory discussed in Chapt. 2. In a periodic medium, the second harmonic wave is a sum of two plane waves $\mathbf{E} = \mathbf{E}_0 e^{i\mathbf{k}_0 \mathbf{r}} + \mathbf{E}_1 e^{i\mathbf{k}_1 \mathbf{r}}$ where $\mathbf{k}_1 = \mathbf{k}_0 + \boldsymbol{\tau}$. Assume also that one of the principal axes of the tensor $\hat{\epsilon}$ is perpendicular to the plane $\mathbf{k}_0, \mathbf{k}_1$. Then the system of two equations can be obtained from Maxwell equations (5.2) for $\mathbf{E}_0, \mathbf{E}_1$ within the approximation of a nondepleted pumping wave (see [7]) by replacing the inhomogeneity related to particle current by $4\pi \left(\frac{2\omega}{c}\right)^2 P^{(2)}(r, 2\omega)$ where

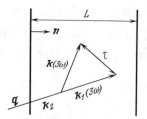

FIGURE 6.1. Geometry of nonlinear frequency transformation in a periodic medium.

$P^{(2)}$ is the nonlinear polarization:

$$(1 - k_0^2/q_q^2)E_0^q + \delta^q E_1^q = -\frac{4\pi}{\epsilon} P_0^{qmn} \delta(\mathbf{k}_0 - \mathbf{k}_m - \mathbf{k}_n)$$

$$\delta^q E_0^q + (1 - k_1^2/q_q^2)E_1^q = -\frac{4\pi}{\epsilon} P_1^{qmn} \delta(\mathbf{k}_0 - \mathbf{k}_m - \mathbf{k}_n). \qquad (6.2)$$

Here $\mathbf{q}_q$ is the wave vector in the homogeneous medium with the dielectric tensor $\hat{\epsilon}$, and P_0 and P_1 are the corresponding zero and first components of the Fourier expansion of the nonlinear polarization $\mathbf{P}^{(2)}_{(r,2\omega)} = \sum_j \mathbf{P}_j e^{i\boldsymbol{\tau}_j \mathbf{r}}$. All quantities in (6.2) are marked by polarization indices q, m, n which are either π or σ, denoting, respectively, the linear polarization in the plane $\mathbf{k}_0, \mathbf{k}_1$ or perpendicular to this plane:

$$\delta^\pi = \delta \cos \mathbf{k}_0 \hat{\mathbf{k}}_1, \quad \delta^\sigma = \delta.$$

The solution of system (6.2) essentially depends on the ratio of δ to the value of frequency dispersion $\eta = (1 - k_0^2/q_q^2)/\delta$ (see the next section for more details), on the geometry of the experiment (see Fig. 6.1), in particular, on the parameter $b = \cos(\hat{\mathbf{k}_1, \mathbf{n}})/\cos(\hat{\mathbf{kn}})$ where $\mathbf{n}$ is the normal to the sample's surface, and on the closeness of the harmonic wave to the diffraction conditions as defined by the parameter

$$\alpha = \frac{1}{2}[\cos \hat{\mathbf{kn}} \cos \hat{\mathbf{k}}\hat{\mathbf{k}}_1 (\eta + 2\nu) + \eta \cos \hat{\mathbf{k}_1 \mathbf{n}}](\cos \hat{\mathbf{kn}} \cos \hat{\mathbf{k}_1 \mathbf{n}})^{-1}$$

where $\nu = -(2\boldsymbol{\tau}\mathbf{q} + \tau^2)/(2\mathbf{q}^2\delta \cos \hat{\mathbf{k}}\hat{\mathbf{k}}_1)$, the vector $\mathbf{q}$ being directed along $\mathbf{k}$, $\mathbf{q}^2 = \left(\frac{2\omega}{c}\right)^2 \epsilon(2\omega)$. Solving the boundary problem for $b < 0$ (Bragg geometry) as was done in Chapt. 5, we obtain the electric field amplitudes for the harmonic wave at the sample's boundary:

$$E_0 = \left[E_0^i + \frac{E_0^i(\xi_+ - \xi_-)e^{i\alpha\ell} + 2iE_1^i \sin \Delta\ell}{\xi_- e^{-i\Delta\ell} - \xi_+ e^{i\Delta\ell}}\right] \cdot D^{-1}$$

$$E_1 = \left[E_1^i + \frac{E_1^i(\xi_+ - \xi_-)e^{i\alpha\ell} + 2i\epsilon_0^i \xi_- \xi_+ \sin \Delta\ell}{\xi_- e^{-i\Delta\ell} - \xi_+ e^{i\Delta\ell}}\right] \cdot D^{-1} \qquad (6.3)$$

where we have put

$$D = \cos \mathbf{k}\hat{\mathbf{k}}_1 \eta (\eta + 2\nu) - 1$$
$$\Delta = [\alpha^2 - D/(\cos \widehat{\mathbf{k}_1 \mathbf{n}} \cos \widehat{\mathbf{k}_1 \mathbf{n}})]^{1/2}$$
$$\ell = \delta q L/2, \quad \xi_{\pm} = \eta + (\alpha \pm \Delta) \cos \widehat{\mathbf{k}\mathbf{n}},$$
$$E_0^i = [\cos \widehat{\mathbf{k}_1 \mathbf{k}} (\eta + 2\nu) P_0 - P_1], \quad E_1^i = (\eta P_1 - P_0)/\delta.$$

In (6.3) and everywhere below, the polarization indices are omitted for the sake of simplicity. So the above formulas describe the generation of the σ-polarized harmonic only. When analyzing other physical situations, the same formulas are used, provided that the corresponding values are marked with the proper polarization indices.

It follows from (6.3) that phase matching occurs when the argument of the delta function (6.2) is zero, i.e., phase-matching conditions can be represented in the form

$$2\alpha = (\eta \cos \widehat{\mathbf{k}\mathbf{n}} \cos \widehat{\mathbf{k}_1 \mathbf{n}})^{-1} (\cos \widehat{\mathbf{k}\mathbf{n}} + \eta^2 \cos \widehat{\mathbf{k}_1 \mathbf{n}}). \tag{6.4}$$

The matching conditions of (6.4) with $k_m(2\omega) = k_q(\omega) + k_\ell(\omega) + s\tau$ are more diverse than in the case of a homogeneous medium. For example, for fixed values of ω and δ, condition (6.4) can be satisfied by varying the direction of the pumping wave or the modulation period. Condition (6.4) being satisfied, the harmonic wave intensities in the directions k_0 and k_1 are described by

$$|E_q|^2 = \eta^2 E_q^{i^2} (\xi_+ - \xi_-) \ell^2 (\cos \widehat{\mathbf{k}\mathbf{n}} + \eta^2 \cos \widehat{\mathbf{k}_1 \mathbf{n}})^{-2} |\xi_- e^{i\Delta \ell} - \xi_+ e^{i\Delta \ell}|. \tag{6.5}$$

It follows from (6.5) that if the ratio $\eta = \left(1 - \frac{\epsilon(\omega)}{\epsilon(2\omega)}\right) / \delta$ is close to the critical value $\eta = \eta_c = 1/\sqrt{|b|}$, i.e., if the harmonic frequency exactly equals the boundary frequency of the forbidden band, then the second harmonic experiences strong diffraction scattering and the generation efficiency increases sharply, compared with that in a homogeneous medium.

In this case, i.e., when $(\delta q L)^{-1} << |\eta - \eta_c| \leq 1$, Eq. (6.5) becomes

$$|E_0|^2 = |\eta E_1|^2 \tag{6.6}$$
$$= \left[\frac{2\pi (P_0 - P_1 \eta) q L}{\epsilon \cos \widehat{\mathbf{k}\mathbf{n}}}\right]^2 \left[(B\eta^2 + 1)^2 - 4B\eta \sin^2 a \left(\frac{b\eta^2 + 1}{q}\right)\right]^{-1},$$

$$a = \ell / \cos \widehat{\mathbf{k}_1 \mathbf{n}}.$$

When η tends to η_c, the radiation intensity oscillates and increases (see Fig. 6.2). The maximum values of the intensities $|E_q|^2$ are reached when η differs slightly from its critical value: $\eta - \eta_c = \pi s/(ab)$, $s = \pm 1, \pm 2, \pm 3$ and are proportional to the fourth power of the sample thickness:

$$|E_0|^2 = |b E_1|^2 = (q L)^4 [\delta (P_1 - P_0 \sqrt{|b|})/(2s\epsilon \cos \widehat{\mathbf{k}\mathbf{n}} \cos \widehat{\mathbf{k}_1 \mathbf{n}})]. \tag{6.7}$$

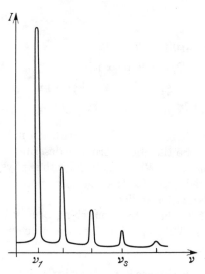

FIGURE 6.2. Frequency dependence of the amplitude of nonlinear harmonic generation for the matching conditions satisfied near the boundary of the selective reflection band: $\nu = \frac{2\omega - \omega_e}{\omega_B}$, $\nu_s = \delta \left(\frac{\pi s}{a}\right)^2 / (4\sqrt{|b|} \cdot \cos \hat{\mathbf{k}}_1 \tau)$.

When the pumping wave direction is fixed, the maximum amplitudes can be expressed in terms of the difference between second harmonic frequency 2ω and that of the forbidden band boundary ω_e:

$$|E_0|^2 = |bE_1|^2 = \left[\frac{\pi q L (P_1 - P_0\sqrt{|b|})}{2\epsilon \cos \hat{\mathbf{k}}\mathbf{n}}\right]^2 \delta\omega_e / \left[\sqrt{|B|}(2\omega - \omega_e)\cos^2 \hat{\mathbf{k}}\boldsymbol{\tau}\right]$$

(6.8)

where

$$2\omega - \omega_e = \omega_B \delta \left(\frac{\pi s}{a}\right)^2 / (4\sqrt{|b|}\cos^2 \hat{\mathbf{k}}\boldsymbol{\tau}), \quad \omega_e = \omega_B \left(1 + \frac{\delta\sqrt{|b|}}{2\cos^2 \hat{\mathbf{k}}\boldsymbol{\tau}}\right)$$

$$\omega_B = -c\tau \left(1 + \frac{\eta}{2}tg\hat{\mathbf{k}}\boldsymbol{\tau}tg\hat{\mathbf{k}}\mathbf{n}\right) / [2\sqrt{\epsilon(2\omega)}\cos \hat{\mathbf{k}}\boldsymbol{\tau}].$$

We will now discuss the conditions under which the enhancement of efficiency of the nonlinear transformation described by (6.7, 6.8) can be at a maximum. In the simplest collinear geometry, for instance, a strong correlation of the frequency dispersion of ϵ and δ is required for the phase-matching conditions to be satisfied. Such correlation is not required in noncollinear geometry. For the maximum efficiency of the nonlinear transformation to be achieved, the linewidth of the pumping wave must be narrow compared with the frequency width of the first maximum of the amplitude oscillations (see Fig. 6.2). This width depends on the sample thickness and is of the order of $\Delta\omega/\omega = \lambda^2/(\delta L^2)$. If the pumping wavelength $\lambda \sim 10^4$ Å and $\delta = 0.01$, $\Delta\omega/\omega = 10^{-4}$ at $L = 0.1$ cm, the maximum enhancement of harmonic intensity (compared with that in a homogeneous sample with constant dielectric permittivity ϵ) must appear even for nanosecond pulses (more so for longer pulses) and be of the order of 10^2, provided the enhancement remains within the framework of the nondepleted pumping wave approximation.

If the pumping wave is not monochromatic and the harmonic linewidth exceeds those of the generation intensity maxima in (6.6), then SHG efficiency is low, compared with that at the maxima of (6.7, 6.8). Harmonic intensity can now be obtained by averaging over the linewidth in accordance with (6.3). If the harmonic linewidth is close to the distance between the maxima in (6.8), then averaging yields the following expression for mean harmonic intensity

$$|E_0|^2 = |bE_1|^2 = \frac{\pi^2(P_1 - P_0\sqrt{|b|})^2(kL)^2}{2\epsilon\cos^2\hat{\mathbf{k}\mathbf{n}}\cos^2\hat{\mathbf{k}\tau}}\left(\frac{\delta\omega_e\sqrt{|b|}}{|2\omega - \omega_e|}\right)^{1/2}. \tag{6.9}$$

Note that the simplifications made here do not affect the above qualitative results and can be removed. For example, we can easily take account of the diffraction of the pumping wave, reflection at the sample's boundaries, etc. The smallness of δ is, however, essential in quantitative analysis, numerical calculations being necessary if this condition is not met. The requirement $b < 0$ is important for the enhancement of efficiency of the nonlinear transformation, which for $b > 0$ does not arise. As for media in which the effect considered can be observed, there are, in addition to artificial heterostructures, incommensurable crystal structures, crystals with helicoidal magnetic ordering, and chiral liquid crystals. The latter crystals seem the most promising because their parameters can be varied easily.

6.2   Third Harmonic Generation in a Cholesteric Liquid Crystal

The pecularities of nonlinear generation of the second harmonic in periodic media discussed in the previous section are very genral, and they also manifest themselves in other nonlinear optical phenomena. Here third harmonic generation (THG) will be considered for cholesteric crystals in collinear geometry, both the pumping wave and harmonic being directed along the cholesteric axis. Cholesteric crystals were chosen since they are used in experiments on harmonic generation [4,8], and the exact solutions of Maxwell equations within linear optics are known for them, unlike the previous section. Hence, the analytical description of THG is not limited to the case of small dielectric anisotropy δ. Besides, the nonlinear optical phenomena in liquid crystals provide important information on the structures of those crystals, revealing their nonlinear susceptibilities. For the sake of brevity, we will not give here a detailed description of this topic that is completely outlined in [8]. We shall only note that in order to observe in liquid crystals the nonlinear optical effects related to the electron mechanism discussed below, pulse lasers (e.g., Q-switched Nd-lasers) as a rule are used [8] in order to provide high intensity of light and to avoid overheating of liquid crystals and the distortion of their structures. The effects of orientational optical nonlinearity, which are not considered here, can be observed with low-power lasers, such as a He-Ne laser, in a continuous mode [19].

Within the approximation of a nondepleted pumping wave of frequency ω which propagates along the cholesteric axis, the generation of the third harmonic is described by an equation of the type of (1.17) which includes the nonlinear polarization vector P as the inhomogeneity:

$$\left[\frac{\partial^2}{\partial z^2} + \left(\frac{3\omega}{c} \right)^2 \hat{\epsilon}(z, 3\omega) \right] \epsilon(z, 3\omega) = -4\pi \left(\frac{3\omega}{c} \right)^2 \mathbf{P}^{(3)}(z, 3\omega) \qquad (6.10)$$

where the nonlinear polarization vector $\mathbf{P}^{(3)}$ is defined from the cubic tensor of nonlinear susceptibility [1,2] as

$$\mathbf{P}^{(3)}(z, 3\omega) = \hat{\chi}(z, 3\omega) \mathbf{E}(z, \omega) \mathbf{E}(z, \omega) \mathbf{E}(z, \omega). \qquad (6.11)$$

All quantities in (6.10) and (6.11) are functions of the coordinate Z and frequency ω, which emphasize their spatial inhomogeneity and the frequency dispersion of the dielectric properties of liquid crystals.

6.2.1 Conditions for Phase Matching

In order to determine the intensity of the third harmonic radiation from a cholesteric sample, as well as its direction and polarization, we must solve the equations (6.10) with regard to boundary conditions. However, before solving the boundary problem, let us find the conditions for the most effective generation of the third harmonic, i.e., phase-matching conditions [1].

We use the fact that the pumping wave in a CLC must satisfy the eqution (1.7); hence, we represent this wave $\mathbf{E}(z, \omega)$ as the sum, as in (1.17), over the eigensolutions of (1.9) and perform Fourier expansion of $\mathbf{P}^{(3)}$ in (6.10).

The particular solution of (6.10) will be looked for, as was done in Chapt. 1, in the form:

$$\epsilon(z, 3\omega) = \left[E^+ \mathbf{n}_+ e^{i\tilde{K}^+(3\omega)z} + E^- \mathbf{n}_- e^{i\tilde{K}^-(3\omega, z)} \right] e^{-3i\omega t} \qquad (6.12)$$

where $\tilde{K}^+ - \tilde{K}^- = \tau$.

After substituting the Fourier expansions of the nonlinear polarization of (6.11) and (6.12) into (6.10), we obtain the following system of equations for the amplitudes E^+ and E^-:

$$\left[1 - \left(\frac{\tilde{K}^+}{q}\right)^2\right] E^+ + \delta E^-$$

$$= \sum_{s\ell mn} \mathcal{P}^{\ell mn}_{+s}(j_1 j_2 j_3)\delta(\tilde{K}^+ - k^m_{j_2} - k^\ell_{j_1} + k^n_{j_3} + s\tau)$$

$$\delta E^+ + \left[1 - \left(\frac{\tilde{K}^-}{q}\right)^2\right] E^-$$

$$= \sum_{s\ell mn} \mathcal{P}^{\ell mn}_{-s}(j_1 j_2 j_3)\delta(\tilde{K}^- - k^\ell_{j_1} - k^m_{j_2} - k^n_{j_3} + s\tau) \qquad (6.13)$$

where the right-hand sides are defined by the Fourier coefficients of the nonlinear polarization vector obtained when substituting into (6.11) the pumping wave field in the form of expansion (1.9, 1.17), where the indices ℓ, m, n take the values $+$ and $-$

$$\mathbf{P}(3\omega z) = -\left(\frac{4\pi}{\epsilon_3}\right)^{-1} \sum_{s,\ell,m,n=\pm} \mathbf{n}_\pm \mathcal{P}^{\ell mn}_{\pm s}(j_1 j_2 j_3)$$

$$\cdot \exp\left\{i\left[(k^\ell_{j_1} + k^m_{j_2} + k^n_{j_3} + s\tau) - 3\omega t\right]\right\}. \qquad (6.14)$$

For the sake of generality, when deriving (6.14), we have assumed that the pumping wave is a superposition of all four eigenwaves of the CLC. Then expressing $K^\pm_j$ through $k^\pm$ and taking $\tilde{K}^\pm = \tilde{k}_3 \pm \tau/2$ (see the notation of Chapt. 1), we find that delta-functions on the right-hand sides of (6.13) can be represented in the form $\delta(\tilde{k}_3 \pm k^\ell \pm k^m \pm k^n + s_+\tau)$ with $s_+ = 3, 2, 1, 0, -1, -2$ for the first equation, and in the form $\delta(\tilde{k}_3 \pm k^\ell \pm k^m \pm k^n + s_-\tau)$ with $s_- = 2, 1, 0, -1, -2, -3$ for the second one; all possible combinations of signs in the arguments of the delta-functions are admissible. Formally, the value of $\tilde{k}_3(\tilde{K}^\pm)$ in the inhomogeneous solution does not necessarily satisfy the dispersion equation (1.13) at the frequency 3ω. However, harmonic generation can actually be effective, provided that $\tilde{k}_3$ satisfies or almost satisfies the dispersion equation. Hence, the quantity $\tilde{k}_3$ is convenient to represent as $\tilde{k}_3 = \Delta k^\pm + k^\pm_3$ where $k^\pm_3$ satisfies the dispersion equation, i.e., it describes the eigenwaves in the CLC at the frequency 3ω. Then the condition that an individual addend of the right-hand side of (6.13) contributes to the solution is

$$\Delta k^\pm_{\ell mn} = k^\ell + k^m + k^n - k^\pm_3 + s_\pm\tau. \qquad (6.15)$$

The indices ℓ, m, n marking $\Delta k^\pm$ in (6.15) characterize the pumping field. Note that (6.15) was obtained using very general suppositions about the linear and nonlinear dielectric properties of CLC.

It is usually assumed (also below), that the local dielectric parameters of a CLC are similar to those of nematic crystals [4,8], i.e., to those of uniaxial crystals. This assumption is a sufficiently good approximation; it means that in (6.15) $s_+ = s_- = 0$.

Without fixing our attention on the explicit dependence of the coefficients $\mathcal{P}_{\pm}^{\ell mn}$ on the tensor of nonlinear susceptibility $\hat{\chi}$, we can easily see that the maximum efficiency of THG is achieved when the right-hand side of (6.15) is zero.

The solution of the inhomogeneous system (6.13) (the particular solution) can be written as

$$E^+ = \frac{D^+}{D^0}, \qquad E^- = \frac{D^-}{D^0} \tag{6.16}$$

where

$$D^0 = \left[1 - \left(\frac{\tilde{K}^+}{q}\right)^2\right]\left[1 - \left(\frac{\tilde{K}^-}{q}\right)^2\right] - \delta^2$$

$$D^+ = \mathcal{P}_+^{\ell mn}\left[1 - \left(\frac{\tilde{K}^-}{q}\right)^2\right] - \delta\mathcal{P}_-^{\ell mn}$$

$$D^- = \mathcal{P}_-^{\ell mn}\left[1 - \left(\frac{\tilde{K}^+}{q}\right)^2\right] - \delta\mathcal{P}_+^{\ell mn}.$$

Thus, expression (6.15) being zero determines the conditions for phase-matched generation, the number of such conditions being much greater than in the case of a homogeneous medium. As follows from (1.9) and (6.12), the conditions do not reduce now to the relationship between the wave vectors of the pumping wave and the harmonic; they may also include the reciprocal lattice vector τ. Formally speaking, relationship (6.15) determines 40 different matching conditions corresponding to various combinations of indices with both positive and negative values of $K^\pm$. However, not every such condition can be achieved in practice. This will be discussed below.

6.2.2   Solution of the Boundary Problem

In order to find THG intensity in a sample with finite dimensions, we have to determine the field $\mathbf{E}$ in the sample, this field being the superposition of the particular solution of (6.12, 6.16) and the eigensolutions of the homogeneous system obtained from (6.13) by cancelling the right-hand sides

$$\mathbf{E}(z, 3\omega) = e^{-3i\omega t}\left[E^+\mathbf{n}_+e^{i\tilde{K}^+z} + E^-\mathbf{n}_-e^{i\tilde{K}^-z}\right.$$

$$\left. + \sum_{j=1}^{4}C_j\left(E_j^+\mathbf{n}_+e^{iK_{3j}^+z} + E_j^-\mathbf{n}_-e^{iK_{3j}^-z}\right)\right]. \tag{6.17}$$

The coefficients C_j in this superposition are determined from the boundary conditions.

When the pumping wave propagates in a CLC along the optic axis, the harmonic propagation direction may be either the same or opposite, or the harmonic may propagate in both directions simultaneously. For the three cases, the boundary conditions reduce to the following:

1. The field of the third harmonic is zero at the front surface of the sample.

2. The field is zero at the exit (with respect to the pumping wave) surface.

3. The field component of the harmonic directed along the pumping wave is zero at the front surface and the component corresponding to the backward propagation harmonic is zero at the exit surface.

Using one of the above conditions, we can find the coefficients C_j in (6.17) and then the field and flux of the third harmonic at the boundary that gives the final solution to the problem in question. Note that the boundary conditions formulated above correspond to the case where $\bar{\epsilon}$ coincides with the dielectric permittivity of the medium outside the sample, i.e., when there is no reflection at the boundaries. The reflection can be easily taken into account, but the corresponding formulae prove to be very cumbersome.

6.2.3 Third Harmonic Generation in the Band of Selective Reflection

So far we have dealt with the general analysis of the third harmonic in CLC using the exact solution of the corresponding linear problem. Nothing was said about specific values of the frequency of pumping wave ω and of 3ω, and about how they are located with respect to the bands of selective reflection.

Let us now consider the situation in which there is a strong selective reflection in CLC of either the pumping wave or the harmonic (other situations are considered in [4,7,12]). The qualitative features of the nonlinear frequency transformation will be shown to appear due to light diffraction and to the specific features of the propagation of electromagnetic waves in the periodic structure, these features having been discussed in Sect. 6.1 for a simple periodic medium. Selective reflection, in particular, can change the efficiency of phase-matched harmonic generation so that the matching conditions which were not effective when there was no selective reflection may now prove otherwise.

Even before solving the boundary problem, we can conclude from the form of the particular solution of the inhomogeneous problem that harmonic generation efficiency increases if the phase-matching conditions are satisfied at the frequency 3ω corresponding to the multiple roots of the dispersion equation (1.13) or near this frequency. This is so because at those frequencies, the determinant in the denominator of the solutions of (6.16) tends to zero as $(\Delta k_{\ell mn})^2$ when approaching the phase-matching conditions, whereas in the absence of multiple roots, it tends to zero as $\Delta k_{\ell mn}$. Since the multiple roots of (1.13) correspond to the boundary frequencies of the selective reflection band, qualitatively new features in harmonic generation must be expected when the phase-matching conditions are satisfied at the frequency 3ω near the boundary of the selective reflection band.

We will now consider the quantitative description of THG under the conditions of selective scattering of the third harmonic. That is, we shall solve the boundary problem formulated above and find the following formulae for the amplitudes of the third harmonic at the front and exit surface of a CLC sample:

$$E^+(3\omega, L) = D^{-1}\left[D^+ + \frac{D^+(\xi_+ - \xi_-)e^{-i\tilde{k}_3 L} + 2iD^- \sin k_3^- L}{\xi_- e^{-ik_3^- L} - \xi_+ e^{ik_3^- L}}\right]e^{i(\tilde{k}_3 + \tau/2)L}$$

$$E^-(3\omega, L) = D^{-1}\left[D^- + \frac{D^-(\xi_+ - \xi_-)e^{i\tilde{k}_3 L} + 2i\xi_+ \xi_- \sin k_3^- L D^+}{\xi_- e^{-ik_3^- L} - \xi_+ e^{ik_3^- L}}\right] \quad (6.18)$$

where $D, D^\pm$ are defined in (6.16), $\xi_\pm = \delta^{-1}\{\kappa_3^{-2}(\tau/2 \pm k_3^-)^2 - 1]$, and ϵ_1, ϵ_3, δ_1, and δ_3 are, respectively, the dielectric permittivity and anisotropy at the frequencies ω and 3ω.

According to our assumption that the polarization of the harmonic corresponds to the diffracting polarization, we find that diffractive scattering of the harmonic may occur for the following phase-matching conditions:

$$3k^+(\omega) = k^-(3\omega)$$
$$3k^-(\omega) = k^-(3\omega)$$
$$2k^+(\omega) + k^-(\omega) = k^-(3\omega)$$
$$k^+(\omega) + 2k^-(\omega) = k^-(3\omega) \quad (6.19)$$

For any given frequency ω, the matching conditions can, in general, be provided by the proper choice of the cholesteric helical pitch whose phase-matched value depends on the dispersion of dielectric permittivity and anisotropy δ. Under this condition, the phase-matching conditions are satisfied, in general, outside the band of selective reflection of the harmonic.

As was shown before, it is of special interest to analyze the situations when phase matching occurs near the selective reflection band. However, in order for the phase-matching conditions (6.19) to be satisfied at the boundary of the selective reflection band or near it, certain relationships between the dispersion of dielectric permittivity and the anisotropy δ must be satisfied. The existence of frequency dispersion of the dielectric permittivity of a CLC means that the corresponding conditions are satisfied (if they can be satisfied at all) for a specific frequency of the pumping wave and for a specific helical pitch.

Assuming that the frequency of the third harmonic coincides with one of the edges of the selective reflection band, i.e., that $k^-(3\omega) = 0$ and

$$\tau = 2q(3\omega)(1 \pm \delta_3)^{1/2} \tag{6.20}$$

we see from (6.19) that only the last of the conditions of that equation can be met.

The condition that phase matching occurs at the boundary of the selective reflection band is

$$k^+(\omega) + 2k^-(\omega) = 0. \tag{6.21}$$

It follows from (6.19) and (6.21) that the pumping field must contain the waves with both diffracting and nondiffracting polarizations.

Substituting the expressions for $k^\pm$ from (1.14) into (6.21) and taking into account (6.20), we find the phase-matching conditions at the boundary of the selective reflection band:

$$d^2 \left[1 - \left(\frac{25}{9}\right)\delta_1^2\right] - 82d + 81 = 0 \tag{6.22}$$

where

$$d = \bar{\epsilon}_1[\bar{\epsilon}_3(1 \pm \delta_3)]^{-1}.$$

Hence,

$$\frac{\epsilon_1}{\epsilon_3} = \frac{1 \pm \delta_3}{1 - 25\delta_1^2/9}\left[41 \pm \sqrt{41^2 - 81\left(1 - \frac{25}{9}\delta_1^2\right)}\right]. \tag{6.23}$$

Assuming the frequency dependence of ϵ and δ to be known, we can find from (6.22) the pumping wave frequency ω and from (6.20) the helical pitch which corresponds to the phase matching at the boundary of the selective reflection band. It follows from (6.22) that the phase-matching condition can be satisfied near or at the boundary of the selective reflection band, provided that the frequency dispersion of the dielectric permittivity $|1 - \epsilon_1/\epsilon_3|$ is of the order of the anisotropy δ_3 at the frequency 3ω. This condition requires that the type of CLC be properly chosen. Nematic-cholesteric mixtures seem to be the most suitable to obtain the needed relationship between the dispersion and anisotropy.

Let us now consider in more detail the expressions (6.18) which yield the third harmonic amplitude in the case when the phase-matching conditions are satisfied at the boundary of the selective reflection band. These expressions show that the third harmonic's amplitude oscillates as a function of the pumping wave frequency or of the difference between the boundary frequency and 3ω.

In order to prove this, we will find the phase-matched values of the amplitudes $E^\pm$, i.e., when the last of the conditions (6.13) is satisfied and the values $\tilde{K}^\pm$ satisfy the dispersion equation [the determinant D of (6.13) is zero]. For this purpose, it is sufficient to discard the indeterminacy $0/0$ for the limit $D \to 0$ in (6.18). As a result, the phase-matched amplitudes of the harmonic exiting the crystal are

$$E^\pm = \pm \frac{iq_3 L(\xi_+ - \xi_-)e^{ik_3^- L}D^\pm}{4(k_3^-/q_3)[(\tau/q_3)^2 + \delta_3^2](\xi_- e^{-ik_3^- L} - \xi_+ e^{ik_3^- L})}. \tag{6.24}$$

As follows from (6.24), the efficiency of the nonlinear frequency transformation can be much greater when phase matching is achieved near the frequency boundary of the selective reflection band: $|k_3^-/q_3| \le \delta_3$. In fact, the amplitudes in (6.24) oscillate strongly as a function of the phase-matching values of k_3^-, these amplitudes being proportional to sample thickness at the minima and to squared thickness at the maxima.

6.2.4 Extreme Efficiency of Transformation

As follows from (6.24), maxima of amplitudes of (6.24) are reached at $k_3^- = s\pi/L$ and minima at $k_3^- = (2s+1)\pi/2L$ where s is an integer. These extreme amplitudes are:

$$E_{\max}^\pm = -\frac{i(q_3 L)^2 D^\pm}{4s\pi} \sim \frac{i\delta(q_3 L)^2 \mathcal{P}}{4\pi s} \tag{6.25}$$

$$E_{\min}^\pm = \frac{i\,\mathrm{sign}\,k_3^- (q_3 L)(\tau + 2k_3^-)q_3^{-1}D^\pm}{4[\frac{1}{2}(\tau/q_3)^2 - \sqrt{(\tau/q_3)^2 + \delta_3^2}]\sqrt{(\delta/q_3)^2 + \delta_3^2}}T \sim i(q_3 L)\mathcal{P}$$

where $\mathcal{P}$ is the characteristic value of nonlinear polarization. It is seen in (6.24, 6.25) that the minimum of at least one of the amplitudes E^+ or E^- is of the order of the amplitude in the absence of diffraction. As for the maximum value, it corresponds to the enhanced efficiency of nonlinear transformation in the range of noticeable diffraction scattering of the harmonics, but outside this region ($|k_3^-/q_3| > \delta_3$), this maximum reduces to the known expressions proportional to $q_3 L$ [1,2]. Also, from (6.24, 6.25), it follows that the generation maxima occur when the parameter $\nu = (\omega_3 - \omega_e)/\omega_e$ describing the location of the harmonic frequency outside the boundaries of the selective reflection band $\omega_e = \tau c/[2\sqrt{\epsilon_3(1 \pm \delta_3)}]$ is

$$\nu = \frac{3\omega - \omega_e}{\omega_e} = \frac{1}{\delta_3\epsilon_3}\left(\frac{c\pi s}{\omega_e L}\right)^2 \approx \frac{1}{\delta_3}\left(\frac{2\pi s}{\tau L}\right)^2. \tag{6.26}$$

It is assumed that when the harmonic frequency 3ω satisfies (6.26), the phase-matching condition is also satisfied. Comparing (6.25) and (6.26), we can see that the maximum intensity of THG decreases as

$$I_{(3\omega)}^{\max} = I_0 \frac{\delta_3\omega_3}{|3\omega - \omega_e|} \tag{6.27}$$

as the phase-matching frequency 3ω goes from the boundary of the selective reflection band, where the frequency 3ω takes discrete values given by (6.26); I_0 is approximately equal to the THG intensity in the same sample but far from the selective reflection band.

The latter formula can be represented in another form if we assume that the frequency of the light is fixed and the helical pitch is variable so that phase matching is achieved at the boundary of the selective reflection band by varying the pitch, i.e.,

$$I_{(\tau)}^{\max} = I_0 \frac{\delta \tau_e(3\omega)}{|\tau - \tau_e(3\omega)|} \qquad (6.28)$$

where $4\pi/\tau_e$ is the helical pitch corresponding to the coincidence of the frequency 3ω with the boundary of the selective reflection band, and the discrete values of τ_e are given by (6.26) if 3ω is replaced by τ and ω_e by τ_e. Note that harmonic amplitude turns out to be proportional to L^2 near the selective reflection band not only when the determinant D of the system (6.13) is exactly zero, in particular, when $\tilde{k}_3$ coincides with k_3^-, but also near these conditions at $\tilde{k}_3 = \pi n/L$ and $k_3^- = \pi s/L$. Hence, the formulae (6.19), (6.21) prove to be the conditions of proportionality of $E^\pm$ to L^2, provided that $\pm \pi n/L$ is added to the right-hand side of (6.19), (6.21) and k_3^- is set equal to $\pi s/L$. In this case, the maximum amplitudes are

$$E^\pm = \frac{(q_3 L)^2 D^\pm}{\pi^2(n^2 - s^2)} \left[1 - e^{i(s\pm n)\pi} \right] \left(\frac{q_3}{k_3^+} \right)^2 \qquad (6.29)$$

where s is an integer, and n is a number with the opposite parity or a half-integer. It is seen from (6.25), (6.26), and (6.29) that the absolute maximum of harmonic intensity is, in fact, at the boundary of the selective reflection band. When the phase-matching condition (6.21) is exactly satisfied, the maximum occurs at $s = 1$ in (6.26), which means that the frequency slightly deviates from the boundary of the selective reflection band $[\nu = \delta^{-1}(\pi s/q_3 L)^2]$. When the deviation of the dispersion $\Delta d = n\pi/q_3 L$ from the boundary value given by (6.22) is small, an absolute maximum occurs at $n = 1$ and $s = 0$ in (6.29), i.e., when the harmonic frequency exactly coincides with the boundary of the selective reflection band. The absolute value of the maximum amplitude in (6.29) is close to that of (6.25); they differ only by factor.

6.2.5   The Conditions for the Enhancement Effect

It is then natural to ask what the conditions are for maximum amplitude of the harmonic, if the dispersion differs from the boundary value of (6.22) by more than $\Delta d = \pi/q_3 L$. Since the cholesteric pitch can be easily varied in an experiment, we will take this parameter τ as the free parameter of the problem while the other parameters, and the frequency ω also, are fixed.

By varying τ, we can, for example, satisfy the phase-matching conditions (6.19) which do not correspond to the boundary of the selective reflection band ω_e, or we can superimpose the harmonic frequency with ω_e, the phase-matching conditions not being satisfied in this case. The former may mean that harmonic amplitude is described by (6.25) with $s > 1$, and the latter that it is described by (6.29) at $s = 0$ and $n > 1$. An analysis of (6.22–6.29) shows that when the value $\nu = |\tau_e - \tau_p|/\tau_e$ is greater than $\delta^{-1}(\pi/q_3 L)$ where τ_p corresponds to the phase-matched pitch and $\tau_e^{-1} = 2(3\omega/c)\sqrt{\epsilon_3(1 \pm \delta_3)}$, the absolute maximum of the harmonic amplitude proportional to $(q_3 L)^2$ is reached under phase-matching conditions and is given by (6.25). The condition of the absolute maximum is the requirement that $q_3 L \sqrt{\delta_3 \nu} = s$ should be an integer. If s is not an integer, then a greater value of amplitude may correspond to a small deviation of τ from τ_p. Whether or not the amplitude is described by (6.29) depends on the compatibility of the considered variations of τ with the conditions for n and s in (6.29).

The maximum value of the phase-matched generation intensity determined by (6.25), (6.27), and (6.29) is related to a strictly monochromatic wave. An effect of the same order can be achieved for a wave with a finite width $\Delta\omega/\omega$ if the linewidth is less than that of the harmonic's maximum, which is about $(\pi s/\delta)^2/(q_3/L)^3$, and this determines the sharpness of phase matching under diffraction conditions.

If the linewidth is of the order of the frequency width of oscillations in (6.24), i.e., $(s\pi/\delta L)^2/\delta_3$, then the generation efficiency proves to be less than the maximum value and is described by averaging the square of the modulus of (6.24) over the linewidth. This efficiency is

$$|E^\pm|^2_{ev} \sim I_0 \sqrt{\frac{\delta_3 \tau_e}{|\tau - \tau_e|}} = I_0 \sqrt{\frac{3\omega\delta}{|3\omega - \omega_e|}} \qquad (6.30)$$

where 3ω now stands for the frequency position of the line's center and can vary continuously. It is seen in (6.30) that the enhancement of the nonlinear frequency transformation occurs when the detuning from ω_e and the width of the pumping line are not greater than δ_3.

Thus, when the phase-matching conditions are satisfied at the boundary of the selective reflection band, the nature of nonlinear generation greatly changes compared with that apart from the selective reflection band [4,7] and the angular distribution of the generated harmonic also changes. (The generations with comparable intensities run both in the direction of the pumping wave and in the opposite direction and generation efficiency is enhanced.)

The possibility of achieving phase matching at the boundary of the selective reflection band depends on certain relationships between the frequency dispersion of dielectric permittivity $\bar{\epsilon}$ and anisotropy δ [see (6.22) and (6.23)]. As mentioned above, these relationships could be satisfied in properly chosen cholesteric crystals, since the dielectric and structure parameters of available cholesteric crystals and cholesteric compositions indicate that such a choice is indeed possible [7]. Up until now, we have completely neglected CLC absorption, which results in a lessening of the effects in question, and this must be taken into account when describing a real experiment (see [13]). So far we have considered only third harmonic generation in a CLC. However, all the results hold, with obvious modifications, for second harmonic generation in chiral smectics also [13].

The specific features of the nonlinear frequency transformation in CLC discussed above are not inherent in the propagation of waves along the cholesteric axis only. They also appear when light propagates at an angle to the axis. We will not consider here the case of oblique incidence and refer the reader to the original papers on the subject [4,15]. We only note that the polarization parameters of generation prove to be very complicated and a theoretical description is more difficult because exact solutions to Maxwell equations in linear optics are not known for this case.

6.3 Phase-Matching Conditions Independent of Frequency Dispersion

In Sect. 6.1 and 6.2, we analyzed the increase in efficiency of the phase-matched nonlinear transformation of frequency under the conditions of diffraction of the generated wave, those conditions requiring certain relationships between the frequency dispersion of dielectric permittivity $\hat{\epsilon}(\mathbf{r})$ and the amplitude of its modulation δ to be fulfilled in the periodic medium. Here we will study the enhancement of efficiency of the nonlinear frequency transformation when the phase-matching conditions do not depend on the frequency dispersion of ϵ [16] but, to be satisfied, require that the pumping field has a special configuration, for example, two oppositely directed pumping waves.

Let us consider second harmonic generation in a periodic nonlinear medium having the shape of a plane-parallel sample with thickness L. For the sake of simplicity, we assume this medium is periodic in one dimension and the direction Z along which the dielectric and nonlinear properties are periodic is perpendicular to the sample surfaces (see Fig. 6.3). The harmonic field is described by the solutions of

$$\mathrm{grad}\,\mathrm{div}\,\mathbf{E} - \Delta\mathbf{E} - \frac{4\omega^2}{c^2}\epsilon(z, 2\omega)\mathbf{E} = \mathcal{P}(z, 2\omega) \qquad (6.31)$$

where ϵ is the dielectric permittivity and $\mathcal{P}$ is the quadratic nonlinear polarization.

Keeping in mind that the efficiency of nonlinear frequency transformation increases under the diffraction of harmonic waves, we assume that the frequency 2ω coincides with or is close to the boundary of the frequency band forbidden for propagating solutions, ω_e. The analysis of (6.31), similar

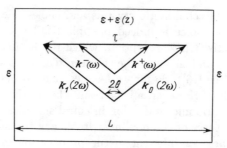

FIGURE 6.3. Pumping geometry which provides the fulfillment of the phase-matching conditions, independent of the dispersion of dielectric permittivity. $\mathbf{k}_{\pm}(\omega)$ and $\mathbf{k}_{0,1}(2\omega)$ are the wave vectors of the pumping wave and harmonic, respectively.

to that given above, indicates that along with the phase-matching conditions depending on the frequency dispersion ϵ and on the amplitude of its modulation δ, there is one more condition which does not depend on ϵ. This condition is realized if the frequency 2ω coincides, practically speaking, with the forbidden band boundary and the nonlinear polarization in (6.31) contains a component modulated with a standing wave, with a period determined by the medium's periodicity.

In order to describe harmonic generation under the above conditions, we consider on the right-hand side of (6.31) only the components with a nonlinear polarization of the form

$$\mathcal{P} \sim e^{-2i\omega t} e^{2i\mathbf{k}_{\perp}(\omega)\mathbf{r}} \cos(\boldsymbol{\tau}\mathbf{r} + \varphi)$$

where φ is the phase determining the positions of nodes of the standing wave modulating the nonlinear susceptibility, τ is the reciprocal lattice vector of the periodic structure, and $\mathbf{k}_{\perp}$ is the vector perpendicular to the periodicity direction. It is evident that such components do exist in the nonlinear polarization, provided that the pumping field (see Fig. 6.3) is the coherent superposition of two plane waves of the type $A_+ e^{i\mathbf{k}_1(\omega)\mathbf{r}} + A_- e^{\mathbf{k}_2(\omega)\mathbf{r}}$, in which the projections of the wave vectors on the periodicity direction have opposite signs (in one particular case, it may be two oppositely directed plane waves).

Let us assume for the sake of simplicity, that the modulation amplitude in the sample is small and that the mean value of $\epsilon(r)$ coincides with the dielectric permittivity of the external medium. Then the solution of (6.31) found within the two-wave approximation of dynamic diffraction theory, neglecting the depletion of the pumping wave (see the sections above) shows that harmonic amplitude at the exit sample surface oscillates as a function of frequency and sharply decreases when 2ω moves away from the forbidden band. This amplitude outside the forbidden band is described by

$$E = \frac{8\pi\Delta\chi A_+ A_-}{\epsilon\delta} \frac{(\alpha e^{i\varphi} - e^{-i\varphi})(1 - \cos\ell\sqrt{\nu}) - i\sqrt{\nu}e^{i\varphi}\sin(\ell\sqrt{\nu})}{(\sqrt{\nu}\cos\ell\sqrt{\nu} + i\alpha\sin\ell\sqrt{\nu})\sqrt{\nu}} \quad (6.32)$$

where $\ell = \delta\omega\sqrt{\epsilon}L(c\sin\theta)^{-1}$; $\nu = 4\sin\theta|2\omega - \omega_e|(\delta\omega_B)^{-1}$ is the parameter describing the shift of harmonic frequency from the forbidden band whose boundaries are $\omega_e = \omega_B[1 \pm \delta(2\sin^2\theta)^{-1}]$ and $\omega_B = c\tau[2\sqrt{\epsilon(2\omega)}\sin\theta]^{-1}$, $\alpha = \pm(1 + \nu/2)$ and $\alpha > 0$, $\alpha < 0$, respectively, for the high- and low-frequency edges of the forbidden band for light polarized in the plane perpendicular to the plane of Fig. 6.3. $\Delta\chi$ is the modulation amplitude of nonlinear susceptibility with the period $4\pi/\tau$ and $\delta\epsilon$ is that of ϵ with the period $2\pi/\tau$.

At the oscillation maxima, which are reached at $\nu = (\pi s/\ell)^2$ with odd s, E takes the values

$$E_{max} = 8\pi\Delta\chi\delta A_+ A_-(\alpha e^{i\varphi} - e^{-i\varphi})\left(\frac{\omega L}{c\sin\theta}\right)^2 (\pi s)^{-2}. \quad (6.33)$$

The frequency width of maxima in (6.33) is of the order of $\omega\delta^{-2}(c/\omega L)^3$, the amplitude of these maxima being proportional to L^2 and the distance between the maxima being of the order of $\omega\delta^{-1}(c/\omega L)$. The intensity of the harmonic at the maxima, as a function of frequency, behaves as $I \sim I_0\left(\frac{c}{\omega L}\right)^2 \omega_e^2/(2\omega - \omega_e)^2$ when 2ω tends to ω_e where I_0 is the phase-matched generation intensity if the diffraction conditions do not hold. The greatest value of intensity defined by (6.33) is reached at $s = 1$, this value being $(\delta L\omega/c)^2$ times greater than I_0 just at the boundary of the forbidden band. This value can be reached only for a sufficiently monochromatic pumping wave $[\Delta\omega/\omega \leq \delta^{-2}(c/\omega L)^3]$. If, on the contrary, the pumping linewidth is wide, SGH enhancement can occur only if $\Delta\omega/\omega < c/\omega L$. If the width of the pumping line is of the order of the distance between maxima, then intensity depends on the distance between harmonic frequency and the forbidden band's boundary

$$I(2\omega) \sim \frac{I_0}{\sqrt{\delta}}\left(\frac{c}{\omega L}\right)^2 \left(\frac{\omega_e}{|2\omega - \omega_e|}\right)^{3/2}$$

where ω now denotes the center of the line and can assume continuous values. The physical reasons for the possibility of the phase matching described are quite clear. For example, in collinear geometry ($\theta = \pi/2$), the eigensolutions of (6.31) at the boundary of the forbidden band are standing waves, with the period determined by medium periodicity. The standing waves of nonlinear polarization with the same period (the pumping field as two oppositely directed waves) are provided by the spatial modulation of the nonlinear susceptibility caused by the pumping field in the form of a standing wave. This is just the basis for the phase-matching conditions independent of the frequency dispersion of ϵ.

In addition to the frequency oscillations of harmonic intensity, there are oscillations connected to small changes of the pumping wave direction, the latter oscillations also being described by the above formulae if the value of ν is set equal to $2(|\Delta\theta|/\delta)\sin 2\theta$ where $\Delta\theta$ describes the deviation of harmonic direction from those defining the angular boundaries of the forbidden band. Since the phase matching under discussion can be achieved in both collinear and noncollinear geometry, it is convenient to use these conditions for experimental studies of the effect of the enhancement of efficiency of a nonlinear frequency transformation in periodic media.

To conclude this section, we note that the phase-matching conditions independent of the frequency dispersion of and the enhancement of efficiency of the nonlinear process in a periodic medium are of a very general nature and their analysis for a simple periodic medium permitting SGH was carried out only for the sake of simplicity. The same effects appear, for example, in a generation of the third harmonic if the pumping field is a coherent superposition of three plane waves, with a zero sum for their wave vectors. Similar effects may surely occur in more complicated periodic media, for example, in three-dimensional periodic media and chiral liquid crystals [7], the latter crystals being a convenient subject for experimental studies.

6.4   Nonlinear Generation of Harmonics in Blue Phase of Liquid Crystals

In Chapt. 4 we considered the optical properties of the blue phases and discussed the possibility of revealing their structure by means of linear optics. In this section, we use as an example the nonlinear generation of the second harmonic in the blue phases [17] to analyze the possibility of nonlinear optics in studying blue phase structure. Since nonlinear harmonic generation in the blue phase is described in a similar way to that outlined above for CLCs and simple periodic media, we will focus our attention on the symmetry analysis of the nonlinear susceptibility tensor of the blue phases and finding restrictions on its components for specific space groups which may describe the symmetry properties of blue phases.

6.4.1   Phase-Matching Condition

As is known from [1,2], nonlinear generation is most effective under phase-matching conditions which have the following well-known form for a homogeneous medium:

$$\mathbf{k}^{\sigma_2}(2\omega) = \mathbf{k}^{\sigma_1}(\omega) + \mathbf{k}^{\sigma_3}(\omega) \tag{6.34}$$

where $\mathbf{k}(m\omega)$ are the wave vectors of the harmonic and pumping waves, and σ_i are the indices of the eigenpolarizations. Because of the optical isotropy of the blue phases, the matching conditions (6.34) cannot be satisfied when there is a frequency dispersion of dielectric permittivity if the waves of harmonic and pumping propagate in the sample as plane waves. In this case, the frequency dispersion in (6.34) cannot be compensated by birefringence.

However, the phase-matching conditions (6.34) may be met in the blue phases, provided that either the pumping wave or the harmonic is subjected to diffraction. In this case, the renormalization of the wave vectors due to diffraction [6,15] can compensate for the frequency dispersion of dielectric permittivity if the latter is not too large. Hence, under phase-matching conditions (6.34), the blue phase can generate the second harmonic if the light in this phase is subject to diffraction. Below, we will assume that only the harmonic, but not the pumping wave, is subjected or almost subjected to diffraction. Note that the phase-matching condition (6.34) corresponds to SGH due to homogeneous component of nonlinear susceptibility. Under diffraction conditions, these phase-matching conditions become [7]:

$$\mathbf{k}^{\sigma_1}(2\omega) = \mathbf{k}^{\sigma_2}(\omega) + \mathbf{k}^{\sigma_3}(\omega) + s\boldsymbol{\tau} \qquad (6.35)$$

where $s = \pm 1, \pm 2, \ldots$, and $\boldsymbol{\tau}$ is the reciprocal lattice vector of the blue phase. The phase-matching conditions (6.34) are easily satisfied, provided that the light within the blue phase experiences diffraction or the diffraction conditions are almost satisfied. Formally speaking, the frequency dispersion in a blue phase with a large period might be compensated in (6.34) by the term $s\boldsymbol{\tau}$.

In order to describe second harmonic generation in a blue phase, it is necessary to solve equations of the type of (6.13) within the two-wave approximation, as was done in [15].

6.4.2 Symmetry Restrictions for the Tensor of Nonlinear Susceptibility

Leaving it to the reader to use the equations presented above to describe SGH in the blue phases, we will instead place our emphasis on an analysis of symmetry restrictions on the quadratic nonlinear susceptibility for BP.

It was mentioned previously that possible spatial structures of the order parameter of BPI and BPII are surely described by enantiomorphic cubic space groups, there being 13 such groups. Hence, the study of nonlinear optical properties of the blue phases, which is conducted next using the example of SHG, must include, as a stage, an analysis of nonlinear susceptibilities for corresponding groups.

The specifics of nonlinear optics of blue phases require that the structure period of the order parameter, and hence of other physical parameters, lie within the optical wavelength band. Unlike ordinary crystals in which spatially inhomogeneous components of the tensor of nonlinear susceptibility are described by the value averaged over the crystal unit cell (in fact, over a much larger volume), in the case of blue phases an essential role is also played by the tensor component varing in space. Since the local symmetry of general locations in a unit cell is, in general, lower than that of the crystal class, the symmetry of inhomogeneous components of the nonlinear susceptibility tensor is, in general, lower. Hence, this symmetry can permit nonlinear processes prohibited by the point group, i.e., by the crystal class of the space group determining the symmetry of the averaged nonlinear susceptibility tensor.

In this connection, the tensor of quadratic nonlinear susceptibility of the blue phase which determines SHG is conveniently expanded into a Fourier series:

$$\hat{\chi}(\mathbf{r}) = \sum_{\tau} \hat{\chi}_{\tau} e^{i\tau\mathbf{r}} \tag{6.36}$$

where τ is the blue phase reciprocal lattice vector. The zero harmonic of this expansion corresponds to the nonlinear susceptibility tensor averaged over a unit cell. Just this quantity is essential to the optics of usual crystals, and its symmetry properties have been well studied [1,2].

Using the results of symmetry analysis [17], we conclude that the averaged component of quadratic nonlinear susceptibility is nonzero only in groups T^i $(i = 1, \ldots, 5)$ and is exactly zero in groups O^i $(i = 1, \ldots, 8)$. For groups with point symmetry 23: $\chi_{123} = \chi_{132} = \chi_{213} \neq \chi_{231} = \chi_{312} = \chi_{321} \neq 0$, and for groups with point symmetry $43m$: $\chi_{123} = \chi_{132} = \chi_{213} = \chi_{231} = \chi_{312} = \chi_{321} \neq 0$. These relationships enable us to distinguish the groups T^1, T^2, T^3, T^4, T^5 and $O^1, O^2, O^3, O^4, O^5, O^6, O^7, O^8$ on the basis of their nonlinear optical properties, such as SHG, by considering whether they have a homogeneous component.

Let us now consider the symmetry restrictions for the spatially inhomogeneous part of the second-order nonlinear susceptibility tensor. An arbitrary tensor $T(\mathbf{r})$ must remain unchanged under symmetry transformations which belong to the space symmetry group of the crystal. Hence, for any space group, the general form of the tensor satisfying all symmetry restrictions for this group can, as mentioned in Chapt. 4, be written as

$$T(\mathbf{r}) = \frac{1}{N} \sum_{a} \hat{T}^{a}(\mathbf{r}) \tag{6.37}$$

where summation is performed over all transformations a included in the crystal symmetry group [18], N is the number of elements in the group, and T^a is the result of the $\hat{a}$ transformation of T

$$T_{ijk}^{a}(\mathbf{r}) = a_{ii'} a_{jj'} a_{kk'} T_{i'j'k'}(\mathbf{r}) \tag{6.38}$$

where $\hat{a}$ is the coordinate-transforming matrix $\mathbf{r}' = \hat{a}(\mathbf{r} - \mathbf{b}_a)$, and $\mathbf{b}_a$ is the translation vector.

The presence of the third-order axis in all cubic groups results in the following restrictions for the tensor $\hat{\chi}(\mathbf{r})$. If we denote

$$\chi_{111}(\mathbf{r}) = d_1(x, y, z), \quad \chi_{123}(\mathbf{r}) = d_4(x, y, z)$$
$$\chi_{112}(\mathbf{r}) = d_2(x, y, z), \quad \chi_{122}(\mathbf{r}) = d_5(x, y, z) \qquad (6.39)$$
$$\chi_{113}(\mathbf{r}) = d_3(x, y, z), \quad \chi_{133}(\mathbf{r}) = d_6(x, y, z)$$

we can easily see that all components of $\chi(\mathbf{r})$ are obtained by a cyclic permutation of the coordinates x, y, z in the functions $d_i(x, y, z)$.

Thus, the second-order nonlinear susceptibility tensor is determined from six arbitrary periodic functions

$$\begin{vmatrix} d_1(x,y,z) & d_5(x,y,z) & d_6(x,y,z) & d_4(x,y,z) & d_3(x,y,z) & d_2(x,y,z) \\ d_6(y,z,x) & d_1(y,z,x) & d_5(y,z,x) & d_2(y,z,x) & d_4(y,z,x) & d_3(y,z,x) \\ d_5(z,x,y) & d_6(z,x,y) & d_1(z,x,y) & d_3(z,x,y) & d_2(z,x,y) & d_4(z,y,x) \end{vmatrix}. \qquad (6.40)$$

Here the third-rank tensor is written as a matrix, the first index of the tensor component χ_{ijk} corresponds to the first index of the matrix elements d, and the other two indices are determined from the second and third indices χ_{ijk} as $1,1 = 1$; $2,2 = 2$; $3,3 = 3$; $2,3 = 4$; $1,3 = 5$; $1,2 = 6$. In (6.40) we introduced additional symmetry restrictions which are not related to the third-order axis, but which assume the tensor is symmetric with respect to the two latter indices (in what follows we will be interested in SHG when the pumping waves have the same frequency [1]).

The rest of the symmetry elements for the space groups T result in additional restrictions on the functions $d_i(\mathbf{r})$, but no relationships between these functions occur (and for the dielectric permittivity tensor as well, see Chapt. 4).

For example, for the space groups T^4 and T^5, the functions d_i are related by

$$d_1(x, y, z) = d_1\left(\tfrac{1}{2} + x, \tfrac{1}{2} - y, \overline{v}z\right) = -d_1\left(\overline{x}, \tfrac{1}{2} + y, \tfrac{1}{2} - z\right)$$
$$= -d_1\left(\tfrac{1}{2} - x, \overline{v}y, \tfrac{1}{2} + z\right)$$
$$d_2(x, y, z) = -d_2\left(\tfrac{1}{2} + x, \tfrac{1}{2} - y, \overline{v}z\right) = d_2\left(\overline{x}, \tfrac{1}{2} + y, \tfrac{1}{2} - z\right)$$
$$= -d_2\left(\tfrac{1}{2} - x, \overline{v}y, \tfrac{1}{2} + z\right)$$
$$d_3(x, y, z) = -d_3\left(\tfrac{1}{2} + x, \tfrac{1}{2} - y, \overline{v}z\right) = -d_3\left(\overline{x}, \tfrac{1}{2} + y, \tfrac{1}{2} - z\right)$$
$$= d_3\left(\tfrac{1}{2} - x, \overline{v}y, \tfrac{1}{2} + z\right) \qquad (6.41)$$
$$d_4(x, y, z) = d_4\left(\tfrac{1}{2} + x, \tfrac{1}{2} - y, \overline{v}z\right) = d_4\left(\overline{x}, \tfrac{1}{2} + y, \tfrac{1}{2} - z\right)$$
$$= d_4\left(\tfrac{1}{2} - x, \overline{v}y, \tfrac{1}{2} + z\right)$$
$$d_5(x, y, z) = d_5\left(\tfrac{1}{2} + x, \tfrac{1}{2} - y, \overline{v}z\right) = -d_5\left(\overline{x}, \tfrac{1}{2} + y, \tfrac{1}{2} - z\right)$$
$$= -d_5\left(\tfrac{1}{2} - x, \overline{v}y, \tfrac{1}{2} + z\right)$$
$$d_6(x, y, z) = d_6\left(\tfrac{1}{2} + x, \tfrac{1}{2} - y, \overline{v}z\right) = -d_6\left(\overline{x}, \tfrac{1}{2} + y, \tfrac{1}{z} - z\right)$$
$$= -d_6\left(\tfrac{1}{2} - x, \overline{v}y, \tfrac{1}{2} + z\right)$$

where $\bar{x} = -x$, etc. In contrast to the T group, the O groups possess a fourth-order axis that results in the following relationships between the functions d_i themselves. For the groups O^6, O^8, and O^4

$$d_3(x,y,z) = -d_2\left(\tfrac{1}{4}-x, \tfrac{1}{4}-z, \tfrac{1}{4}-y\right) = -d_2\left(\tfrac{3}{4}+x, \tfrac{3}{4}-z, \tfrac{1}{4}+y\right)$$
$$= d_2\left(\tfrac{3}{4}-x, \tfrac{1}{4}+z, \tfrac{3}{4}+y\right) = d_2\left(\tfrac{1}{4}+x, \tfrac{3}{4}+z, \tfrac{3}{4}-y\right)$$
$$d_5(x,y,z) = -d_6\left(\tfrac{1}{4}-x, \tfrac{1}{4}-z, \tfrac{1}{4}-y\right) = d_6\left(\tfrac{3}{4}+x, \tfrac{3}{4}-z, \tfrac{1}{4}+y\right)$$
$$= -d_6\left(\tfrac{3}{4}-x, \tfrac{1}{4}+z, \tfrac{3}{4}-y\right) = d_6\left(\tfrac{1}{4}+x, \tfrac{3}{4}+z, \tfrac{3}{4}-y\right) (6.42)$$

and for the groups O^1, O^2, and O^3,

$$d_3(x,y,z) = -d_2(\bar{x}, \bar{z}, \bar{y}) = d_2(x, \bar{z}, y)$$
$$= d_2(\bar{x}, z, y) = -d_2(x, z, \bar{y})$$
$$d_6(x,y,z) = -d_5(\bar{x}, \bar{z}, \bar{y}) = d_5(x, \bar{z}, y)$$
$$= -d_5(\bar{x}, z, y) = d_5(x, z, \bar{y}).$$

As a result, of the 18 components of the O groups (any third-rank tensor symmetric in its two indices has 18 components), 12 components (with two coinciding indices) are similar to those of the corresponding T groups: T^1 and O^1, O^2; T^2 and O^3, O^4; T^3 and O^5; T^4 and O^6; T^5 and O^8. Components χ_{xxx} and χ_{xyz} with all possible cyclic permutations of the indices in the O groups differ from those in the T groups. Thus, we obtained the restrictions for the third-rank tensor of the blue phase. As was mentioned in our discussion of the nonlinear optics of blue phases, the Fourier components of the nonlinear susceptibility are important. The above relationships for the components of $\hat{\chi}(\mathbf{r})$ can lead to the restrictions for the Fourier components χ_τ.

6.4.3 Fourier Components of the $\hat{\chi}(\mathbf{r})$ Tensor

The Fourier components of the tensor $\hat{\chi}(r)$ are determined as follows:

$$\hat{\chi}_\tau = \frac{1}{V} \int \hat{\chi}(\mathbf{r}) e^{i\boldsymbol{\tau}\mathbf{r}} d^3\mathbf{r} \qquad (6.43)$$

where integration is carried out over the unit cell volume V.

Calculating the Fourier components of $\chi(\mathbf{r})$ using (6.43) and the relationships presented in the previous chapter, we can find the most general form of $\hat{\chi}_\tau$ which follows from the symmetry restrictions of the space groups considered above (see Table 6.1).

In Table 6.1, R_i are arbitrary real numbers, I_i are imaginary numbers, and C_i are complex numbers, these numbers being different for different reciprocal lattice vectors. For any reciprocal lattice vector, the conditions

TABLE 6.1. Fourier components of quadratic nonlinear susceptibility tensor χ in cubic groups.

T^4, T^5		χ_{111}	χ_{222}	χ_{333}	χ_{122}	χ_{233}	χ_{311}	χ_{133}	χ_{211}	χ_{322}	χ_{123}	χ_{231}	χ_{312}	χ_{131}	χ_{212}	χ_{323}	χ_{312}	χ_{223}	χ_{331}	T^1, T^2, T^3
$h00$	$h=2n$	I_1	0	0	I_2	0	0	I_3	0	0	R_1	R_2	R_3	0	I_4	0	0	0	I_5	$h00$
	$h=2n+1$	0	R_1	I_1	0	R_2	I_2	0	R_3	I_3	0	0	0	I_4	0	R_4	R_5	I_5	0	
$hk0$	$h=2n$	I_1	I_2	R_1	I_3	I_4	R_2	I_5	I_6	R_3	R_4	R_5	R_6	R_7	I_7	I_8	I_9	R_8	I_{10}	$hk0$
	$h=2n+1$	R_1	R_2	I_1	R_3	R_4	I_2	R_5	R_6	I_3	I_4	I_5	I_6	I_7	R_7	R_8	R_9	I_8	R_{10}	
hhh	—	C_1	C_1	C_1	C_2	C_2	C_2	C_3	C_3	C_3	C_4	C_4	C_4	C_5	C_5	C_5	C_6	C_6	C_6	hhh

TABLE 6.1 (continued)

O groups

O^2	O^4,O^6,O^8	χ_{111}	χ_{222}	χ_{333}	χ_{122}	χ_{233}	χ_{311}	χ_{133}	χ_{211}	χ_{322}	χ_{123}	χ_{231}	χ_{312}	χ_{131}	χ_{212}	χ_{323}	χ_{312}	χ_{223}	χ_{331}	O^1,O^3,O^5	O^7
$h00$ $h=2n$	$h=4n$	I_1	0	0	I_2	0	0	I_2	0	0	0	R_1	R_1	0	I_3	0	0	0	I_4	$h00$	$h=4n$
	$h=4n+2$	0	0	0	I_1	0	0	I_1	0	0	R_1	R_2	R_2	0	I_2	0	0	0	$-I_2$		$h=4n+2$
$h=2n+1$	$h=4n\pm1$	0	R_1	$\pm iR_1$	R_2	R_2	$\pm iR_3$	0	R_3	$\pm iR_2$	0	0	0	$\pm iR_4$	0	R_5	R_4	$\pm iR_5$	0		$h=4n+1$
$hk0$ $h=n$	$h=2n$	I_1	I_2	R_1	I_3	I_4	R_2	I_5	I_6	R_3	R_4	R_5	R_6	R_7	I_7	I_8	I_9	R_8	I_{10}	$hk0$	$h=2n$
	$h=2n+1$	R_1	R_2	I_1	R_3	R_4	I_2	R_5	R_6	I_3	I_4	I_5	I_6	I_7	R_7	R_8	R_9	I_8	R_{10}		$h=2n+1$
$hh0$ $h=n$	$h=2n$	I_1	I_1	0	I_2	I_3	R_1	I_3	I_2	$-R_1$	R_2	$-R_2$	0	R_3	I_4	I_5	I_4	$-R_3$	I_5	$hh0$	$h=2n$
	$h=2n+1$	R_1	R_1	0	R_2	R_3	I_1	R_3	R_2	$-I_1$	I_2	$-I_2$	0	I_3	R_4	R_5	R_4	$-I_3$	R_5		$h=2n+1$
hhh $h=2n$	$h=4n$	I_1	I_1	I_1	C_1	C_1	C_1	$-C_1^*$	$-C_1^*$	$-C_1^*$	I_2	I_2	I_2	C_2	C_2	C_2	I_2	$-C_2^*$	$-C_2^*$	hhh	$h=4n$
	$h=4n+2$	R_1	R_2	R_1	C_1	C_1	C_1	C_1^*	C_1^*	C_1^*	I_2	R_2	R_2	C_2	C_2	C_2	R_2	C_2^*	C_2^*		$h=4n+2$
$h=2n+1$	$h=4n\pm1$	$(1\mp i)R$	$(1\mp i)R$	$(1\mp i)R$	C_1	C_1	C_1	$\mp iC_1^*$	$\mp iC_1^*$	$\mp iC_1^*$	$(1\mp i)R_1$	$(1\mp i)R_1$	$(1\mp i)R_1$	C_2	C_2	C_2	$(1\mp i)R_1$	$\mp iC_2^*$	$\pm iC_2^*$		$h=4n\pm1$

for a Fourier component to be nonzero are $h + k + l = 2n$ for the body-centered lattices T^3, T^5, O^5, O^8 and the same parity of h, k, l for the face-centered lattices T^2, O^3, O^4. The components of a general type $(h \neq k \neq l)$ are determined in the general case by 18 complex numbers, without any relationship between χ_{pqs} with different subscripts for both the T and O groups.

Here we do not touch upon the calculations of intensity and polarization of the harmonics because they are similar to those given in previous sections of this chapter (see also [17]). We only illustrate how the above symmetry restrictions can be used in a study of blue phases. The symmetry analysis of quadratic nonlinear susceptibility shows that the phase-matched SHG due to the homogeneous component $\hat{\chi}_0$ (T groups) is rigorously forbidden if the pumping wave is directed along [100] and [110]. Since (hhh) reflections were not observed experimentally in the blue phases, there should not also be phase-matched SHG due to $\hat{\chi}_0$ for the [111] direction because the dielectric tensor frequency dispersion cannot be compensated by the diffraction effects. A corresponding symmetry analysis of the phase-matched SHG due to the spatially inhomogeneous components of $\hat{\chi}$ showed that it is not forbidden for the [100] direction only for groups T^4, O^6, and O^7. Therefore, for example, if SHG in colinear geometry for the [100] direction was observed, this same fact would allow us to distinguish these groups from the above-mentioned 13 enantiomorphic groups.

References

1. S.A. Akhmanov, R.V. Khokhlov: *Problems of Nonlinear Optics* (Moscow, VINITI, 1964) (in Russian).

2. N. Bloembergen: *Nonlinear Optics* (New York, W.A. Benjamin Inc., 1965).

3. N. Bloembergen, A. Sielvers: *Appl. Phys. Lett.* **17**, 483 (1970).

4. J.W. Shelton, Y.R. Shen: *Phys. Rev. A* **5A**, 1867 (1972).

5. V.A. Belyakov, V.N. Shipov: *Phys. Lett.* **86**, 94 (1981).

6. A.A. Maier, A.P. Sukhorukov, R.N. Kuz'min: *Zh. ETF* **77**, 1282 (1979).

7. V.A. Belyakov, A.S. Sonin: *Optics of Cholesteric Liquid Crystals* (Moscow, Nauka, 1982) (in Russian).

8. S.M. Arakelyan, Yu. S. Chiligaryan: *Nonlinear Optics of Liquid Crystals* (Moscow, Nauka, 1984) (in Russian).

9. V.A. Belyakov, V.E. Dmitrienko, V.P. Orlov: *Pis'ma Zh TF* **1**, 978 (1975).

10. N.V. Shipov, V.A. Belyakov: *Zh ETF* **75**, 1589 (1978) [*Sov. Phys. - JETP* **48**, 802 (1978)].

11. S.K. Saha, G.K. Wong: *Appl. Phys. Lett.* **34**, 423 (1979).

12. N.V. Shipov, V.A. Belyakov: *ZhTF* **50**, 205 (1980).

13. S.V. Shiyanovskii: *UFZh* **27**, 361 (1982).

14. V.A. Belyakov, V.E. Dmitrienko, S.M. Osadchii: *ZhETF* **83**, 585 (1982) [*Sov. Phys. - JETP* **56**, 322 (1982)].

15. V.A. Belyakov, N.V. Shipov: *ZhETF* **82**, 1159 (1982) [*Sov. Phys. - JETP* **55**, 674 (1982)].

16. V.A. Belyakov, N.V. Shipov: *Pis'ma ZhTF* **9**, 22 (1983).

17. V.A. Belyakov, V.E. Dmitrienko: *Optics of Chiral Liquid Crystals*, ed. by I.M. Khalatnikov, (London, Soviet Scientific Reviews, Sect. A, 1989).

18. *Atlas of Space Groups of the Cubic System* (Moscow, Nauka, 1980).

19. B.Ya. Zel'dovich, N.V. Tabiryan: *Uspekhi Fiz. Nauk* **147**, 633 (1985).

20. V.Eh. Pozhar, L.A. Chernozatonskii: *Fizika Tv. Tela* **27**, 682 (1985).

7

Dynamic Scattering of Thermal Neutrons in Magnetically Ordered Crystals

In the previous chapters we considered the scattering and emission of electromagnetic radiation in crystals. Now we will discuss the coherent scattering of particles in crystals using as an example thermal neutrons. In general, the coherent scattering of neutrons in a crystal is qualitatively similar to that of electromagnetic radiation. However, an elementary act of interaction of neutrons with crystal atoms is essentially different. For example, the energy of neutrons used in diffraction experiments is of the order of the characteristic energy of phonons in the crystal, and hence, it is easy to distinguish the scattering processes which are accompanied with emission (absorption) of phonons and to study crystal dynamics in this way. Until recently, such a separation of elastic and inelastic processes in the case of X-ray diffraction was not possible; a certain success has now been achieved with the help of Mössbauer radiation. The other difference, which is the main subject of the discussions that follow and is very essential to neutron scattering studies, is connected to the magnetic interaction of neutrons with atoms.

It is widely known how important neutron diffraction methods are in studying the structure and dynamics of condensed media [1–5]. These methods are very essential to studies of crystal magnetic structure. In fact, magnetic neutron diffraction is the only method presently available which permits the direct determination of crystal magnetic structure (for a discussion of Mössbauer radiation and X-rays, see Chapt. 3 and 9). Although the magnetic scattering of neutrons has been studied for about 30 years, its theoretical description is mainly related to the kinematic approximation [1,2,4,5]. This is so because until recently there was no real need to apply the dynamic theory to experiments. On the other hand, the dynamic description of coherent scattering of neutrons in magnetically ordered crystals turns out to be much more complicated than the kinematic one and the dynamic description of neutron scattering in paramagnetic crystals. The application of the kinematic approximation to analyze the experimental results was justified by the fact that the interaction of neutrons with the crystal is much weaker than that of an X-ray; hence, crystals which are thick (thus requiring the dynamic approach) for X-rays prove to be thin for neutrons.

Recently, perfect single crystals with sizable dimensions became available [6,7]. This stimulated experimental studies of dynamic scattering of neutrons in perfect crystals [8–16]. It is precisely why the dynamic theory of neutron scattering is now attracting great attention. The dynamic theory of neutron scattering in paramagnetic crystals may be considered complete, although this is not the case with the theory of scattering in magnetically ordered crystals. This is apparently because the neutron scattering in paramagnetic crystals is to a great extent similar to X-ray scattering whose dynamic theory is well developed and may be directly applied to neutron scattering. Until now, there has been no dynamic theory which could be used to describe neutron scattering in full detail in magnetically ordered crystals.

The experimental conditions of magnetic scattering of neutrons correspond, as a rule, to the kinematic approximation of the diffraction theory [1–5]. This is due to the small dimensions of the regions ($\leq 10^{-3}$ cm) in which the samples exhibit perfect crystal and magnetic structure. However, much larger magnetic single crystals as big as 0.1 cm are available at present (see, e.g., [6,7]). These crystals consist of one or several magnetic domains. In order to describe the coherent scattering of neutrons in such samples, we must use the results of dynamic diffraction theory which takes into account multiple scattering and attenuation of the incident neutron beam in the sample.

The successful experiments on dynamic effects in neutron diffraction in paramagnetic crystals [8–12] and the first experiments on dynamic effects in magnetically ordered crystals [14–15,51] show that the experimental studies of dynamic scattering in perfect magnetically ordered crystals have become quite important. These experiments are of interest because, in particular, the dynamic effects, such as pendulum beats of diffracting beams, can assist in revealing exact quantitative data on the coherent amplitudes of neutron scattering. This question was addressed in [11–13] for the case of paramagnetic crystals. In what concerns the magnetically ordered crystals, there is no theory which includes a dynamic description of neutron scattering in crystals with an arbitrary type of magnetic ordering and which could be used to describe quantitatively experiments in an arbitrary geometry. Several specific cases of magnetic ordering were considered in [18–20,44–48], as well as the case of a magnetic structure for a general type of sample whose magnetization is parallel to its surface [21,41,42]. The first works devoted to dynamic diffraction of neutrons and to neutron topography in magnetically ordered crystals are surveyed in [49,50]. In this chapter we will discuss the two-wave approximation of the dynamic theory of neutron diffraction in crystals with an arbitrary magnetic structure. In general, this theory proves to be similar to the dynamic theory of Mössbauer diffraction in magnetic crystals (see Chapt. 3). However, the elementary acts of scattering of a neutron and a gamma-quantum on an atom differ to the point that the solutions of the dynamic system and the boundary problem prove to be very different. Since neutron scattering amplitude depends on orientation of the atom (nucleus) magnetic momentum in a different way, the description of dynamic neutron scattering appears to be simpler than that for gamma-quanta. For example, the problem of dynamic scattering of neutrons in crystals with compensated magnetic momentum of a unit cell is solved analytically in its general form [21], but from another viewpoint, this description turns out to be more complicated. In the general case, the coefficients of the dynamic system depend on the sample shape, which introduces the difficulties unknown in the diffraction of gamma-quanta. Below we present the analytical solution to the problem of diffraction for the class of magnetic structures with compensated magnetic momentum in a unit cell, and for some other magnetic structures which are of physical interest. Also, the dynamic scattering of neutrons in magnetically ordered crystals is analyzed qualitatively in the general case. The qualitative difference between the dynamic scattering of neutrons in magnetic and paramagnetic crystals is discussed. The difference between the results obtained for neutron diffraction within the dynamic and kinematic approach is analyzed in brief.

7.1 Scattering Amplitude of Thermal Neutrons

Before considering the neutron diffraction due to elastic scattering in perfect magnetically ordered crystals, we would like to touch briefly on the physics of interaction of a neutron with an individual atom of the crystal. This interaction determines, in fact, all the specifics of elastic scattering of thermal neutrons in crystals.

It is known [1] that the neutron wave function ψ in a crystal satisfies the Schrödinger equation:

$$\frac{1}{k^2}\Delta\psi + \left[1 - \frac{2mV(\mathbf{r})}{\hbar^2 k^2}\right]\psi = 0 \qquad (7.1)$$

where $\mathbf{k}$ is the wave vector of the neutron wave, m is the neutron mass, and $V(\mathbf{r})$ is the potential of neutron interaction with the crystal lattice. The $V(\mathbf{r})$ potential can be represented as a sum of the potentials of interaction with individual atoms:

$$V(\mathbf{r}) = \sum_a U_a(\mathbf{r} - \mathbf{r}_a) \qquad (7.2)$$

where $\mathbf{r}_a$ determines the location of the ath atom. In its turn, the potential of interaction with an atom is a sum of two terms

$$U_a = u_{a_N} + u_{E_M} \tag{7.3}$$

which describe the nuclear and electromagnetic interaction of neutrons.

The nuclear component of the potential for thermal neutrons can be represented in the form of a Fermi pseudopotential [1,22]

$$u_{a_N} = \frac{2\pi\hbar^2}{m}(b - ib' + B\mathbf{I}\sigma)\delta(\mathbf{r} - \mathbf{r}_a) \tag{7.4}$$

where $B = (B_+ - B_-)/(2I + 1)$, $b' = \sigma_t/2\lambda$, b is the nuclear scattering length, b' is the imaginary part of the nuclear scattering amplitude, I is the nuclear spin, σ is the Pauli matrix, b_+ and b_- are the nuclear scattering lengths for the spin states $I + \frac{1}{2}$ and $I - \frac{1}{2}$ of the system nucleus-neutron, respectively, σ_t is the total cross section of interaction of neutron with the nucleus, and λ is the neutron wavelength.

The electromagnetic part of the interaction potential can be presented in the form

$$u_{a_{EM}} = -\mu\sigma\mathbf{H} - \frac{\mu\sigma(\mathbf{E} \times \mathbf{P})}{mc} - \frac{\mu\hbar}{2mc}(\nabla\mathbf{E}) \tag{7.5}$$

where μ is the magnetic moment of a neutron, $\mathbf{P}$ is its momentum, and $\mathbf{E}$ and $\mathbf{H}$ are the electric and magnetic fields. The first term in (7.5) is the well-known magnetic dipole interaction, and the second one describes the interaction of neutron magnetic momentum with electric field, or Schwinger interaction [1,23], and the third term describes the specific interaction of a neutron with an electron which is due to the relativistic jitter of the neutron, the so-called Foldy interaction [1,24].

It is known that Swinger and Foldy interactions are very weak as compared with nuclear and magnetic dipole interactions and can be observed only in specially arranged experiments [1,17]. Hence, below we will only take into account nuclear and magnetic dipole interactions.

As was mentioned above (see, e.g., Chapt. 3), the expressions describing diffraction include coherent scattering amplitudes. In the case involved, this means that the amplitudes of elastic neutron scattering corresponding to the interaction potentials of (7.4) and (7.5) must be averaged over the ensemble of scattering centers. The averaging procedure is very similar to that described in Chapt. 3. Just as in the case of gamma-quanta, the thermal motion of crystal atoms and isotopic and spin incoherence lead to a decrease in the coherent interaction of neutrons with the crystal. In particular, due to spin incoherence, the nuclear magnetic interaction described by the last term in (7.4) does not usually manifest itself. For this interaction to contribute to the coherent amplitude, it is necessary that the nuclear spins in the crystal be polarized. The polarization of nuclear spins, as is known, can be achieved by special measures exerted on the crystal— namely, either by cooling it to hyperlow temperatures ($T \lesssim 10^{-3}$ K) or by dynamic methods of polarization of nuclei [25]. Therefore, if the opposite is not specified, we take into account only the spin-independent part of the nuclear interaction and magnetic dipole interaction and use the coherent amplitude of elastic neutron scattering corresponding to those interactions.

The amplitude of magnetic scattering of a neutron on an individual atom is usually written [1] as

$$f_{s'm'sm} = 2r_e \gamma P(q) \langle s'm' | (\mathbf{Ss}_n) - (\mathbf{es}_n)(\mathbf{Se}) | sm \rangle \qquad (7.6)$$

where r_e is the classical radius of electron, γ is the gyromagnetic ratio of neutron ($\gamma = -1.91$), and $q = \mathbf{k}_1 - \mathbf{k}_0$ where $\mathbf{k}_0$ and $\mathbf{k}_1$ are the neutron wave vector before and after scattering, $\mathbf{e} = q/|q|$, $P(q)$ is the magnetic form factor of the atom, $\mathbf{s}_n$ and $\mathbf{S}$ are the spins of the neutron and atom, and s, m, s', m' are, respectively, the quantum numbers describing the spin states of the neutron and atom before and after scattering. The structure amplitude of magnetically ordered crystal which enters the system of dynamic equations (see below) is the sum of two terms, nuclear and magnetic. The expression for the nuclear terms of the structure amplitude of magnetically ordered crystals is the same as for paramagnetic crystals, provided that nonpolarized nuclei are considered [1]. The magnetic term of the structure amplitude is obtained from (7.6) by a proper averaging of the elastic amplitude of individual atoms and summing over the magnetic atoms inside a unit cell (see Chapt. 3) with regard to phase factor. Since the explicit expressions for the structure amplitude and the eigenpolarizations of neutron waves (for which the structure amplitudes are of the most interest) differ for different magnetic structures, we do not give here their general form and discuss only the corresponding quantities when solving the dynamic problem for every specific type of magnetic structure.

7.2   System of Dynamic Equations

Let us consider the coherent scattering of slow neutrons in a perfect magnetically ordered crystal when the Bragg conditions are fulfilled. Within the two-wave approximation, the system of dynamic equations describing neutron wave function in the crystal is

$$
\begin{aligned}
(k_0^2/q^2 - 1)\psi_1 &= F_{00}\psi_0 + F_{01}\psi_1 \\
(k_1^2/q^2 - 1)\psi_2 &= F_{10}\psi_0 + F_{11}\psi_1.
\end{aligned}
\qquad (7.7)
$$

Here $\mathbf{k}_0, \mathbf{k}_1$ are the wave vectors and ψ_0, ψ_1 are the spin functions of the primary and secondary neutron waves propagating in the crystal in the incidence direction and at the Bragg angle, respectively, $\mathbf{k}_1 - \mathbf{k}_0 = \boldsymbol{\tau}$, $\boldsymbol{\tau}$ is the reciprocal lattice vector of the magnetic structure, $\mathbf{q}$ is the wave vector of the primary neutron wave in vacuum, $\hbar^2 q^2 = 2mE$, m is neutron mass, E is neutron energy, F_{ij} ($i, j = 0, 1$) is the spin operator describing the scattering of the neutron wave with the wave vector $\mathbf{k}_j$ into the wave with the wave vector $\mathbf{k}_i$.

The elements of the matrix F_{ij} are related to the structure amplitude of neutron scattering [1] as

$$F_{ij}^{\sigma'\sigma} = \frac{4\pi}{q^2 V} F(\mathbf{k}_j, \sigma; \mathbf{k}_i, \sigma') \tag{7.8}$$

where $F(\mathbf{k}_j, \sigma; \mathbf{k}_i, \sigma')$ is the structure amplitude corresponding to the neutron scattering from the state with the wave vector $\mathbf{k}_j$ and spin σ into the state with $\mathbf{k}_i, \sigma'$. The explicit form of the matrices of operators F_{ij} is given below when considering the specific types of magnetic structures.

In order to describe the reflection of the neutron beam from the sample and its transmission through the sample, we should supplement the dynamic system with the boundary conditions at the sample surface. Because magnetic scattering is determined by long-range dipole interaction, the matrices F_{jj} $(j = 0, 1)$ describing the scattering to a nonzero angle in the dynamic system (7.7) are shown to depend on the average magnetic field in the crystal and hence on the sample shape [26]. Therefore, in contrast to diffraction of Mössbauer radiation, the explicit form of the dynamic system depends not only on the type of magnetic ordering in the crystal, but also on the shape of the sample. In this connection, the stage of solving the dynamic diffraction problem in which the eigensolution is looked for (in particular, eigenpolarizations) proves to be more complicated than in the case of Mössbauer radiation. Even at this stage, we must take into accout boundary conditions in determining the coefficients of the dynamic system. Thus, the parameters of the eigensolutions of the dynamic system are not determined uniquely by the type of magnetic ordering (as was the case in Mössbauer radiation), but are directly connected to sample shape.

Because of this, the general analysis of the dynamic system (7.7) is very limited, as compared with that in the case of Mössbauer radiation. The problem of dynamic scattering of neutrons will be considered below for specific types of magnetic structures and for some cases of physical interest. We only note that in the case of magnetically ordered crystals, the polarization properties of the eigensolutions of the dynamic system, and the polarization properties of diffracted neutron beams as well, depend on magnetic structure and sample shape and vary within the range of diffraction scattering.

7.3   Structures with Compensated Magnetic Moment of a Unit Cell

In this section we consider within the two-wave approximation the dynamic neutron scattering in magnetically ordered crystals with compensated magnetic momentum of a unit cell. It was shown that because of the spin dependence of magnetic scattering, the dynamic theory equations for such crystals, in particular antiferromagnets, coincide with the known equations for nonmagnetic crystals [27] only for certain (eigen) polarizations of neutrons. These eigenpolarizations are determined, and the scattering and its polarization parameters for both polarized and nonpolarized neutron beams and for both Laue and Bragg geometries are considered.

The dynamic consideration yields for the polarization of neutrons results which are qualitatively different from those of kinematic theory [4,28]. The scattered beam polarization becomes dependent not only on the crystal magnetic structure, but also on its thickness, the polarization experiencing beats as the thickness varies.

7.3.1   Matrix of the Dynamic System

In structures with compensated magnetic momentum, the coefficients in the dynamic system (7.7) are not connected to the boundary problem; this makes the analysis of dynamic neutron scattering easier. In this case, the average magnetic field in the sample is zero. Hence, the forward scattering amplitude does not depend on neutron polarization and the form of the matrices F_{ii} in (7.7) does not depend on sample shape. Due to this, the dynamic theory of diffraction for antiferromagnets proves to be much more advanced than for structures with noncompensated magnetic momentum. In particular, the zero-angle scattering in our case does not depend on neutron polarization and on wave vector $\mathbf{k}_j$ ($j = 0, 1$); hence, $F_{00} = F_{11} = F_n^0 E^0$ where E^0 is the two-row unit matrix, and F_n^0 is the coherent amplitude of forward scattering, which is

$$F_n = \frac{4\pi}{q^2 V} \sum_\rho b_\rho \qquad (7.9)$$

b_ρ is the nuclear coherent amplitude, and V is the volume of a magnetic unit cell over which the summing is performed. In the elements $F_{10}^{\sigma'\sigma}$ and $F_{01}^{\sigma\sigma'}$ of the scattering operator matrix, the spin projections in the primary σ and secondary σ' waves, in general, correspond to different directions of quantization axes $\mathbf{Z}_0$ and $\mathbf{Z}_1$. If the quantization axes for the primary and secondary waves are the same ($\mathbf{Z}_0 = \mathbf{Z}_1 = \mathbf{Z}$), then the matrix elements F_{10} are expressed with the structure amplitude of nuclear scattering $F_n(\boldsymbol{\tau})$ and the vector $M(\boldsymbol{\tau})$, which characterizes the crystal magnetic structure as

$$F_{10}^{\sigma\sigma} = F_n(\boldsymbol{\tau}) + \frac{\sigma}{|\sigma|}\hat{\mathbf{Z}}\mathbf{M}(\tau)$$

$$F_{10}^{-\sigma\sigma} = \sqrt{2}\hat{n}_+ \mathbf{M}(\tau), \qquad F_{10}^{\sigma-\sigma} = \sqrt{2}\hat{n}_- \mathbf{M}(\tau)$$

$$F_n(\boldsymbol{\tau}) = \frac{4\pi}{q^2 V} \sum_\rho b_\rho e^{i\tau\rho}$$

$$M(\boldsymbol{\tau}) = \frac{4\pi}{q^2 V} \sum_{\rho} f_\rho(\boldsymbol{\tau})[\hat{\tau} \times (\hat{\mathbf{S}}_\rho \times \hat{\tau})]e^{i\tau\rho} \qquad (7.10)$$

where $\hat{n}_\pm = \frac{1}{\sqrt{2}}(\hat{\mathbf{x}} \pm i\hat{\mathbf{y}})$ are circular unit vectors, $\hat{\mathbf{x}} \perp \hat{\mathbf{y}}$ are unit vectors perpendicular to the Z-axis, $f_\rho(\boldsymbol{\tau})$ is the coherent amplitude of magnetic scattering at a single atom [1,3], and $\mathbf{S}_\rho$ is the atom spin where the vector ρ determines the atom location within a magnetic unit cell and the index ρ labels the values related to this atom. Note that the vector $\mathbf{M}(\boldsymbol{\tau})$ coincides, apart from its factor, with the projection of the τ-component of Fourier expansion of crystal magnetization on the scattering plane. In the absence of absorption in the crystal, i.e., when the amplitudes $f_\rho(-\boldsymbol{\tau}) = f_\rho^*(\boldsymbol{\tau})$ and f_ρ are real, the matrix elements F_{01} are related to the matrix elements F_{10} as $F_{01}^{\sigma\sigma'} = (F_{10}^{\sigma\sigma'})^*$. Using the formulae of spin transformation and the above expressions, we can find the F_{10} and F_{01} matrices for any arbitrary orientation of the quantization axes $\mathbf{Z}_0$ and $\mathbf{Z}_1$.

7.3.2   Eigenpolarizations

To simplify the dynamic system, it is convenient to take such axes $\mathbf{Z}_0$ and $\mathbf{Z}_1$, as the quantization axes, in which the operators F_{10} and F_{01} are diagonal. Then the system of dynamic equations consisting of four equation splits into two independent second-order systems. In one of these systems, the amplitudes of the primary and secondary waves are connected only to projections of spin $1/2$ in the directions Z_0 and Z_1. The second system relates the wave amplitudes with a neutron spin projection of $-1/2$ in the same directions. The spin states (polarizations) defined in this way will be referred to as eigenstates, and we can say that under the foregoing choice of quantization axes, the polarizations will be separated in the dynamic system. Note that a neutron wave with eigenpolarization yields a wave with initial polarization after being twice scattered at an angle. Hence to find the eigenpolarization, it is sufficient to diagonalize the matrix defined by the products $F_{01}F_{10}$ and $F_{10}F_{01}$. We will give explicit expressions for the directions Z_0 and Z_1 which provide the separation of polarizations (see Fig. 7.1):

$$\mathbf{Z}_0 = \mathbf{m} + \mathbf{O}, \qquad \mathbf{Z}_1 = \mathbf{m} - \mathbf{O}$$
$$\mathbf{m} = \mathrm{Re}(F_n^*\mathbf{M}), \qquad \mathbf{O} = (\mathrm{Im}\,\mathbf{M}) \times (\mathrm{Re}\,\mathbf{M}). \qquad (7.11)$$

Thus, when the quantization axes are chosen as in (7.11), only the diagonal elements $F_{10}^{\frac{1}{2}\frac{1}{2}} = F_+$ and $F_{10}^{-\frac{1}{2}-\frac{1}{2}} = F_-$ of the matrix $F_{10}(F_{01})$ are other than zero where

$$F_\pm = F_n/\cos\beta \pm \hat{m}\mathbf{M}$$
$$|F_\pm|^2 = |F_n|^2 + |M|^2 \pm 2|\mathbf{m} + \mathbf{O}| \qquad (7.12)$$

and 2β is the angle between $\mathbf{Z}_0$ and $\mathbf{Z}_1$.

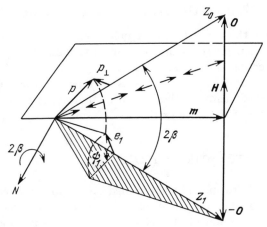

FIGURE 7.1. Schematics for determining the orientation of the quantization axes Z_0, Z_1 and for transforming the polarizations in scattering of neutrons in an antiferromagnetic crystal.

As is seen in (7.11), the quantization directions Z_0 and Z_1 are determined by the reflection (τ) and by the structure of a magnetic unit cell. Note that in many antiferromagnetic structures, the vectors Z_0 and Z_1 in (7.11) are zeros; hence, the direction of one of the quantization axes determining the eigenpolarizations, say, Z_0, can be chosen arbitrarily. Then the direction of Z_1 is obtained from Z_0 by rotating around M by the angle φ defined by $\cos\varphi = (|F_n|^2 - |M|^2)/(|F_n|^2 + |M|^2)$. Such a situation takes place in structures in which the module of the structure amplitude does not depend on the orientation of the neutron spin ($|F_+| = |F_-|$). In such structures, in particular, a nonpolarized beam remains nonpolarized after scattering. In this case, the solution to the dynamic problem is simplified and the polarization properties coincide with those of the kinematic approximation [4,28].

Below we consider the diffraction of neutrons in structures in which the strength of interaction with neutrons depends on spin orientation ($|F_+| \neq |F_-|$). Such dependence can exist, for example, in collinear antiferromagnetics in which spatial inversion connects the equivalent sites with similarly directed spins of magnetic atoms. For such structures, the direction of the quantization axes (7.11) is uniquely determined. In the case of a collinear antiferromagnetic, the directions Z_0 and Z_1 coincide and go along the projection of the antiferromagnetic axis on the scattering plane T, i.e., along m in Fig. 7.1. The polarization dependence of scattering in this instance differs from that of the above case [4], in particular, a nonpolarized beam becomes polarized after scattering.

7.3.3   Solution to the Boundary Problem

Let us consider the diffraction of a polarized neutron beam in a plane-parallel antiferromagnetic sample of finite thickness. Using the results of the previous section and the known boundary conditions [27] for the wave vectors and amplitudes of neutron waves, we represent the amplitude $\psi(\mathbf{k}_0)$ of the wave passed through the crystal in the primary direction and the amplitude $\psi(\mathbf{k}_1)$ of the wave exiting the crystal in the Bragg direction in the following form:

$$\psi(\mathbf{k}_j) = a^+ C_j^+ \psi^+(\mathbf{k}_j) + a^- C_j^- \psi^-(\mathbf{k}_j), \qquad (j = 0, 1) \qquad (7.13)$$

where a^+ and a^- are uniquely determined by the polarization vector of the incident beam $\mathbf{p}$, they are the coefficients of the expansion of spin functions of the incident wave in the eigen spin functions $\psi^+(\mathbf{k}_0)$ and $\psi^-(\mathbf{k}_0)$ for the primary direction, and $\psi^\pm(\mathbf{k}_1)$ are the eigen spin functions for the secondary direction. The coefficients $C_j^\pm$ characterize the reflection (transmission) of the waves with eigenpolarization; they are described by the formulae known from the dynamic theory of X-ray diffraction [29] (see also Chapt. 3 and 4). Within Laue and Bragg geometry, the expressions for $C_j^\pm$ (see Chapt. 2) are:

$$C_j^\pm(\Delta\theta)_L = e^{i\delta_j qL_0} \left(-\frac{\gamma_0}{\gamma_1} F_\pm^* \right)^j \frac{1}{4W_\pm} \left[(\delta + 2W_\pm)^{1-j} e^{iW_\pm qL_0} \right.$$
$$\left. - (\delta - 2W_\pm)^{1-j} e^{iW_\pm \kappa L_0} \right]$$

$$C_j^\pm(\Delta\theta)_B = e^{i(1-j)\delta_1 qL_0} \left(\frac{\gamma_0}{\gamma_1} F_\pm \right)^j (4W_\pm)^{-j} \left(e^{iW_\pm qL_0} \right.$$
$$\left. - e^{-iW_\pm qL_0} \right)^j \left[(\delta + 2W_\pm) e^{iW_\pm qL_0} - (\delta - 2W_\pm) e^{iW_\pm qL_0} \right]^{-1}$$

$$(2W_\pm)^2 = \delta^2 + \frac{\gamma_0}{\gamma_1} |F_\pm|^2, \qquad 2\delta = \frac{\gamma_0}{\gamma_1} \alpha + F_n^0 \left(1 - \frac{\gamma_0}{\gamma_1} \right)$$

$$4\delta_j = (2j-1)\gamma_0 \left(\frac{\alpha}{\gamma_1} \right) + F_n^0 \left(1 + \frac{\gamma_0}{\gamma_1} \right), \qquad (j = 0, 1) \qquad (7.14)$$

where γ_j is the cosine of the angle between $\mathbf{k}_j$ and the inner normal to the entrance surface of the sample ($\gamma_0 \geq 0$, $\gamma_1 < 0$ in Bragg geometry and $\gamma_1 > 0$ in Laue geometry); $\alpha = (\hat{\mathbf{q}} + \boldsymbol{\tau}/q)^2 - 1 \approx 2\sin 2\theta_b \cdot \Delta\theta$ is the parameter determining the deviation from the exact Bragg condition ($\Delta\theta = \theta - \theta_b$ is the deviation of the incidence angle from the Bragg one θ_b); $L_0 = L/\gamma_0$; L is the sample thickness. The coefficients $C_j^\pm(\Delta\theta)$ determine the intensity $I_j(\Delta\theta)$ and polarization [the polarization vector $\mathbf{p}_j'(\Delta\theta)$] of scattered neutrons for unit amplitude of the incident wave by the expressions

$$I_j(\Delta\theta) = \frac{1}{2} \left[|C_j^+|^2 + |C_j^-|^2 + p_\parallel (|C_j^+|^2 - |C_j^-|^2) \right]$$

$$I_j(\Delta\theta)\mathbf{p}'_j(\Delta\theta) = \frac{1}{2}\left[|C_j^+|^2 - |C_j^-|^2 + p_\parallel(|C_j^+|^2 + |C_j^-|^2)\right]\hat{\mathbf{Z}}_j$$

$$+ p_\perp|C_j^+C_j^-|(\cos\psi_j\hat{\mathbf{e}}_j + \sin\psi_j\hat{\mathbf{e}}_j \times \hat{\mathbf{Z}}_j), \quad (j = 0,1) \qquad (7.15)$$

where $p_\parallel = \mathbf{p}\hat{\mathbf{Z}}_0$, $p_\perp\hat{\mathbf{e}}_1 = \hat{\mathbf{p}}_\perp$ is the component of the polarization vector $\mathbf{p}$ perpendicular to Z,

$$p_\perp\hat{\mathbf{e}}_1 = \left[\mathbf{p}(\hat{\mathbf{Z}}_0 \times \hat{\mathbf{N}})\right](\hat{\mathbf{Z}}_1 \times \hat{\mathbf{N}}) + (\mathbf{p}\hat{\mathbf{N}})\hat{\mathbf{N}}, \quad \mathbf{N} = \hat{\mathbf{Z}}_1 \times \hat{\mathbf{Z}}_0$$

is the normal to the plane $(\mathbf{Z}_0, \mathbf{Z}_1)$, and $e^{j\psi_j} = C_j^+C_j^*/|C_j^+C_j^-|$.

Equation (7.15) describes the rotation which transforms the initial polarization vector $\mathbf{p}$ into the polarization vector $\mathbf{p}'_j$ of the wave exiting the crystal. In particular, for the wave exiting the crystal in the primary direction, the vector $\mathbf{p}'_0$ lies in the plane $(\mathbf{p}'_0, \mathbf{Z}_0)$ which is produced by rotating $(\mathbf{p}, \mathbf{Z}_0)$ by the angle ψ_0 around $\mathbf{Z}_0$. Orientation of $\mathbf{p}'_0$ in this plane depends on the direction $\mathbf{p}$ and on the ratio $|C_j^+/C_j^-|$; this orientation can be easily found from (7.15). For the wave leaving the crystal in the secondary direction, the polarization vector $\mathbf{p}'_1$ lies in the plane $(\mathbf{p}'_1, \mathbf{Z}_1)$ which is obtained by rotating the plane $(\mathbf{p}, \mathbf{Z}_0)$ by the angle $\mathbf{Z}_0\mathbf{Z}_1 = 2\beta$ around $\mathbf{N}$ and then by the angle ψ_1 around $\mathbf{Z}_1$ (see Fig. 7.1).

As is seen in (7.14), for the structures in question $|F_+| \neq |F_-|$ and $C_j^+ \neq C_j^-$; hence, according to (7.15), the intensities I_j $(j = 0,1)$ depend on the incident beam polarization and the polarization vectors $\mathbf{p}'_j$ do not coincide with $\mathbf{p}$. It is essential that the dependence of I_i and $\mathbf{p}'_j$ on the polarization of the primary beam be directly connected to the detailed magnetic structure of the crystal.

If, in contrast to the routinely considered case of $|F_+| = |F_-|$, $|F_+| \neq |F_-|$, then not only intensities I_i exhibit beats as functions of the crystal thickness but also the polarizations of neutron beams do so, two beat periods taking place. It was mentioned in [20] that polarization beats in an antiferromagnet are possible, but the conditions for this effect were not analyzed there.

7.3.4 Averaged Scattering Parameters

Formulas (7.15) were obtained for a completely polarized primary beam $(|\mathbf{p}| = |\mathbf{p}'_j| = 1)$. When a partially polarized beam is scattered, the intensity and polarization are also described by (7.15), but with $|\mathbf{p}| = P < 1$ where P is the polarization degree of the incident beam. In general, the polarization degree $P'_j = |\mathbf{p}'_j|$ does not coincide with P. In the specific case of a nonpolarized incident beam, the scattered neutrons are partially polarized. For scattering along the Bragg direction, the corresponding polarization vector $\mathbf{p}'$ can be presented using (7.14) and (7.15) as

$$\mathbf{p}' = \frac{R^+ - R^-}{R^+ + R^-}\hat{\mathbf{Z}}_1 \qquad (7.16)$$

where $R^{\pm}(\Delta\theta)$ are the reflection coefficients of eigenpolarization.

On the basis of (7.14–7.15), we can obtain the integrated over the incidence angle $\Delta\theta$ intensity

$$I_j^i = \int I_i(\Delta\theta)w(\Delta\theta)\,d(\Delta\theta)$$

and the polarization vector

$$\mathbf{p}_j'^i = (I_j^i)^{-1}\int \mathbf{p}_j'(\Delta\theta)w(\Delta\theta)\,d(\Delta\theta)$$

where $w(\Delta\theta)$ is the angular intensity distribution in the incident beam. In particular, for a nonpolarized incident beam the integral polarization vector of the scattered beam $\mathbf{p}^{i'}$ is determined by (7.16), in which the values $R^{\pm}(\Delta\theta)$ must be substituted by the corresponding integral reflection coefficients $R_i^{\pm}$ [29]. As is seen from (7.16), the beam of scattered neutrons is completely polarized ($|\mathbf{p}^{i'}| = 1$), provided that the reflection coefficient of one of the eigenpolarizations is zero, for example, when one of the values $F_{\pm}$ becomes zero due to mutual canceling of nuclear and magnetic scattering. In this case, the diffraction on an antiferromagnetic crystal can be used to produce polarized neutrons. It is known [1,30] that the similar cancellation of nuclear and magnetic scattering is already used to produce highly polarized neutron beams by diffraction at ferromagnet.

Now we give the expressions for the integrated intensity I_1^i and polarization vector $\mathbf{p}_1^i$ of the beam scattering along the Bragg direction when the incident beam is arbitrarily polarized

$$I_1^i = \frac{1}{2}(R_i^+ + R_i^-)[1 + P^{i'}(\mathbf{p}\hat{\mathbf{Z}}_0)]$$
$$\mathbf{p}_1^i = (1 + P^{i'}p_{\parallel})^{-1}[(P^{i'} + p_{11})\hat{\mathbf{Z}}_1$$
$$+ p_{\perp}(\operatorname{Re}G_i\hat{\mathbf{e}}_2 + \operatorname{Im}G_i\hat{\mathbf{e}}_2 \times \hat{\mathbf{Z}}_1)] \tag{7.17}$$

where

$$G_i = 2(R_i^+ + R_i^-)^{-1}\int C_1^+(\Delta\theta)C_1^{-*}(\Delta\theta)w(\Delta\theta)\,d(\Delta\theta), \quad |G_i| \leq 1.$$

As is seen from (7.17), the integral parameters, as well as the differential parameters discussed above, depend on the incident beam polarization, and the specific form of this dependence is determined by the details of the crystal structure [because, in particular, the orientation of $\mathbf{Z}_0$ and $\mathbf{Z}_1$ is connected to magnetic structure; see (7.11)]. Therefore, in addition to the information obtained using nonpolarized beams without polarization measurements, we can obtain the information on magnetic structure (orientation of $\mathbf{Z}_0$) in experiments with polarized neutrons, restricting ourselves to measuring scattered beam intensity. To do this, as follows from (7.17),

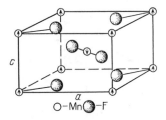

FIGURE 7.2. Structure of MnF$_2$ [31].

either the incident beam polarization should be varied or the sample should be rotated around the normal to the reflecting surface in measurements of some reflection. Similar information can be obtained by polarization measurements of the scattered beam if the incident beam is unpolarized.

7.3.5   Observed Dynamic Effects

The above consideration of dynamic scattering in antiferromagnetic crystals indicate that dynamic effects are essential for perfect crystals in describing not only the intensities of neutron beams but also their polarization properties. Hence, the analysis of polarization of a scattered neutron beam can be used to obtain information about the magnetic structure of a crystal with compensated magnetic momentum of a unit cell. We must keep in mind that the dynamic expressions for the polarization parameters may differ strongly from corresponding kinematic expressions (see also the corresponding expression for Mössbauer diffraction in Chapt. 3). However, the polarization measurements were not used until now for studying magnetic structures [5].

Note that there is a qualitative difference in the pendulum solution for an antiferromagnetic crystal from that for a paramagnetic crystal. In fact, in the general case of an antiferromagnet, there are two beat periods of intensity and polarization as a function of the sample thickness, whereas there is only one such period in a paramagnetic crystal.

To conclude this section, we give several examples of antiferromagnetic structures in which the results obtained are applicable in a description of neutron diffraction. First of all, it is the well-known collinear antiferromagnetic MnF$_2$ (Fig. 7.2) in which nuclear and magnetic scattering may interfere [31]. There are also other antiferromagnets with the same type of lattice [32]. In these structures, $|F_+| \neq |F_-|$ only in crystal reflections [where $F_n(\tau) \neq 0$]. The other example is noncollinear antiferromagnets in which the antiferromagnetic ordering has no center of symmetry or antisymmetry; in particular, CrSe (Fig. 7.3) is an umbrella-type antiferromagnet [33] in which scattering depends on the incident neutron polarization both in crystal and magnetic reflexes.

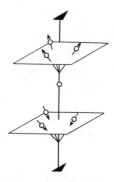

FIGURE 7.3. Magnetic structure of CrSe [33].

7.4   Dynamic Scattering in Magnetically Ordered Structures of the General Type

In the previous section, we considered the structure with compensated magnetic moment which permits an analytical solution of the problem of dynamic scattering of thermal neutrons in the general case. The analytic expressions thereby obtained allowed us to analyze easily both the qualitative and quantitative features of dynamic scattering of neutrons for this class of magnetic structures. The results indicate that the current state of experiments on dynamic scattering on nonmagnetic crystals ([8–10,11,12,34–37]) and references therein) and the recent studies on magnetic crystals [14–16] will enable us to also study dynamic magnetic scattering. In this connection, it is important to elaborate on the theory of dynamic diffraction in regard to magnetic structures of the general type and to reveal the conditions which are optimal for experiments. Some specific cases of dynamic scattering in magnetic crystals were considered in [38–40].

In this section we consider the dynamic theory of diffraction in a magnetically ordered structure of the general type within the framework of the two-wave approximation. For the general case, we present the solution of the boundary problem for scattering at plane-parallel samples and the expressions for the intensities and polarizations of scattered beams [41]. The solution and its polarization properties are analyzed qualitatively; the peculiarities of dynamic scattering in the general-type magnetically ordered structures have been pointed out. It is shown, in particular, that in the pendulum solution six beat periods can occur and that the Bragg reflection curve possesses a fine angular structure. The analytical description of dynamic scattering for both Laue and Bragg symmetric cases is presented for the cases permitting simplification of general formulae due to transformation of the secular equation of the dynamic system into a biquadratic one

[42]. The case of thick crystals, in particular, those with strong magnetic birefringence, is considered in detail.

7.4.1 Eigensolutions to the Dynamic System

Let us consider the scattering of slow neutrons in a perfect and sufficiently thick magnetic crystal. The system of dynamic equations which describe diffraction has the form of (7.7) in the two-wave approximation. For scattering at the Bragg angle $(i \neq j)$, the form of the operator F_{ij} coincides with that for antiferromagnets (7.10). For zero-angle scattering $(i = j)$, the operator F_{ii} depends on the orientation of neutron spin with respect to the average magnetic field (magnetization) in the crystal. In general, magnetization is not uniform and depends on sample shape. Hence, zero-angle scattering amplitudes also depend on sample shape. For the sake of simplicity, we consider neutron scattering in a plane-parallel sample whose magnetization is homogeneous and parallel to the sample surface.

For this shape, we can write explicitly all the matrices of spin operators F_{ij} for a crystal with an arbitrary magnetic structure. The system (7.7) becomes simplified if we choose the magnetization direction as the axis of quantization of neutron spins. In this case, $F_{00} = F_{11}$, the corresponding operators are diagonal, and their diagonal elements are

$$F_{00} = F_n^0 + \frac{\sigma}{|\sigma|}|\mathbf{M}_0|, \qquad \mathbf{M}_0 = \frac{4\pi}{q^2}r_e\gamma\sum_\rho \mathbf{S}_\rho'' \qquad (7.18)$$

where F_n^0 is the structure amplitude of forward nuclear scattering (7.9), $\mathbf{S}_\rho''$ is the component of atomic spin parallel to the sample surface, and $\sigma = \pm\frac{1}{2}$ is the neutron spin projection on the direction $\mathbf{M}_0$ which coincides with the magnetization direction. The matrices of the operators F_{ij} and F_{ji} $(i \neq j)$ in the absence of absorption are Hermitian-conjugated. At a chosen direction of the quantization axis, the matrix elements are determined by (7.10), in which $\hat{\mathbf{Z}}$ is substituted by $\hat{\mathbf{M}}_0$. To be more precise, the direction of the unit vector $\hat{\mathbf{x}}$ in (7.10), which is perpendicular to the quantization axis $\hat{\mathbf{M}}_0$, is taken to be parallel to the sample surface.

To determine the eigensolution to the system (7.7), it is necessary to find the roots ϵ_j $(j = 1, 2, 3, 4)$ of its secular equation with respect to ϵ $(\epsilon = k_0^2/q^2 - 1)$, $k_1^2/q^2 - 1 = (\epsilon\gamma_1/\gamma_0) + 2\alpha$, $\alpha = \tau(\tau + 2\mathbf{q})/2q^2$, $\gamma_0/\gamma_1 = \cos\mathbf{k}_0\mathbf{n}/\cos\mathbf{k}_1\mathbf{n}$, and $\mathbf{n}$ is the inner normal to the sample input surface. The eigensolution of the system (7.7) ϕ_j corresponding to the root ϵ_j is a superposition of two plane waves:

$$\psi_j = \sum_{q=0,1} e^{i\mathbf{k}_{qj}\mathbf{r}}(a_{qj}^+|\uparrow\rangle + a_{qj}^-|\downarrow\rangle)$$

$$\mathbf{k}_{qj} = \mathbf{q} + \frac{q}{2\gamma_0}\epsilon_j\hat{\mathbf{n}} + \mathbf{q}\tau, \qquad (q = 0, 1) \qquad (7.19)$$

where $|\uparrow\rangle$ and $|\downarrow\rangle$ are the spin functions of neutrons with spin projections at $\mathbf{M}_0$ being $1/2$ and $-1/2$, respectively, $\mathbf{k}_{qj}$ are the wave vectors of two eigenwaves in the crystal corresponding to the root ϵ_j, four coefficients $(a_{0j}^{\pm}, a_{1j}^{\pm})$ are the eigensolutions to the system (7.7) corresponding to the root ϵ_j (see also Chapt. 3), those solutions determining the eigenpolarizations of neutron waves in the crystal and depending, in general, on the incidence angle (on α).

The explicit expression for the eigensolutions $(a_{0j}^{\pm}, a_{1j}^{\pm})$ of the dynamic system (7.7) is

$$a_{0j}^{+} = M_{+}^{*}(F_n^0 - M_0 - \epsilon_j)(F_n^0 + M_0 - \epsilon_j - 2\alpha) + M_{-}(F_{+}^{*}F_{-}^{*} - M_{+}^{*}M_{-}^{*})$$

$$a_{0j}^{-} = F_{-}^{*}(F_n^0 - M_0 - \epsilon_j)(F_n^0 + M_0 - \epsilon_j - 2\alpha) - F_{+}(F_{+}^{*}F_{-}^{*} - M_{+}^{*}M_{-}^{*})$$

$$a_{1j}^{+} = -F_{-}^{*}M_{-}(F_n^0 + M_0 - \epsilon_j) - F_{+}M_{+}^{*}(F_n^0 - M_0 - \epsilon_j)$$

$$a_{1j}^{-} = |F_{+}|^2(F_n^0 - M_0 - \epsilon_j) - (F_0^n - M_0 - \epsilon_j) \qquad (7.20)$$

$$\cdot[(F_n^0 - M_0 - \epsilon_j)(F_n^0 + M_0 - \epsilon_j - 2\alpha) - |M_{-}|^2], \quad (j = 1, 2, 3, 4)$$

where $F_{\pm} = F_n \pm \hat{\mathbf{M}}_0 \mathbf{M}$ and $\mathbf{M}_{\pm} = (\hat{\mathbf{x}} \pm i\hat{\mathbf{y}})\mathbf{M}$ are the structure amplitudes of scattering of the primary wave at the Bragg angle without and with spin-flip from the state with the spin projection $\pm 1/2$.

7.4.2   Solution to the Boundary Problem

Let a neutron wave $e^{i\mathbf{qr}}(a_{\uparrow}|\uparrow\rangle + a_{\downarrow}|\downarrow\rangle)$ fall on a plane-parallel crystal plate at an angle close to the Bragg one where $a_{\uparrow}$ and $a_{\downarrow}$ are the components of the spin function directed along the quantization axis and opposite to it. As a result of diffraction, two waves leave the crystal: the wave transmitted in the primary direction (index t) and the wave reflected at the Bragg angle (index r). The amplitudes of the waves leaving the crystal may be presented as

$$\psi_g = a_g^{+}|\uparrow\rangle + a_g^{-}|\downarrow\rangle, \qquad g = t, r. \qquad (7.21)$$

The coefficients a_g^{+} and a_g^{-} determine the intensity and polarization of neutron waves outside the crystal ($g = t$ for the transmitted wave and $g = r$ for the Bragg reflected wave). These coefficients are expressed through the solutions of the dynamic system (7.7).

The explicit expressions for $a_g^{\pm}$ are different in the Bragg and Laue cases. In the former case, the values $a_g^{\pm}$, $g = t, r$, are

$$a_{gB}^{\pm} = -\frac{\Delta_{gB}^{\pm}}{D_B}, \qquad D_B = \begin{vmatrix} a_{01}^{+} & a_{02}^{+} & a_{03}^{+} & a_{04}^{+} \\ a_{01}^{-} & a_{02}^{-} & a_{03}^{-} & a_{04}^{-} \\ \eta_1 a_{11}^{+} & \eta_2 a_{12}^{+} & \eta_3 a_{13}^{+} & \eta_4 a_{14}^{+} \\ \eta_1 a_{11}^{-} & \eta_2 a_{12}^{-} & \eta_3 a_{13}^{-} & \eta_4 a_{14}^{-} \end{vmatrix} \qquad (7.22)$$

$$\eta_i = \exp\left(i\frac{q}{2\gamma_0}\epsilon_j L\right),$$

$$\Delta_B = \begin{vmatrix} a_{01}^+ & a_{02}^+ & a_{03}^+ & a_{04}^+ & a_\uparrow \\ a_{01}^- & a_{02}^- & a_{03}^- & a_{04}^- & a_\downarrow \\ \eta_1 a_{11}^+ & \eta_2 a_{12}^+ & \eta_3 a_{13}^+ & \eta_4 a_{14}^+ & 0 \\ \eta_1 a_{11}^- & \eta_2 a_{12}^- & \eta_3 a_{13}^- & \eta_4 a_{14}^- & 0 \\ \xi_1 & \xi_2 & \xi_3 & \xi_4 & 0 \end{vmatrix}$$

where the determinants $\Delta_{tB}^\pm$ and $\Delta_{rB}^\pm$ are determined by the formula for Δ_B in which $\xi_i = \eta_i a_{0i}^\pm$ and $\xi_j = a_{1j}^\pm$, respectively, and L is the sample thickness.

In the Laue case, the values $a_g^\pm$ $(g = t, r)$ are

$$a_{rL}^\pm = -\exp\left(i\frac{q}{\gamma_1}\alpha L\right)\frac{\Delta_{rL}^\pm}{D_L}, \qquad a_{tL}^\pm = -\frac{\Delta_{tL}^\pm}{D_L} \qquad (7.23)$$

where the determinants D_L and Δ_{gL} are given by (7.22) for D_B and Δ_B in which we must take $\eta_j = 1$, and $\xi_j = \eta_j a_{0j}^\pm$ and $\xi_j = \eta_j a_{1j}^\pm$ for $\Delta_{tL}^\pm$ and $\Delta_{rL}^\pm$, respectively. The formulae (7.22) and (7.23) give a solution to the problem of dynamic diffraction of neutrons in a plane-parallel crystal of an arbitrary magnetic structure. The above expressions are valid for a completely polarized incident beam. For a partially polarized beam (with the polarization vector $\mathbf{p}$), the formulas (7.22) and (7.23) yield the intensities I_g $(g = t, r)$ of diffracted beams in the form of

$$I_g = I_{0g} + p_\parallel I_{1g} + p_\perp I_{2g}$$

$$I_{0g} = \frac{1}{2}(|C_{g\uparrow}^+|^2 + |C_{g\uparrow}^-|^2 + |C_{g\downarrow}^+|^2 + |C_{g\downarrow}^-|^2)$$

$$I_{1g} = \frac{1}{2}(|C_{g\uparrow}^+|^2 + |C_{g\uparrow}^-|^2 - |C_{g\downarrow}^+|^2 - |C_{g\downarrow}^-|^2)$$

$$I_{2g} = \operatorname{Re} e^{i\varphi_\perp}(C_{g\downarrow}^+ C_{g\uparrow}^{+*} + C_{g\downarrow}^- C_{g\uparrow}^{-*}) \qquad (7.24)$$

where $p_\parallel \hat{\mathbf{M}}_0 + \mathbf{p}_\perp = \mathbf{p}$, $p_\parallel = \mathbf{p}\hat{\mathbf{M}}_0$ is the projection of the polarization vector $\mathbf{p}$ on the quantization axis M_0, $p_\perp = |\mathbf{p}_\perp|$ is the length of its component perpendicular to the quantization axis, and $\varphi_\perp$ is the azimuthal angle of the $\mathbf{p}_\perp$ vector with respect to the X-axis parallel to the sample surface. The values $C_{g\uparrow}^\pm$ and $C_{g\downarrow}^\pm$ $(g = t, r)$ are given by (7.22) and (7.23) as $C_{g\uparrow}^\pm = a_g^\pm$ at $a_\uparrow = 1$, $a_\downarrow = 0$, and $C_{g\downarrow}^\pm = a_g^\pm$ at $a_\uparrow = 0$, $a_\downarrow = 1$. The polarization vector of diffracted beams $\mathbf{p}_g$ $(g = t, r)$ is expressed from $C_{g\uparrow}^\pm$ and $C_{g\downarrow}^\pm$ by virtue of the following formulae:

$$\mathbf{P}_g = \mathbf{P}_{g0} + p_\parallel \mathbf{P}_{g1} + p_\perp \mathbf{P}_{g2}$$

$$I_g \mathbf{P}_{g0} = \frac{1}{2}(|C_{g\uparrow}^+|^2 + |C_{g\downarrow}^+|^2 - |C_{g\uparrow}^-|^2 - |C_{g\downarrow}^-|^2)\hat{\mathbf{M}}_0$$

$$+ \operatorname{Re}(\hat{\mathbf{x}} + i\hat{\mathbf{y}})(C_{g\uparrow}^+ C_{g\uparrow}^{-*} + C_{g\downarrow}^+ C_{g\downarrow}^{-*})$$

$$I_g \mathbf{P}_{g1} = \frac{1}{2}(|C_{g\uparrow}^+|^2 + |C_{g\downarrow}^-|^2 - |C_{g\downarrow}^+|^2 - |C_{g\uparrow}^-|^2)\hat{\mathbf{M}}_0$$
$$+ \operatorname{Re}(\hat{\mathbf{x}} + i\hat{\mathbf{y}})(C_{g\uparrow}^+ C_{g\uparrow}^{-*} - C_{g\downarrow}^+ C_{g\downarrow}^{-*})$$
$$I_g \mathbf{P}_{g2} = \operatorname{Re} e^{i\varphi_\perp} \left[(\hat{\mathbf{x}} - i\hat{\mathbf{y}}) C_{g\downarrow}^- C_{g\uparrow}^{+*} + (\hat{\mathbf{x}} + i\hat{\mathbf{y}}) C_{g\uparrow}^- C_{g\downarrow}^{+*} \right]. \qquad (7.25)$$

Formulae (7.24) and (7.25) describe also the scattering of nonpolarized incident beams at $p_\parallel = p_\perp = 0$.

As it follows, in particular, from (7.22) and (7.25), there are some qualitative difference between neutron scattering in magnetically ordered and paramagnetic crystals. The difference is connected to the polarization properties, the pendulum solution, and the shape of the reflection curve. That is, it follows from (7.22) and (7.23) that the polarization of diffracting neutron beams (at a given initial polarization) depends on the detailed magnetic structure of the crystal, the crystal thickness L, and the incidence angle (parameter α).

The peculiarity of the pendulum solution for magnetic crystals is that there are six beat periods of intensity in the general case. The beats are also present in the polarization parameters. These six beat periods are due to the fact that in a magnetically ordered crystal (in contrast to a nonmagnetic one), all four eigenwaves determined by (7.7) interfere. Note that for unlimitedly thick crystals, the general expressions (7.22) and (7.23) in the Bragg case become simplified because the waves increasing inwardly toward the crystal are not excited in this case. Hence, we can analyze in general form the dependence of the reflection coefficient on small variations of the incident angle near the Bragg value, i.e., the shape of the reflection curve. Such analysis shows that the reflection curve may consist of three separate angular ranges of diffraction reflection with different polarization parameters in each of these regions. Namely, in one range that corresponds to four complex roots of the secular equation, a neutron wave with any polarization is reflected. In two other ranges in which there are two complex and two real roots, only a wave whose polarization coincides with that of an eigensolution decaying inside the crystal is reflected.

Since the roots of the secular equation depend on the parameters of the problem in a complicated way, general analysis of the dynamic scattering is difficult. Below we consider the cases in which the secular equation is simplified and an analytic description of the dynamic scattering becomes possible.

7.4.3   The Cases of Analytic Solution of the Dynamic Problem

It is known that the dynamic system becomes simplified when polarization separation occurs. Then the problem of dynamic scattering of neutrons in magnetically ordered crystals is reduced to that in nonmagnetic crystals.

Note that polarizations can always be separated (both in the Bragg and

Laue cases), provided that in the amplitude of magnetic scattering at an angle the quantity $\mathbf{M}(\tau)\|\mathbf{M}_0$, i.e., it is parallel to the sample magnetization direction. We should point out that the condition $\mathbf{M}(\tau)\|\mathbf{M}_0$ does not assume necessarily that the atomic spins are parallel to the sample surface (we should remind the reader that $\mathbf{M}_0$ is parallel to the sample surface if the sample is a plane-parallel plate). For example, in the Bragg symmetric case, this condition is fulfilled for a collinear structure with arbitrary orientations of spins.

The solution is also simplified if the secular equation of the dynamic system (7.7) is biquadratic. It is easy to show that this can occur only in the case of symmetric diffraction ($\gamma_1/\gamma_0 = \pm 1$). In the Bragg case ($\gamma_1/\gamma_0 = -1$), the secular equation is biquadratic for an arbitrary magnetic structure, whereas in the Laue case ($\gamma_1/\gamma_0 = 1$), it is so only for reflections in which the nuclear and magnetic scatterings do not interfere.

Let us now consider in more detail the cases of the biquadratic secular equations for the Laue and Bragg geometries. In the Laue case, for the secular equation of (7.7) to become biquadratic, the condition $|F_+| = |F_-|$ should be satisfied which is equivalent to the absence of interference in the nuclear and magnetic scattering within the reflection involved. Then the roots ϵ_j of the secular equation in this case are given by

$$(\epsilon_j + \alpha - F_n^0)^2 = \alpha^2 + M_0^2 + |F_n|^2 + |M|^2 \pm 2\big[(M_0\alpha + |\mathrm{Re}\,M \times \mathrm{Im}\,M|)^2$$
$$+ (\mathrm{Im}\,F_n^* M)^2 + M_0^2|F_n|^2 + |M_0 M_2|^2\big]^{1/2}. \tag{7.26}$$

This formula, together with (7.22) and (7.25), provides an analytical description of the dynamic neutron scattering for the symmetric Laue case. Note that (7.26) yields, in particular, the periods of the intensity beats in the pendulum solution, the number of these periods being four for the biquadratic secular equation.

In the Bragg case, the conditions for the biquadratic secular equation are $\gamma_1/\gamma_0 = -1$ and $|M_+| = |M_-|$. According to the analysis completed, these conditions are fulfilled for an arbitrary magnetic structure in the symmetric Bragg case. The roots ϵ_j of the corresponding secular equation are given by

$$(\epsilon_j - \alpha)^2 = (\alpha - F_n^0)^2 + \mathbf{M}_0^2 - |F_n|^2 - |\mathbf{M}|^2$$
$$\pm 2\big\{[\mathbf{M}_0(\alpha - F_n^0) + \mathrm{Re}(F_n^*\hat{\mathbf{M}}_0\mathbf{M})]^2 + |\mathrm{Re}\,\mathbf{M} \times \mathrm{Im}\,\mathbf{M}|^2$$
$$+ (\mathrm{Re}\,F_n^*\mathbf{M}_\perp)^2 - \mathbf{M}_0^2|\mathbf{M}_\perp|^2\big\}^{1/2} \tag{7.27}$$

where $\mathbf{M}_\perp = \mathbf{M} - (\hat{\mathbf{M}}_0\mathbf{M})\hat{\mathbf{M}}_0$ is the component of the vector $\mathbf{M}(\tau)$ which is perpendicular to the quantization axis, in this case, $|\mathbf{M}_\perp| = |\mathbf{M}_\pm|$. Formula (7.27) together with (7.22–7.25) yields the analytical solution to the problem of dynamic scattering of neutrons in magnetically ordered crystals of arbitrary thickness in the symmetric Bragg case. In both the Laue and

Bragg cases, the intensities and polarizations of transmitted and reflected beams experience beats with the crystal thickness. However, in the latter case, the beat amplitudes decrease much faster with an increase in thickness because in this case, diffraction damping takes place. The analysis of dynamic reflection is essentially simplified for thick crystals.

7.4.4   Bragg Reflection from Thick Crystals

For thick samples $[|\mathrm{Im}(\mathbf{k}_{0j}\hat{\mathbf{n}})L| \gg 1]$, the problem is simplified because the waves growing in the depth of the crystal are not excited. Hence, the shape of the reflection curve can be easily analyzed. It follows from (7.22) that the character of dynamic scattering may be different because of the dependence on the incidence angle (on α) for different ranges of the parameter, and the reflection curve may consist of several disconnected ranges of strong diffraction reflection (see Fig. 3.9). In the range of incidence angles in which all four roots of the secular equation are complex, the reflection coefficient is unity for any polarization. When two roots are complex and two are real, the reflection coefficient is unity only for the incident polarization coinciding with the eigenpolarization determined by the eigensolution damping in the depth of a crystal. When the four roots are real, the reflection coefficient for any wave is less than unity. For the angles which strongly deviate from the Bragg value, the four roots are real and the reflection coefficient tends to zero as α increases.

In the symmetric Bragg case, the reflection curve for a thick crystal for a nonpolarized beam looks like that in Fig. 3.9. In this case, there may be two disconnected ranges of α, (α_1, α_2) and (α_5, α_6), in which two roots are real and two roots are complex, and one range (α_3, α_4) in which all roots are complex. Dependent on the specific magnetic structure and the reflection involved, the locations of the ranges of complete (α_3, α_4) and selective (α_1, α_2), (α_5, α_6) reflection can change. In particular, one or two ranges of selective reflection can either adjoin the total reflection band or be completely absent. Note that the polarization of the beam reflected in (α_1, α_2) and (α_5, α_6) varies with α, whereas in the band of total reflection the scattered radiation remains nonpolarized.

This consideration indicates that diffraction scattering and its polarization properties possess qualitative differences for magnetically ordered and nonmagnetic crystals. The quantitative results are illustrated below using the example which permits the analytical description of the reflection curve. The explicit expressions for $\alpha = \alpha_n$ $(n = 1, 2, \ldots, 6)$ which separate the bands with different numbers of complex roots of the secular equation can be obtained in the symmetric Bragg case in the absence of interference between nuclear and magnetic scatterings $\mathrm{Re}(F_n^* \hat{\mathbf{M}}_0 \mathbf{M}) = 0$. In this case, the secular equation, and the equation for the boundaries α_m as well, are biquadratic and the reflection curve is symmetric with respect to the middle of the total reflection band $\alpha_0 = F_n^0$. The boundaries of the bands α_m

are determined by the formulas

$$(\alpha_m - F_n^0)^2 = \mathbf{M}_0^2 + |F_n|^2 + |\mathbf{M}|^2 \pm 2\big[\mathbf{M}_0^2|F_n|^2 + |\mathbf{M}_0\mathbf{M}|^2 \tag{7.28}$$
$$+ (\operatorname{Re} F_n^*\mathbf{M}_\perp)^2 + |\operatorname{Re}\mathbf{M} \times \operatorname{Im}\mathbf{M}|^2\big]^{1/2} \quad (m = 1, 2, 5, 6)$$
$$(\alpha_{3,4} - F_n^0)^2 = \xi = |\mathbf{M}_\perp|^2 - \mathbf{M}_0^{-2}[(\operatorname{Re} F_n^*\mathbf{M}_\perp)^2 + |\operatorname{Re}\mathbf{M} \times \operatorname{Im}\mathbf{M}|^2].$$

For the existence of three disconnected ranges (humps) of strong reflection (see Fig. 3.9), it is necessary that the inequalities $\xi > 0$ and $\xi > |F_n|^2 + |M|^2 - \mathbf{M}_0^2$ be simultaneously fulfilled where ξ is defined in (7.28) ($\sqrt{\xi}$ is the half-width of the total reflection band in the case of a tree-hump curve). If $\xi > 0$ and the second inequality is substituted by equality, the bands of total and selective reflection merge. If $\xi > 0$ and the second inequality is wrong, the reflection curve does not split into separate ranges and the half-width of the total reflection band is greater than that in the case of the three-hump curve, this half-width being $(\alpha_5 - F_1^0)$, as defined by the first formula in (7.28). When $\xi = 0$, the band of total reflection disappears. At $\xi < 0$, the reflection curve is always a one-hump one and the width of the total reflection band is $(\alpha_5 - F_n^0)$, i.e., it is smaller than for the three-hump curve.

7.4.5   Reflection in the Case of Strong Birefringence

The description of dynamic scattering is simplified in the presence of strong birefringence, i.e., when the coherent amplitude of zero-angle magnetic scattering is much greater than the amplitude of the scattering for the Bragg angle $M_0/[|F_n|^2 + |M|^2]^{1/2} = m >> 1$. This can occur when scattering runs into magnetic reflexes at large Bragg angles and when the amplitudes of magnetic scattering f_ρ in (7.10) are small because the atomic magnetic form factors are small. It also takes place in samples with large magnetization. Using formulas (7.22) and (7.25) with the limit of infinitely thick crystals, we can obtain for the case in question the explicit expressions (with accuracy m^{-1}) for intensity and polarization of the reflected beam at any point of the ranges (α_1, α_2), (α_3, α_1), and (α_5, α_6) and for any polarization (polarization vector p) of the incident beam. Thus, in the bands of selective reflection, (α_1, α_2) and (α_5, α_6), the reflected beam is always completely polarized no matter what its initial polarization was, the output polarization being directed along magnetization for one band and against magnetization for another band. The corresponding polarization vector $\mathbf{p}'$ and reflection coefficient R are

$$\mathbf{p}' = \mp\hat{M}_0, \qquad R = \frac{1}{2}(1 - p_\parallel) \tag{7.29}$$

where the minus sign corresponds to (α_1, α_2) and the plus sign to (α_5, α_6), and $p_\parallel = \mathbf{p}\hat{\mathbf{M}}$. In the total reflection band (α_3, α_4) for any incident polarization, the polarization vector $\mathbf{p}'$ is determined by

$$\mathbf{p}' = p_x\hat{\mathbf{x}} - p_y\hat{\mathbf{y}} - p_\parallel\hat{\mathbf{M}}_0 \tag{7.30}$$

where $p_x = \mathbf{p}\hat{\mathbf{x}} = p_\perp\cos\varphi_\perp$; $p_y = p_\perp\sin\varphi_\perp$. Thus, when scattering in this band, the polarization degree does not change $|\mathbf{p}'| = |\mathbf{p}|$, but the polarization vector changes its orientation. The polarization vector of the reflected beam $\mathbf{p}'$ is obtained from $\mathbf{p}$ by rotating it by 180 around the axis $\hat{\mathbf{x}}$, i.e., the axis parallel to the sample surface and perpendicular to magnetization. The half-widths of the bands of selective reflection, (α_1, α_2) and (α_5, α_6), are $|F_-|$ and $|F_+|$, respectively, and the half-width of the total reflection band is $|M_\perp|$, the centers of the selective reflection bands, $\frac{1}{2}(\alpha_2 - \alpha_1)$ and $\frac{1}{2}(\alpha_6 - \alpha_5)$, are shifted from $\alpha_0 = F_n^0$ by M_0.

Formulae (7.24) and (7.25) describe the differential values (in the incident angle or α) of the intensity $I(\alpha)$ and polarization vector $\mathbf{p}'(\alpha)$ of scattered beams. Using these formulas, we can obtain the integrated intensities and polarization vectors which are usually measured in experiments. In the general case, the corresponding expressions are very cumbersome, and hence, we give these expressions for the case of strong birefringence only ($m \gg 1$). Neglecting weak reflection outside the bands (α_1, α_2), (α_3, α_7), and (α_5, α_6) and using (7.29) and (7.30), we obtain the following expressions for the integrated intensity I^i and polarization vector $\mathbf{p}^i$ of the reflected beam:

$$I_i = \int I(\alpha)d\alpha = |F_+| + |F_-| + 2|M_+| + p_\parallel(|F_+| - |F_-|)$$

$$I^i\mathbf{p}^i = \int I(\alpha)p(\alpha)d\alpha = [|F_+| - |F_-| + p_\parallel(|F_+| + |F_-| - 2|M_\perp|)]\hat{\mathbf{M}}_0$$
$$+ 2p_\perp|M_\perp|(\cos\varphi_\perp\hat{\mathbf{x}} - \sin\varphi_\perp\hat{\mathbf{y}}). \tag{7.31}$$

It follows from (7.30) that a nonpolarized incident beam ($p_\parallel = p_\perp = 0$), in the case of Bragg reflection, becomes partially polarized parallel to the quantization axis, and its degree of polarization is

$$P^i = ||F_+| - |F_-||/(|F_+| + |F_-| + 2|M_+|).$$

The reflected beam may, in particular, be completely polarized ($P^i = 1$) if there is no band of total reflection, i.e., $\alpha_4 - \alpha_3 = 2|M_+| = 0$ and only one of the widths $2|F_\pm|$ of selective reflection is nonzero. The latter issue holds no matter how strong birefringence is. It is related to magnetic structures in which one of the polarizations, directed along or against magnetization, is absent in scattering.

7.4.6 Concluding Remarks

As was mentioned above, the dynamic scattering of neutrons in magnetically ordered crystals has several qualitative peculiarities, as compared with that in nonmagnetic crystals. Note that the accuracy of current diffraction experiments [14–16,36,37,49,50] has been sufficient for observing those pecularities, such as the three-hump reflection curve and the presence of more than one (up to six) beating periods in the pendulum solution. In fact, the above formulae show that the beats periods and angular dimensions of the fine structure are of the order of the corresponding values measured in experiments with nonmagnetic crystals [34–37], since the amplitude of magnetic scattering is of the same order as the amplitudes of nuclear scattering. Observations about magnetic pendulum beats in a perfect single crystal of yttrium ferrit garnet were recently reported [51]. It is useful to keep in mind that in magnetically ordered crystals there are additional possibilities for studying the dependence of dynamic effects on experimental conditions [13,33,43]. These possibilities arise from the fact that the magnetic properties of a crystal may be influenced by external actions. For example, pendulum beats may depend on not only neutron energy and sample thickness, but also magnetization direction [36].

To conclude with, we should note that the results discussed in this chapter can be used to describe the scattering of neutrons in crystals with ordered nuclear spins. This covers not only the known case of collinear nuclear structures [20], but also noncollinear structures in which dynamic scattering is similar to that considered above.

After the first observation of magnetic dynamical diffraction [51] there was a significant break in the publication of experimental works on this item. However, quite recently several papers were published containing the results of detailed experimental studies on dynamical neutron diffraction of perfect magnetically ordered crystals [55–60]. The authors of the above-mentioned papers observed different types of pendelosung beats in α-Fe_2O_3 and $FeBO_3$ single crystals [55] (see Fig. 7.4), beginning with well-known from X-ray diffraction beats of intensities of diffracting beams with a change of sample thickness [56], up to completely new types of pendelosung beats manifesting themselves with a change of the magnetic field orientation [57] and sample temperature [58]. The last two types of pendelosung beats are connected to the changes of the coherent scattering amplitude of neutrons due to its dependence on the spin orientation and Debay–Waller factor, respectively. The interesting dependence of neutron dynamical diffraction on the orientational phase transition in α-Fe_2O_3 was observed in [59]. The results of the cited investigations show that relative to nuclear scattering, the α-Fe_2O_3 sample behaves itself as if perfect one below and above the point of Morin transition (the temperature of transition from an antiferromagnetic phase to weak ferromagnetic phase), but relative to magnetic scattering, the sample looks perfect only in the weak ferromagnetic phase (Fig. 7.5). The imperfection manifests itself in the disappearance of the pendulum beats in the antiferromagnetic phase for a magnetic reflection. The authors propose to explain the observed phenomenon by a magnetic domain structure which exists in the antiferromagnetic phase and results

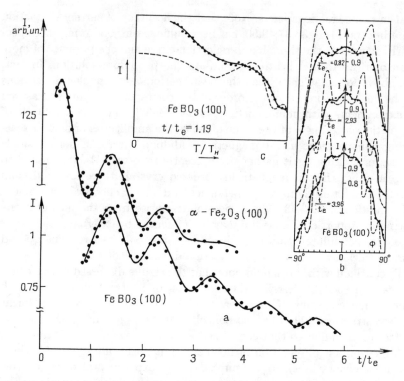

FIGURE 7.4. Experimental (points and solid line) and theoretical (dashed line) magnetic neutron scattering intensity as a function of (a) crystal thickness t, (b) magnetic field rotation angle φ, and (c) temperature [55].

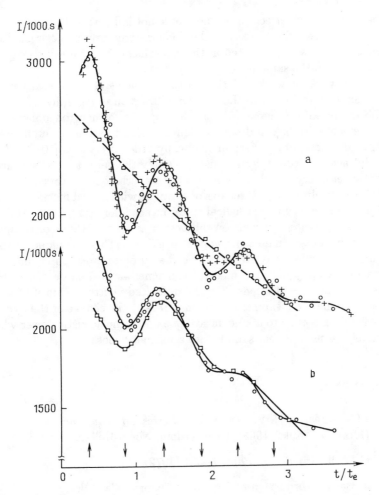

FIGURE 7.5. The intensity of magnetic [reflection (100)] and nuclear [reflection (110)] neutron scattering on the thickness of $\alpha - Fe_2O_3$ sample (expressed in the extinction length). The experiments were carried out above (○), below (□), and again above (+) the Morin temperature. The arrows show the theoretical positions of the pendulum beats' maxima and minima.

in the magnetic imperfection of the sample not influencing the crystal perfection of the sample. Note that the domain structure of the weak ferromagnetic phase was eliminated in the experiment due to a weak magnetic field applied to the sample.

The other recent works relating to the problems of dynamical neutron scattering are connected to the neutron diffraction topography and observation of chirality domains [61–66], development of neutron polarization measurements [67–69], and observation of magneto-acoustic resonance in neutron diffraction of perfect magnetically ordered crystals [70–72]. Neutron diffraction topography was used to study the perfection of (in particular magnetic) crystals, and by means of this technique, so-called chirality domains were observed, the regions of a sample with helical magnetic structure and opposite senses of helical spiral in neighboring regions [61]. Very important to dynamical neutron diffraction are the methods of complete polarization measurements developed in [67–69]. This technique, in particular, was used already to determine the absolute direction of magnetic moments in Cr_2O_3 [67]. And finally, the magneto-acoustic resonance which leads to an increase of intensity in the diffracted neutron beam if acoustic power is applied to a sample is explained [70] by the change of the dynamical scattering regime to a kinematical one under the acoustic power which effectively turns a perfect sample into a magnetic mosaic.

References

1. I.I. Gurevich, L.V. Tarasov: *Physics of Low-Energy Neutrons* (Moscow, Nauka, 1965) [(Amsterdam, North-Holland, 1968)].

2. G. Will: *Annales de Phys.* **7**, 371 (1972).

3. J. Bacon: *Diffraction of Neutrons* (Oxford, Clarendon Press, 1955).

4. Yu.A. Izyumov, R.P. Ozerov: *Magnetic Neutronography* (Moscow, Nauka, 1966) (in Russian).

5. Yu.A. Izyumov, V.E. Naish, R.P. Ozerov: *Neutronography of Magnetics* (Moscow, Atomizdat, 1981).

6. A.N. Balbashov, A.Ya. Chervonenkis, A.V. Antonov, V.E. Bachteuzov: *Izvestiya*, AN SSSR (ser. fiz.) **35**, 1243 (1972).

7. H. Ronningen, B.V. Mil', V.I. Sokolov: *Kristallografiya* **19**, 361 (1974).

8. S.M. Shil'shtein, V.I. Marukhin, V.A. Somenkov, L.A. Sysoev: *Pis'ma v ZhETF* **12**, 80 (1970).

9. S.M. Shil'shtein, V.A. Somenkov, V.P. Dokashenko: *Pis'ma v ZhETF* **13**, 301 (1970).

10. C.G. Shull: *Phys. Rev. Lett.* **21**, 1585 (1968).

11. C.G. Shull, J.A. Oberteuffer: *Phys. Rev. Lett.* **29**, 871 (1972).

12. C.G. Shull: *J. Appl. Cryst.* **6**, 257 (1973).

13. G. Smidt: *Acta Cryst.* **A39**, 679 (1983).

14. G. Smidt: *Acta Cryst.* **A39**, 682 (1983).

15. F. Schlenker: *J. de Physique* **43**, Colloq. C7, 101 (1982).

16. F. Schlenke: *J. de Physique* **43**, Colloq. C7, 107 (1982).

17. C. Stassis, J.A. Oberteuffer: *Phys. Rev. B* **10**, 5192 (1974).

18. G.P. Felcher: *Solid State Commun.* **12**, 1167 (1973).

19. A.M. Zaitseva, L.N. Korennaya, A.K. Lokhov: *Vestsi AN BSSR* (ser. fiz.-mat. nauk) **3**, 106 (1972).

20. V.G. Baryshevskii: *Pis'ma v ZhETF* **20**, 575 (1974).

21. V.A. Belyakov, R.Ch. Bokun: *FTT* **17**, 1758 (1975).

22. E. Fermi: *Recerca Sci.* **1**, 13 (1936).

23. J. Schwinger: *Phys. Rev.* **73**, 407 (1948).

24. L.L. Foldy: *Rev. Mod. Phys.* **30**, 471 (1958).

25. A. Abraham: *Principals of Nuclear Magnetism* (Oxford, Clarendon Press, 1961).

26. O. Halpern: *Phys. Rev.* **88**, 1003 (1952).

27. Yu. Kagan, A.M. Afanas'ev: *ZhETF* **49**, 1504 (1985).

28. M. Blume: *Phys. Rev.* **130**, 1670 (1963).

29. Z.G. Pinsker: *Dynamic Scattering of X-Rays in Ideal Crystals* (Moscow, Nauka, 1974) (in Russian).

30. M.A. Clark, J.M. Robson: *Canad. J. Phys.* **39**, 1 (1961).

31. H.A. Alperin, P.J. Brown, R. Nathans: *Phys. Rev. Lett.* **8**, 237 (1968).

32. J.B. Goodenough: *Magnetism and the Chemical Bond* (New York, John Wiley and Sons, 1963).

33. L. Corliss et al.: *Phys. Rev.* **122**, 1402 (1961).

34. D. Sippel et al.: *Phys. Lett.* **14**, 174 (1965).

35. C.G. Shull: *Phys. Rev.* **21**, 1585 (1968).

36. S.Sh. Shil'shtein, V.A. Somenkov: *Kristallografiya* **20**, 1096 (1975).

37. S. Kikuta, I. Ishikava, K. Kohra, S. Hoshino: *J. Phys. Soc. Japan* **39**, 471 (1975).

38. V.G. Baryshevskii, L.N. Korennaya: Preprint OIYaI, PA-4202 (Dubna, 1968).

39. J. Sivardiere: *Acta Cryst.* **A31**, 340 (1975).

40. H.H. Schmidt, P. Diemel: *Acta Cryst.* **A31**, 53, 243 (1975).

41. V.A. Belyakov, R.Ch. Bokun: *FTT* **18**, 2399 (1976).

42. V.A. Belyakov, R.Ch. Bokun: *Trudy VNIIFTRI* (Moscow, 1976, vyp. 30/60, p. 86).

43. R.Ch. Bokun: *ZhTF* **49**, 1303 (1976).

44. S.K. Mendiratta, M. Blume: *Phys. Rev. B* **14**, 144 (1976).

45. H.H. Schmidt, P. Deimel, H. Daniel: *J. Appl. Cryst.* **8**, 128 (1975).

46. G.P. Guigay, M. Schlenker: in *Neutron Interferometry*, ed. by U. Bonse and H. Rauch (Oxford University Press, 1979).

47. H.H. Schmidt, P. Deimel: *Phys. Stat. Sol. B* **73**, 87 (1976).

48. J.P. Guigay, M. Schlenker, J. Baruchel: *J. de Physique* **43**, Colloq. C7, No. 12, 107 (1982).

49. J.P. Guigay, M. Schlenker, J. Baruchel: in *Application of X-Ray Topographic Methods to Material Science*, ed. by S. Weiss and J.F. Petroff (New York, Plenum Press, 1984, p. 171).

50. J. Baruchel, M. Schlenker: in *Application of X-Ray Topographic Methods to Material Science*, ed. by S. Weissman and J.F. Petroff (New York, Plenum Press, 1984, p. 182).

51. J. Baruchel, J.P. Guigay, C. Mazuve-Espejoc, M. Schlenker, J. Schweizer: *J. de Physique* **43**, Colloq. C7, No. 12, 101 (1982).

52. A.G. Gukasov, V.A. Ruban: *FTT* **17**, 2967 (1975).

53. J. Baruchel, C. Patterson, J.P. Guigay: *Acta Cryst.* **A42**, 47 (1986).

54. M. Schlenker, J. Baruchel: *Physica* **137B**, 309 (1986).

55. V.V. Kvardakov, V.A. Somenkov, S.Sh. Shilstein: *Material Science* **27/28**, 221 (1988).

56. M.V. Zalepukhin, V.V. Kvardakov, V.A. Somenkov, S.Sh. Shilstein: *ZhETF* **95**, 1530 (1989).

57. V.V. Kvardakov, V.A. Somenkov: *Kristallografiya* **35**, 1051 (1990).

58. V.V. Kvardakov, V.A. Somenkov, S.Sh. Shilstein: *Fiz. Tverd. Tela* **32**, 1879 (1990).

59. V.V. Kvardakov, V.A. Somenkov, S.Sh. Shilstein: *Fiz. Tverd. Tela* **32**, 2149 (1990).

60. V.V. Kvardakov, V.A. Somenkov: *Acta Cryst.* (in print).

61. J. Baruchel, S.B. Palmer, M. Schlenker: *J. Physique* **42**, 1279 (1981).

62. C. Patterson, S.B. Palmer, J. Baruchel, and Y. Ishikawa, *Solid State Comm.* **55**, 81 (1981).

63. M. Schlenker, J. Baruchel: *Physica* **137B**, 309 (1986).

64. J. Baruchel, S.B. Palmer, C. Patterson: *J. Physique* **49**, Colloq. C8, C8-1893 (1988).

65. J. Baruchel, M. Schlenker: *Physica* **156/157B**, 666 (1989).

66. D. Sillou, J. Baruchel: *Physica* **156/157B**, 581 (1989).

67. F. Tasset, P.J. Brown, J.B. Forsyth: *J. Appl. Phys.* **63**, 3606 (1988).

68. F. Tasset: *Physica* **156/157B**, 627 (1989).

69. F. Tasset: *Neutron Diffraction Newsletter* (Institute Laue-Langevin, Grenoble, Autumn 88, p. 7).

70. V.V. Kvardakov, V.A. Somenkov: *Fiz. Tverd. Tela* **31**, 235 (1989).

71. V.V. Kvardakov, V.A. Somenkov, A.B. Tiugin: *Pis'ma v ZhETF* **48**, 396 (1990).

72. V.V. Kvardakov, V.A. Somenkov: *Pis'ma v ZhETF* **52**, 901 (1990).

8

Polarization Phenomena in X-Ray Optics

X-ray optics has undergone a veritable renaissance in recent years because of improvement in instrumentation and methods of measurement, which has been partly but not entirely due to the availability of synchrotron sources of radiation. One of the interesting phenomena in the development of X-ray optics has been the study of polarization effects. As far back as 1906, the polarization of X-rays scattered through 90° was used to demonstrate their electromagnetic character [1]. Classical phenomena, such as birefringence, dichroism, and rotation of polarization [2–6], have been observed in X-ray optics, and there has been discussion on the possibility of producing quarter-wave plates capable of transforming linear into circular polarization and vice versa [3,7–9].

It is important to emphasize that, whereas in ordinary optics dealing with visible radiation, the wavelength is much greater than interatomic separation, in X-ray optics, the phenomenon of Bragg diffraction has a significant influence on the optical characteristics of crystal. For example, the contribution of Bragg diffraction to dichroism, birefringence, and the transformation of polarization can sometimes be the dominant factor. In particular, it is interesting to note that diffractive birefringence, essentially a spatial dispersion effect, is observed not only in low-symmetry crystals, but even in cubic crystals. Diffractive polarization phenomena are already being used to investigate imperfection in crystals [10–14], where the complete description of such phenomena necessitates the use of the transfer equation for the polarization tensor of X-ray beams [15] by analogy with the tensors used in the theory of radiation transfer in random media [16–19].

It is important to note that polarization phenomena due to the anisotrophy of the X-ray susceptibility of crystals are relatively weak and mostly observed near the absorption edges, i.e., in regions in which effects associated with the chemical binding of electrons in atoms are significant. Very recently, it has been shown that the anisotropy of X-ray susceptibility leads to a qualitatively new effect, namely, the appearance of additional diffraction maxima with unusual polarization properties, containing information about the structure of crystal and chemical bonds of atoms in crystals [20–23]. Particularly interesting and promising is the magnetic scattering of

X-rays [24–28], which is due to the weak dependence of X-ray scattering amplitude on the magnetic moment of the atom and will be discussed in detail in the following chapter. The reason to mention it here is the unusual polarization properties of X-ray magnetic scattering.

This chapter consist of two sections devoted to the theory of polarization effects and a section devoted to applications, although many of the latter are discussed in the first two sections as well. In the chapter are analyzed and generalized an extensive range of factual material scattered among numerous publications (see also Ref. [4]). Special attention is paid to the possible applications of X-ray polarization methods to the study of the structure and properties of solids. There is no doubt that this subject will advance in the next few years, especially with the advent of sources of synchrotron radiation.

8.1   Polarization Phenomena in Diffraction

X-ray diffraction by crystals is one of the basic methods of studying the crystalline structure of matter and is widely used in X-ray optics. For example, diffraction monocromators and spectrometers are used to produce, transform, and analyze X-ray beams. In Sect. 8.1 the change in the polarization of primary and diffraction beams during diffraction will be examined in detail. The topics of current interest, e.g., diffractive birefringence and dichroism, and change of polarization state on diffraction in perfect and imperfect crystals, are examined first. In contrast to Sect. 8.2, a traditional approach will be adopted and we will assume that X-ray susceptibilities are isotropic in order to isolate polarization phenomena that originate in pure Bragg diffraction.

8.1.1   Diffraction in Perfect Crystals

The diffraction of X-ray by perfect crystals has been discussed in an enormous number of papers (see, e.g., Refs. [29–32]), so it will be sufficient to confine ourselves to a brief description of the basic results, with particular reference to the polarization characteristics of diffracted beams. The dielectric tensor for X-rays in accordance with (2.22) is conveniently presented in the form

$$\hat{\epsilon}(\mathbf{r}) = 1 + \hat{\chi}(\mathbf{r}) = 1 + \sum \hat{\chi}_\tau \exp(i\boldsymbol{\tau}\mathbf{r}) \qquad (8.1)$$

where $\hat{\chi}(\mathbf{r})$ is the susceptibility of the crystal in the X-ray range. If the susceptibility is isotropic, all the $\hat{\chi}_\tau$ are proportional to a unit tensor: $(\chi_\tau)_{ik} \equiv \chi_\tau \delta_{ik}$ where

$$\chi_\tau = -\frac{r_e \lambda^2}{\pi V} F_\tau \qquad (8.2)$$

and F_τ is the dimensionless structure amplitude [compare with (3.22) where Mössbauer amplitude is connected to susceptibility]. In (8.2), $r_e = e^2/mc^2$ is the classical radius of the electron, λ is the wavelength, and V is the volume of the unit cell. For $\lambda \sim 1\text{Å}$, $|\hat{\chi}_\tau|$ is small and usually does not exceed 10^{-5}.

Below the polarization properties of X-ray diffraction will be described in the two-wave approximation in dynamic and kinematic theory according to the approach presented in Chapt. 2 and the chapters that follow it. The characteristic feature of the two-wave approximation for X-rays is the independent diffraction scattering of σ- and π-polarized waves. The characteristic parameter that can be used to distinguish thick from thin perfect crystals is the primary extinction length

$$L_e^I = \frac{\lambda}{|\chi_\tau|}.$$

The kinematic theory is valid when the distances traversed in the crystal by both the incident and diffracted waves are much smaller than the primary extinction length ($L_e^I \geq 10\,\mu$m for $\lambda \sim 1\text{Å}$). The polarization properties of diffraction are found to be particularly simple in the kinematic theory, in which the scattering amplitudes for σ- and π-polarized waves differ by the polarization factor $\cos(2\theta)$ where 2θ is the scattering angle [see also Sect. (3.3.1)] and

$$E_\sigma^d = AF_\tau E_\sigma^i$$
$$E_\pi^d = AF_\tau E_\pi^i \cos 2\theta \qquad (8.3)$$

in which $E_{\sigma,\pi}^i$ and $E_{\sigma,\pi}^d$ are the components of the incident and diffracted waves, respectively, and the factor A depends on the diffraction geometry, wavelength, and deviation of the angle of incidence from the Bragg angle θ_B. The significant point is that the factor A does not affect the polarization properties of kinematic diffraction (the specific form of A may be found from the dynamic theory in the limit of a very small crystal [29]). Since the angle θ changes by a very small amount $\Delta\theta \leq 10^{-4}$ in the diffraction region, the polarization factor in (8.3) can be replaced by $\cos 2\theta_B$. The expressions given by (8.3) are conveniently written in the vector form $\mathbf{E}^d = \hat{R}_k \mathbf{E}^i$ where the scattering matrix is given by

$$\hat{R}_k = AF_\tau \begin{pmatrix} 1 & 0 \\ 0 & \cos 2\theta_B \end{pmatrix} = AF_\tau \hat{K}. \qquad (8.4)$$

The particular feature of the kinematic case is that the polarization of the diffracted wave is determined by the ratio E_π^d/E_σ^d, which is constant throughout the diffraction region. In particular, σ- and π-polarized incident beams produce σ- and π-polarized diffracted beams: Linearly polarized incident beams produce linearly polarized diffracted beams independently of the angle of incidence; the sign of elliptic polarization changes on diffraction if $\cos 2\theta < 0$, but does not change when $\cos 2\theta > 0$; for scattering through $90°$, the diffracted wave is σ-polarized.

The polarization properties are much more complicated in the dynamic rather than kinematic theory. The physical reason for this is that the width of the diffraction region is different for σ- and π-polarization, and moreover, the relative phase of the σ- and π-components of the diffracted waves undergoes a change in the diffraction region. It is well known [29–32] that four Bloch waves (two σ-polarized and two π-polarized) are produced in a perfect crystal, and each has its own wave vector. The characteristic difference between these wave vectors is of the order of $1/L_e^I$, and interference between the Bloch waves gives rise to nontrivial polarization properties. The solution of the dynamic diffraction problem for a plane-parallel plate leads to the following expressions for the amplitudes of diffracted and transmitted waves:

$$E_\sigma^d = R_{\sigma\sigma} E_\sigma^i, \qquad E_\pi^d = R_{\pi\pi} E_\pi^i \qquad (8.5)$$

$$E_\sigma^t = T_{\sigma\sigma} E_\sigma^i, \qquad E_\pi^t = T_{\pi\pi} E_\pi^i. \qquad (8.6)$$

The coefficients $R_{\gamma\gamma}$ and $T_{\gamma\gamma}$ ($\gamma = \sigma, \pi$) in (8.5, 8.6) are given by the following expressions. In the Bragg case (see Fig. 2.5a):

$$R_{\gamma\gamma} = \chi_\tau C_\gamma (\alpha + i\Delta_\gamma \mathrm{ctgl}_\gamma)^{-1}$$

$$T_{\gamma\gamma} = (\cos l_\gamma - i\alpha \Delta_\gamma^{-1} \sin l_\gamma)^{-1} \exp\left[i\mathbf{q}_0^2 L(\chi_0 - ab)(2\mathbf{q}_0 s)^{-1}\right]. \qquad (8.7)$$

In the Laue case (see Fig. 2.5b):

$$R_{\gamma\gamma} = \chi_\tau C_\gamma \Delta_\gamma^{-1} \sin l_\gamma \, \exp\left[i\mathbf{q}_0^2 L(\chi_0 - \alpha b)(2\mathbf{q}_0 s)^{-1}\right]$$

$$T_{\gamma\gamma} = (\cos l_\gamma + i\alpha \Delta_\gamma^{-1} \sin l_\gamma)^{-1} \exp\left[i\mathbf{q}_0^2 L(\chi_0 - \alpha b)(2\mathbf{q}_0 s)^{-1}\right] \qquad (8.8)$$

where

$$a = \frac{\tau^2 + 2\mathbf{q}_0 \tau}{2\mathbf{q}_0^2} + \frac{\chi_0(1 - b)}{2b}$$

$$\Delta_\gamma = \left(\alpha^2 + \frac{C_\gamma^2 \mathbf{q}_\tau \mathbf{q}_{-\tau}}{b}\right)^{1/2}$$

$$l_\gamma = \frac{\Delta_\gamma \mathbf{q}_0^2 L}{2\mathbf{q}_\tau s}, \qquad b = \frac{\mathbf{q}_0 s}{\mathbf{q}_\tau s},$$

$C_\sigma = 1$, $C_\pi = \cos 2\theta_B$. The parameter α [see also (2.30)] can be expressed in terms of the deviation of the angle of incidence from the Bragg angle:

$$\alpha = (\theta_B - \theta) \sin 2\theta_B + \frac{\chi_0(1 - b)}{2b}. \qquad (8.9)$$

Formulas (8.5–8.9) can be used to perform a full analysis of polarization phenomena in perfect crystal. In particular, it follows from them that the amplitude and relative phase of diffracted waves with σ- and π-polarizations undergo variations within the diffraction region (Figs. 8.1a and 8.1b). Usually, the reflection coefficient $|R_{\gamma\gamma}|^2$ for σ-polarization is greater than that for π-polarization, but the reverse situation is also possible because of the so-called pendellosung that arises as a consequence of interference between the two Bloch functions associated with each polarization state.

The amplitude and phase of the waves transmitted by the crystal (see Figs. 8.1b and 8.2b) also undergo changes within the diffraction region, which can be described as the manifestation of an effective diffraction birefringence and dichroism (see Sect. 8.1.3). It should be emphasized that this dichroism is related to not only the difference between the absorption of radiation of different polarizations (as in ordinary optics) but also the difference between the diffracted intensities of σ- and π-polarized waves. As far as true absorption is concerned, this can be significant for dichroism under the conditions of anomalaous absorption (Borrmann effect), which is also used to produce polarized beams (Sect. 8.3.2).

8.1.2   Polarization Tensor and Integrated Polarization Parameters

In practice, one frequently has to deal with the description of partially polarized beams. X-ray sources usually produce either partially polarized or completely unpolarized radiation. It will be seen below that a change in the polarization state occurs when the polarization parameters are averaged over the diffraction region, and also in the case of diffraction by imperfect crystals. The complete description of the polarization properties and intensities of such beams is accomplished with the aid of the polarization tensor $\hat{\mathbf{J}}$ [15–19], defined by

$$J_{\beta\gamma} = \overline{E_\beta E_\gamma^*} \tag{8.10}$$

where the bar indicates averaging (with respect to time in classical language [17] or over the photon ensemble in quantum-mechanical language [33]). In a beam with polarization tensor $\hat{\mathbf{J}}$, the intensity of a component with arbitrary polarization presented by the unit vector $\mathbf{e}$ is given by the expression $(\mathbf{e}^*\hat{\mathbf{J}}\mathbf{e})$. Throughout the discussion presented below, the σ- and π-polarization vectors will be used as polarization basis vectors. The physical meaning of the elements of the tensor $\hat{\mathbf{J}}$ on this basis is as follows: $J_{\sigma\sigma}$ and $J_{\pi\pi}$ are the intensities of the σ- and π-components, respectively, 2Re $J_{\sigma\pi}$ is the intensity difference between the components that are linearly polarized at $\pm45°$ to σ and 2Im $J_{\sigma\pi}$ is the intensity ratio of components with right and left circular polarization (a total of four independent elements). The polarization tensor is convenient because when incoherent beams are combined, their polarization tensors are added. The expressions relating

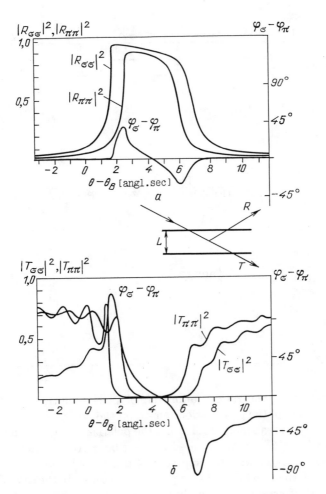

FIGURE 8.1. (a) Reflection coefficients $|R_{\sigma\sigma}|^2$, $|R_{\pi\pi}|^2$ and the difference $\varphi_\sigma - \varphi_\pi$ between diffraction corrections to the σ- and π-polarized waves in the symmetric Bragg case; 200 reflection in Si, semiinfinite crystal, CuK$_\alpha$ radiation. (b) Transmission coefficients $|T_{\sigma\sigma}|^2$, $|T_{\pi\pi}|^2$ and the difference $\varphi_\sigma - \varphi_\pi$ between diffraction corrections to the phase of σ- and π-polarized transmitted waves in the symmetric Bragg case; 220 reflection in Si, $L = 10\,\mu$m, CuK$_\alpha$ radiation; dashed line corresponds to the transmission coefficient in the absence of diffraction. Insert shows the diffraction geometry.

TABLE 8.1. Polarization parameters of beams ($z = E_\pi/E_\sigma$).

Parameter	Polarization	
	Total	Partial
1. Intensity I	$\lvert E_\sigma\rvert^2 + \lvert E_\pi\rvert^2$	$J_{\sigma\sigma} + J_{\pi\pi}$
2. Degree of polarization P	1	$\dfrac{[(J_{\sigma\sigma}-J_{\pi\pi})^2+4\lvert J_{\sigma\pi}\rvert^2]^{1/2}}{J_{\sigma\sigma}+J_{\pi\pi}}$
3. Ratio of axes of polarization ellipse b_e	$\dfrac{-2\,\mathrm{Im}\,z}{1+\lvert z\rvert^2+[(1-\lvert z\rvert^2)^2+4(\mathrm{Re}\,z)^2]^{1/2}}$	$\dfrac{2\,\mathrm{Im}\,J_{\sigma\pi}}{PI+[(J_{\sigma\sigma}-J_{\pi\pi})^2+4(\mathrm{Re}\,J_{\sigma\pi})^2]^{1/2}}$
4. Angle of rotation ψ of the semi-major axis of the ellipse relative to σ-polarization	$\tfrac{1}{2}\mathrm{arctg}\dfrac{2\,\mathrm{Re}\,z}{1-\lvert z\rvert^2}$	$\tfrac{1}{2}\mathrm{arctg}\dfrac{2\,\mathrm{Re}\,J_{\sigma\pi}}{J_{\sigma\sigma}-J_{\pi\pi}}$

the intensity and polarization parameters of a beam, on the one hand, and the components of the polarization tensor, on the other, are given in Table 8.1. The Stokes parameters [16–18] and the polarization density matrix [33] (see also Chapt. 3) are widely used to describe partially polarized beams and, of course, agree with the approach of this section.

The transformation for the beam polarization tensor in the case of diffraction by perfect crystals can be obtained from the above formulas for the field, as given by (8.3–8.8). By constructing the quadratic combinations (8.10) for the diffracted and transmitted wave fields, one obtains

$$\hat{J}^d = \hat{R}\hat{J}^i\hat{R}^*$$
$$\hat{J}^t = \hat{T}\hat{J}^i\hat{T}^* \tag{8.11}$$

where the diagonal components of matrices $\hat{R}$ and $\hat{T}$ are given by (8.7, 8.8), whereas the off-diagonal components of the matrices are all zero. The diagonal form of the matrices $\hat{R}$ and $\hat{T}$ leads to the fact that each of the components of the polarization tensors $\hat{\mathbf{J}}^d$ and $\hat{\mathbf{J}}^t$ depends only on the corresponding components of $\hat{\mathbf{J}}^i$:

$$J^d_{\beta\gamma} = R_{\beta\beta}R^*_{\gamma\gamma}J^i_{\beta\gamma}$$
$$J^t_{\beta\gamma} = T_{\beta\beta}T^*_{\gamma\gamma}J^i_{\beta\gamma} \tag{8.12}$$

where $\beta = \sigma, \pi$, $\gamma = \sigma, \pi$ (no summation over repeated indices!)

Formulas (8.11, 8.12) readily explain the change in polarization and in the degree of polarization of the beam on diffraction. For example, one can use the expressions listed in Table 8.1 to show that if $J_{\sigma\sigma}^i / J_{\pi\pi}^i \leq |R_{\pi\pi}|^2 / |R_{\sigma\sigma}|^2$, then for any incident-beam polarization, the degree of polarization of the diffracted beam does not increase, i.e., $P^d \leq P^i$. The change in the intensity and polarization of the beam on successive reflection from a number of crystals is also conveniently described by the successive application of (8.11, 8.12). This procedure becomes nontrivial if the crystals are rotated relative to one another so that the σ- and π-polarizations for them are not the same. It is necessary to transform the polarization tensor from one basis (σ, π) to another (σ', π'), where the latter is rotated relative to the former by ψ':

$$\hat{J}' = \hat{R}_{\psi'} \hat{J} \hat{R}_{\psi'}^{-1} \tag{8.13}$$

in which the rotation matrix $\hat{R}_{\psi'}$ is given by

$$\hat{R}_{\psi'} = \begin{pmatrix} \cos\psi' & \sin\psi' \\ -\sin\psi' & \cos\psi' \end{pmatrix}. \tag{8.14}$$

The polarization tensor is also convenient in finding the integrated (over the diffraction region) polarization parameters. These are given by the integrated polarization tensor:

$$\overline{J}_{\beta\gamma}^d = \int J_{\beta\gamma}^i R_{\beta\beta} R_{\gamma\gamma}^* \, d\theta$$

$$\overline{J}_{\beta\gamma}^t = \int J_{\beta\gamma}^i T_{\beta\beta} T_{\gamma\gamma}^* \, d\theta. \tag{8.15}$$

It follows from (8.15) that diffraction by a thick perfect crystal, even for a completely polarized incident beam, gives rise to the degree of depolarization, provided only that the incident beam is neither σ- nor π-polarized. Physically, this is related to the fact that the polarization of the diffracted and transmitted waves undergoes a change in the different region (see Figs. 8.1 and 8.2), and the incoherent superpositions of beams with different polarizations result in partially polarized beams.

The integrated reflection coefficients and the integrated polarization parameters (Fig. 8.3) are oscillating functions of thickness that are related to the well-known pendelosung solutions for waves diffracted in crystal. Note that for certain particular thicknesses corresponding to Re $\overline{R_{\sigma\sigma} R_{\pi\pi}^*} = 0$, the diffracted beam is circularly polarized if the incident beam is linearly polarized at the angle $\psi^i = \arctan\left(|\overline{R_{\sigma\sigma}}|^2 / |\overline{R_{\pi\pi}}|^2\right)^{0.5}$ to the direction of σ-polarization. However, the degree of circular polarization is then shown by Fig. 8.3 and Table 8.1 to be relatively small, $P \approx 0.25$ for $L \approx 33\,\mu\text{m}$.

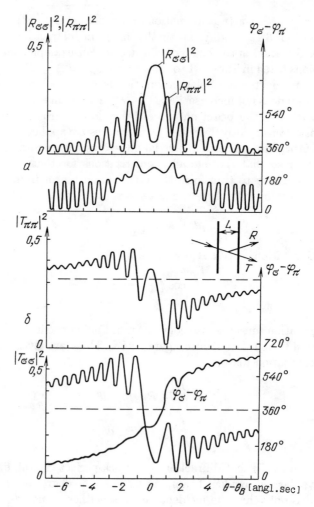

FIGURE 8.2. Same as Fig. 8.1 but for the symmetric Laue case. $L = 72\,\mu$m.

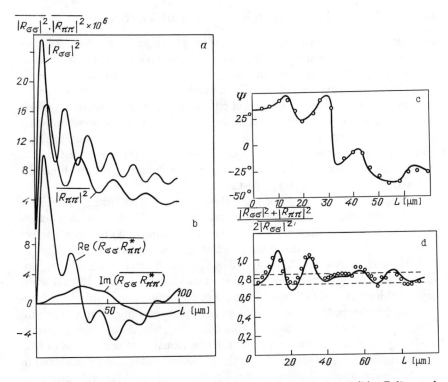

FIGURE 8.3. Integral reflection coefficients for (a) diagonal and (b) off-diagonal components of polarization tensor as function of crystal thickness. (c) and (d), taken from Ref. [2], show the corresponding polarization parameters, i.e., (c) rotation of the plane of polarization and (d) ratio of integral reflection coefficients for σ- and π-polarizations. Solid curves are calculated from dynamical theory, points are experimental values for linear polarization at $45°$ to the σ and π vectors; dashed lines correspond to $(1 + \cos 2\theta_B)/2$ and $(1 + \cos^2 2\theta_B)/2$.

8.1.3   Diffractive Birefringence and Dichroism

A monochromatic wave falling on a crystal produces in the diffraction region two σ- and π-polarized Bloch waves, each with its own wave vector. As they propagate in the crystal, these waves interfere with one another, so that the polarization parameters of the waves transmitted and diffracted by the crystal are very complicated functions of the crystal thickness, which has often been observed experimentally (Refs. [2,3,36]). However, in many cases, only one Bloch component is significant for each of the polarizations, and this simpler situation is analogous to ordinary optics, except that the X-ray birefringence is diffractive in origin. For example, one such case is diffraction by thick absorbing crystals [29].

Two similar cases are discussed below—namely, "off-Bragg" diffractive birefringence and birefringence by mosaic crystals. The former case occurs for sufficiently large deviations from the Bragg angle, so one can neglect the intensity of the diffracted wave, which decreases relatively rapidly (as $|\Delta\theta|^{-2}$) with increasing $|\Delta\theta|$ [see (8.5–8.8)]. It will be clear below that, under these conditions, birefringence decreases slowly (as $|\Delta\theta|^{-1}$), and when crystal thickness is significant enough, a considerable phase difference is established between the σ- and π-polarized waves emerging from the crystal (Fig. 8.2b). Moreover, because of this nth slow reduction of birefringence, off-Bragg birefringence may include the contribution of many reflections.

The diffraction correction Δn to the refractive index can be obtained in the two-wave approximation either directly from the Maxwell equations, using perturbation theory in which it is assumed that the amplitude of the diffracted wave is much smaller than that of the incident wave [(8.7–8.9)], or from (8.6) which yields the amplitude of the transmitted wave in the limit as $|\Delta\theta| \to \infty$. The final expression for Δn that is valid in both Bragg and Laue cases is

$$\Delta n_\gamma = -\frac{C_\gamma^2 \chi_\tau \chi_{-\boldsymbol{\tau}}}{4\Delta\theta \sin 2\theta_B}, \qquad (\gamma = \sigma, \pi). \tag{8.16}$$

The refractive index difference δn is

$$\delta n = \Delta n_\sigma - \Delta n_\pi = -\frac{\chi_\tau \chi_{-\boldsymbol{\tau}} \sin 2\theta_B}{4\Delta\theta}. \tag{8.17}$$

The range of validity for this expression is defined by the inequality $|\Delta\theta| \gg |\chi_\tau| / \sin 2\theta_B$.

The real part of δn determines birefringence and the imaginary part the dichroism (i.e., different absorption) of σ- and π-polarized rays in crystal.

It should be emphasized that the dichroism that follows from (8.17) is displayed by absorbing crystals only, i.e., if Im $\chi_\tau \chi_{-\boldsymbol{\tau}} \neq 0$. Physically, it is related to the difference between the Borrmann effects for σ- and π-polarized waves [29–31]. Note that the usual conditions in an X-ray region are described by $|\mathrm{Re}\,\delta n| \gg |\mathrm{Im}\,\delta n|$ (provided we are not too close to the absorption edge). Only the relative phase of the σ- and π-polarized waves is then found to change during propagation.

In the many-wave case, the solution of the Maxwell equations can also be obtained by perturbation theory [5,7,9]. The solution is sought in the form of a Bloch wave (in this case, it is more convenient to use the induction $\mathbf{D}$):

$$\mathbf{D}(\mathbf{r}) = \left(\mathbf{D}_0 + \sum_{\tau \neq 0} \mathbf{D}_\tau e^{i\mathbf{T}\mathbf{r}}\right) e^{i\mathbf{k}_0\mathbf{r}}. \tag{8.18}$$

Substituting this into Maxwell equations, one obtains, as usual [31] (see also Sect. 4.1.4), the set of equations multiwave dynamic theory of diffraction. Since it cannot be known in advance which polarization will be intrinsic, let us write this set in vector form in the case of direct wave $\mathbf{D}_0$:

$$a) \quad \left(1 - \frac{q_0^2}{\mathbf{K}_0^2} - \chi_0\right) \mathbf{D}_0 = \sum_{\tau \neq 0} \chi_{-\tau} \left[\mathbf{D}_\tau - \frac{\mathbf{K}_0(\mathbf{D}_\tau \mathbf{K}_0)}{\mathbf{K}_0^2}\right]$$

$$b) \quad \left(1 - \frac{q_0^2}{\mathbf{K}_\tau^2} - \chi_0\right) \mathbf{D}_\tau = \sum_{G \neq \tau} \chi_{\tau - G} \left[\mathbf{D}_G - \frac{\mathbf{K}_\tau(\mathbf{D}_G \mathbf{K}_\tau)}{\mathbf{K}_\tau^2}\right] \tag{8.19}$$

($\tau \neq 0$, $\mathbf{K}_\tau = \mathbf{K}_0 + \tau$). Assuming that all the $\mathbf{D}_\tau$ are small and retaining only $\mathbf{D}_0$ on the right-hand side of (8.19), one finds that

$$\mathbf{D}_\tau = \chi_\tau \left[\mathbf{D}_0 - \frac{\mathbf{K}_\tau(\mathbf{D}_0 \mathbf{K}_\tau)}{\mathbf{K}_\tau^2}\right] \left(1 - \frac{q_0^2}{\mathbf{K}_\tau^2} - \chi_0\right)^{-1}. \tag{8.20}$$

Substituting $\mathbf{D}_\tau$ in (8.19a), one obtains the following equation for the wave vector amplitudes in the crystal:

$$\left(1 - \frac{q_0^2}{\mathbf{K}_0^2} - \chi_0 + 2\hat{\delta}^d\right) \mathbf{D}_0 = 0 \tag{8.21}$$

where the tensor $\hat{\delta}^d$ is given by (see also [5,9])

$$\hat{\delta}_{ik}^d = \frac{1}{4} \sum_{\tau \neq 0} \frac{\chi_\tau \chi_{-\tau}}{\alpha_\tau q_0^2} (\mathbf{K}_\tau^2 \delta_{ik} - \tau_i' \tau_k'), \tag{8.22}$$

and the quantity $\alpha_\tau = [\tau^2 + 2(\mathbf{k}_0\tau)]/2\mathbf{k}_0^2$ determines the deviation of the direction of propagation of the wave $\mathbf{D}_0$ from the Bragg direction for the reflection where $\tau' = \tau - \mathbf{k}_0(\mathbf{k}_0(\tau)/\mathbf{k}_0^2$.

By diagonalizing the tensor $\hat{\delta}^d$, one can find the eigenvectors $\mathbf{e}_m$ ($m = 1, 2$) and the eigenvalues $\hat{\delta}^d$ of (8.21), which determine the polarization and wave vector of eigenwaves in the crystal, respectively. The refractive index difference between the eigenwaves is given by $\delta n = \delta_1^d - \delta_2^d$.

Note that, for absorbing crystal, the eigenpolarization can be elliptical and nonorthogonal, i.e., $(\mathbf{e}_1^* \mathbf{e}_2) \neq 0$, and the quantities δ_m^d are complex.

It also follows from (8.12) that, in this particular approximation, there is neither birefringence nor dichroism for a beam propagating along threefold or higher-order symmetry axes.

As an example, let us consider the propagation of X-rays at a right angle to the (110) plane in a cubic crystal. A consideration of symmetry then shows immediately that the eigenpolarizations are linear and one of them ($\mathbf{e}_1$) is parallel to the [001] direction, whereas the other ($\mathbf{e}_2$) is parallel to the [110] direction. Once one knows $\mathbf{e}_m$, δ_m^d can be readily determined since

$$\delta_m^d = (\mathbf{e}_m^* \hat{\delta}^d \, \mathbf{e}_m) \tag{8.23}$$

($\mathbf{e}_m$ are unit vectors). For the present case, (8.22) then leads to the following expression:

$$\delta n = \sum_{h^2+k^2+l^2\neq 0} \frac{[(h-k)^2 - 2l^2]\chi_{hkl}\chi_{\bar{h}\bar{k}\bar{l}}}{4[h^2+k^2+l^2-(\sqrt{2}a/\lambda)(h+k)]} \tag{8.24}$$

where a is the size of the unit cell and h, k, l are the Miller indices.

Let us now analyze the application of the general relations given by (8.22, 8.24) and the resulting estimate of the effects for the special case of propagation of $CuK_{\alpha 1}$ radiation in silicon. Since in this case, $\sqrt{2}a/\lambda = 4.9855 \approx 5$, the main contribution to δn is provided by the 620 and 260 reflections, since the denominator in (8.24) is very small for these reflections. When these reflections alone are taken into account, one has $\delta n = (1.9 + i0.24) \times 10^{-9}$, whereas the inclusion of remaining reflections yields $\delta n = (2.5 + i0.27) \times 10^{-9}$, i.e., distant reflections provide a relatively appreciable contribution. When the crystal thickness is $L \approx 0.04$ cm, the rotation $\Delta\Psi$ of the plane of polarization and the ratio b of the axes of the polarization ellipse are found to be measurable by existing experimental techniques [3,5] (e.g., $\Delta\Psi = 3.8'$ and $b_e = 0.02$ for an incident wave polarized at 45° in the σ- and π-directions).

An experimental study [37] of the above case of the propagation of $CuK\alpha_1$ radiation in silicon showed that both birefringence and dichroism were substantially greater than the theoretical values indicated above. This discrepancy may be due to the fact that appreciable 260 and 620 reflections were produced during the course of collimation of the beam employed in [37]. Diffraction and the Borrmann effect associated with these reflections could have given rise to a substantial increase in observed dichroism and birefringence, as compared with the theoretical predictions. These suggestions are consistent with the fact that, when measurements were performed on the same crystal but with special care to exclude the 260 and 620 reflections [5], this gave rise to significantly smaller effects than those reported in [37] (see also [38]).

It therefore may be concluded that crystals exhibit appreciable birefringence and dichroism in the X-ray region, even in directions well away

from those of strong diffraction scattering. In contrast to the case of strong
diffraction scattering, the propagation of X-rays in such cases can be described by analogy with the ordinary optics of anisotropic media, provided
the diffraction terms discussed above are taken into account in the refractive index. Specific estimates (Sect. 8.3.3) show that off-Bragg diffractive
birefringence can be used as the basis for the transformation of linearly polarized waves into circularly polarized waves and vice versa. On the other
hand, diffractive birefringence and dichroism may impede the observation
of the true anisotropy of X-ray susceptibility. Diffraction correction to the
refractive index can also be very significant in precision measurements of
the refractive index (using X-ray interferometers, etc.), in which a relative
precision of order of 10^{-9}–10^{-10} has already been attained.

8.1.4   Diffraction in Imperfect Crystals

The imperfection of a crystal, i.e., departure from the regular crystal lattice, ensures that waves diffracted at different points in the crystal acquire
an additional phase difference. When these irregularities are random, the
waves become partially incoherent because of the randomness of the phase
difference. It is clear that this incoherence will affect, in the first instance,
the polarization properties of diffracted waves and, in particular, can give
rise to changes in the polarization state. There is, at present, no theory that
can describe diffraction by a crystal with arbitrary imperfections. Existing
theoretical approaches apply to slightly or completely imperfect crystal. In
this section, we examine only one of the simplest models, i.e., that of mosaic crystal, which will nevertheless exhibit many polarization effects that
are specific to imperfect crystals. In this model, crystal imperfection is assumed to be so strong that waves diffracted by different blocks of the mosaic
are completely incoherent, and the kinematical approximation presented in
general form in Chapt. 2 is applied to each such block [29,39–41].

Diffraction by mosaic crystal is traditionally discussed on the basis of
Darwin transfer equations for the intensities of the σ- and π-polarized components [39–41]. For beams of arbitrary polarization, the equations for the
intensities must be replaced with the transfer equation for the polarization tensor $\hat{J}$ defined above. It will be seen below that this generalization
is equivalent to adding to the Darwin equations, which describe the diagonal components of polarization tensors, a further set of equations for
the off-diagonal components. Their general solution provides a complete
description of the polarization properties of beams diffracted by mosaic
crystals [15].

The derivation of the transfer equation for the polarization tensor can
be illustrated by the following simple argument. Consider the evolution
of the polarization tensor of the direct wave $\hat{J}_0$ as its propagates in the
crystal, i.e., let us find the derivative $\partial \hat{J}^0 / \partial s_0$ (s_0 is the distance along
the direction of propagation of the direct wave). First we must take into

account absorption that is present even outside the diffraction region, whose contribution to the derivative has the form $\mu \hat{J}^0$ where μ is the absorption coefficient. Diffraction of the diffracted wave back into the direct wave is described in precisely the same way as the diffraction of the direct wave [see (8.3, 8.4, 8.11)] and leads to a contribution of the form $\sigma_{0\tau}\hat{K}\hat{J}^\tau\hat{K}$ to the required derivative where $\hat{J}^\tau$ is the polarization tensor of the diffracted wave and $\sigma_{0\tau}$ is the mean Bragg scattering cross section per unit volume of the crystal for a given angular deviation $\Delta\theta = \theta - \theta_B$ from the Bragg angle:

$$\sigma_{0\tau} = \frac{\pi^2}{\lambda \sin 2\theta_B}|\chi_\tau|^2 W(\Delta\theta) \equiv QW(\Delta\theta). \qquad (8.25)$$

The function $W(\Delta\theta)$ in (8.25) describes the orientation distribution of the mosaic blocks. In deriving (8.25), it is assumed that the characteristic size L_b of these blocks is much less than L_e^I, so that the kinematic approximation is valid for the individual blocks and the characteristic block disorientation angle $\Delta\theta$ is much less than λ/L_B (the so-called type I mosaic structure [40]).

The terms describing the change in the polarization tensor of the direct beam due to loss by diffraction have the most nontrivial form. Apart from the obvious diffractive reduction in the beam, there is an unavoidable change in the phase velocity (refractive index); they are mutually connected by the dispersion relations [15]. Here, the situation is analogous to the case of ordinary resonance absorption that necessarily leads to a change in the real part of the refractive index. We therefore make a small digression and consider the determination of the diffraction correction to the refractive index in an imperfect crystal.

The most systematic approach is to evaluate Δn_γ in terms of the forward scattering amplitude of an individual mosaic block (of arbitrary shape) [42], followed by averaging over block orientations. If, as assumed above, $L_b \ll L_e^I$ and $\Delta\theta \gg \lambda/L_B$, we need not consider the dependence on the shape and dimension of blocks, and we obtain the following expression for Δn:

$$\Delta n = \frac{\pi C_\gamma^2 \chi_\tau \chi_{-\tau}}{4 \sin 2\theta_B}[\tilde{W}(\Delta\theta) + iW(\Delta\theta)] \qquad (8.26)$$

where $C_\sigma = 1$, $C_\pi = \cos 2\theta_B$, and the function $\tilde{W}(\Delta\theta)$ is related to $W(\Delta\theta)$ by the dispersion relation

$$\tilde{W}(\Delta\theta) = \frac{1}{\pi}\int_{-\infty}^{+\infty}\frac{W(\Delta\theta)}{x - \Delta\theta}. \qquad (8.27)$$

This result is unusual because it involves an integral with respect to the angle of incidence and not the radiation frequency. This replacement is possible because the frequency and the angle are proportional to one another by virtue of the Bragg condition (with the exception of one case, $\theta_B = 90°$!).

The relation given by (8.26) can be obtained in other ways as well. For example, if one neglects absorption, it is possible to conclude from the law of conservation of energy that $\text{Im}(\Delta n_\gamma) = \lambda \sigma_{0\tau} C_\gamma^2 / 4\pi$ [see (8.25)], i.e., whatever was lost from one beam was gained by another, and then obtained $\text{Re}(\Delta n)$ with the aid of the dispersion relations. Another method consists of evaluating Δn from the amplitude of the wave transmitted by the crystal [see (8.6)] in the limit of very thin flat mosaic blocks, followed by averaging over the orientations of the individual blocks. The dispersion relations given by (8.27) are obtained automatically in this approach.

Note that (8.26) enables one to correctly take into account the mutual effects of absorption and diffraction. In particular, it can be shown [42] that partial suppression of absorption (Borrmann effect) will also occur in mosaic crystal in the diffraction region. However, in contrast to perfect crystal, the Borrmann effect is now relatively weak, and we shall not consider it when we examine polarization phenomena in mosaic crystal, i.e., we shall assume that $\text{Im}(\chi_\tau \chi_{-\tau}) = 0$.

The function $W(\Delta\theta)$ is commonly regarded as Gaussian or Lorentzian. The corresponding graphs of $W(\Delta\theta)$ and $\tilde{W}(\Delta\theta)$ are shown in Fig. 8.4. The very significant point is that $W(\Delta\theta)$ decreases slowly with increasing $|\Delta\theta|$: $\tilde{W}(\Delta\theta) \approx -(\pi\Delta\theta)^{-1}$, and whatever the distribution of the mosaic blocks, the refractive index (8.26) reaches the universal form $\Delta n_\gamma \approx -C_\gamma^2 \chi_\tau \chi_{-\tau} / (4\Delta\theta \sin 2\theta_B)$ for $|\Delta\theta| \gg \Delta\theta_B$. Hence, diffractive birefringence $\Delta n_\sigma - \Delta n_\pi$ is appreciable even outside the strong diffraction region. The quantity Δn has identical asymptotic behavior in perfect crystals. The fact that for the large $|\Delta\theta|$, the diffraction correction Δn_γ does not depend on the degree of perfection of the crystal can be understood qualitatively from the following considerations. For large $|\Delta\theta|$, the significant point is that the crystal must be perfect over distances of the order of $\lambda/|\Delta\theta|$ since for large distances, the phase difference between the diffracted waves is greater than π and on the average, the waves tend to cancel one another. This means that when the block size is $L_b \gg \lambda/|\Delta\theta|$, the radiation does not "feel" the imperfection of the crystal.

Let us return now to the examination of the terms in the transfer equation that describe the evolution of the polarization tensor of the direct beam. Since the refractive index is given by (8.26) for each of the polarization components, it can be shown [15,43] that terms describing the change in the polarization tensor of the direct beam by diffraction take the form $-QW(\Delta\theta)(\hat{K}^2 \hat{J}^0 + \hat{J}^0 \hat{K}^2) + iQ\tilde{W}(\Delta\theta)(\hat{K}^2 \hat{J}^0 - \hat{J}^0 \hat{K}^2)$. This finally leads to the equation for $\partial \hat{J}^0 / \partial s_0$ and the analogous equation for $\partial \hat{J}^\tau / \partial s_\tau$:

$$\frac{\partial \hat{J}^0}{\partial s_0} = -\mu \hat{J}^0 - QW(\Delta\theta)(\hat{K}^2 \hat{J}^0 + \hat{J}^0 \hat{K}^2) + iQ\tilde{W}(\Delta\theta)(\hat{K}^2 \hat{J}^0 - \hat{J}^0 \hat{K}^2)$$
$$+ QW(\Delta\theta)\hat{K}\hat{J}^\tau \hat{K}$$

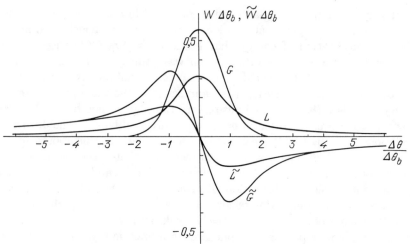

FIGURE 8.4. Graphs of the functions $W(\Delta\theta)$ and $\tilde{W}(\Delta\theta)$ for the Gaussian (G and $\tilde{G}$) and Lorentzian (L and $\tilde{L}$) distributions normalized so that $\int W_L(x)\,dx = \int W_G(x)\,dx = 1$.

$$\frac{\partial \hat{J}^{\tau}}{\partial s_{\tau}} = -\mu\hat{J}^{\tau} - QW(\Delta\theta)(\hat{K}^2\hat{J}^{\tau} + \hat{J}^{\tau}\hat{K}^2) - iQ\tilde{W}(\Delta\theta)(\hat{K}^2\hat{J}^{\tau} - \hat{J}^{\tau}\hat{K}^2)$$

$$+ QW(\Delta\theta)\hat{K}\hat{J}^0\hat{K} \qquad (8.28)$$

where s_{τ} is the distance measured along the direction of propagation of the diffracted wave. The set of equations given by (8.28) provides a complete description of the polarization properties in imperfect crystal with type I mosaic structure. It splits into four sets of equations for each component of the tensor $\hat{J}^0$ and $\hat{J}^{\tau}$. The equations for diagonal elements are identical to the Darwin equations for the intensities of the σ- and π-polarized beams, and the only new feature is the presence of the equations for the off-diagonal elements. The solution of these equations for a crystal in the form of a plane-parallel plate presents no difficulties. A detailed analysis of the results has been given in [15], so that we shall confine our attention to a brief review of the basic results.

The main difference between the X-ray optics of imperfect and perfect crystals is that even for a fixed angle θ_B, the diffracted beam becomes partially polarized whenever the polarization of the incident beam differs from σ- and π-polarization. This can be understood qualitatively from the following considerations. The incident wave polarization varies as the radiation propagates through the crystal, so that waves diffracted at different depths have different polarizations and add up incoherently, giving rise to a partially depolarized wave. An appreciable change in the polarization properties of diffracted beams occurs in mosaic crystal within the secondary extinction length $L_e^{II} = 2\pi\Delta\theta_b/Q$. For type I mosaic structures,

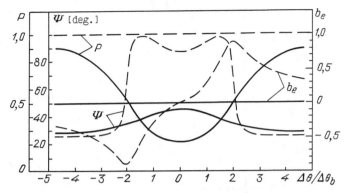

FIGURE 8.5. Degree of polarization P, ratio of axes b_e, and rotation Ψ of the polarization ellipse for a beam diffracted by a mosaic crystal as functions of the ratio $\Delta\theta/\Delta\theta_B$; Laue geometry ($\Delta\theta$, deviation from the Bragg conditions). Solid line, symmetric Laue case $b = 1$, $\cos\theta_B = \sqrt{3}/2$, dashed lines, asymmetric Laue case $b = 2$, $\cos\theta_B = \sqrt{3}/2$. The incident beam was linearly polarized at 45° to the plane of scattering, $W(\Delta\theta) = W_L(\epsilon)$, crystal thickness $L = 2.5L_e^{II} = 5\pi\Delta\theta_B/Q$.

$L_e^{II} \gg L_e^{I}$, but L_e^{II} can be comparable with the absorption length μ^{-1} i.e., both $\mu L_e^{II} \gg 1$ and $\mu L_e^{II} \ll 1$ can occur. The effect of depolarization is enhanced still further for quantities integrated over the diffraction region.

Apart from depolarization, diffraction birefringence of the diffracted and transmitted beams ensures that they become elliptically polarized (Fig. 8.5). For a linearly polarized incident beam, the elliptic polarization is right- or left-handed, depending on the sign of $\Delta\theta$, and ellipticity practically disappears from the integrated parameters (if anomalous absorption can be neglected). Figures 8.5 and 8.6 show examples of differential and integral polarization parameters, and Table 8.2 compares the polarization parameters of perfect and imperfect crystals.

Experimental studies of the polarization properties of imperfect crystals are only just beginning, but they have already led to significant results. In addition to early work [10,12,44–46], particular attention should be paid to the series of experimental studies by N.M. Olekhnovich et al., who investigated the polarization properties for the case of real crystals with different dislocation densities [14]. They examined in detail the integrated [47–49] and differential [11,13,50,51] reflection coefficient for σ- and π-polarizations and have also observed diffractive birefringence and depolarization. Although so far experimental studies have been largely confined to σ- and π-polarizations, they have demonstrated that the Darwin theory of diffraction by mosaic crystals is frequently incapable of providing a quantitative description of experimental data. For example, the ratio of the coefficients of reflection for σ- and π-polarized radiation (the so-called polarization coefficient [11,13,50,51]) reaches a plateau in the diffraction region (Fig. 8.7). This behavior can probably be explained by assuming that

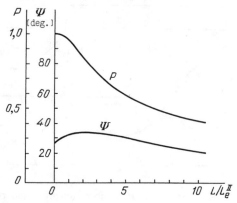

FIGURE 8.6. Integral polarization parameters as a function of crystal thickness. Asymmetric Laue case ($b = 2$), linearly polarized incident beams with polarization at $45°$ to the vector σ (Ref. [15]).

TABLE 8.2. Polarization properties of diffracted wave for a polarized incident wave.

Properties	Crystals		
	Kinematic	Dynamic	Mosaic
1. Different width of the diffraction regions for σ and π components	No	Yes	Yes
2. Phase difference between σ and π components	No	Yes	Yes
3. Depolarization (for a plane monochromatic wave)	No	No	Yes
4. Depolarization integrated over the diffraction region	No	Yes	Yes

the dimensions of the individual blocks are so large that primary extinction becomes significant in each block [50,51]. However, some of the experimentally observed features have not received even a qualitative explanation. For example, it has been reported [53] that, for a certain dislocation density, the integrated reflection coefficient for π-polarization is greater than that for σ-polarization (Fig. 8.8).

In view of the foregoing, it is clear that further advances are necessary in the theory of diffraction by imperfect crystals, and polarization measurements have to be made for the verification of theoretical predictions, since such measurements are the most informative. In particular, the Kato extinction theory [54] and its more developed versions [55,56] could be generalized to the case of arbitrary polarization.

The following general question has arisen: What is the maximum number

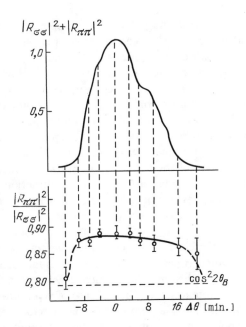

FIGURE 8.7. Reflection coefficients $|R_{\sigma\sigma}|^2 + |R_{\pi\pi}|^2$ for unpolarized radiation and the ratio $|R_{\pi\pi}|^2/|R_{\sigma\sigma}|^2$ as functions of $\Delta\theta$ for a dislocation density of $3 \cdot 10^5$ mm^{-2}. Symmetric Bragg case, Ge(III), CuK$_\alpha$ radiation [50].

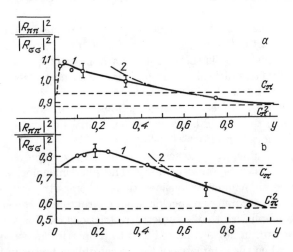

FIGURE 8.8. Ratio of integral reflection coefficients for σ- and π-polarized beams as a function of extinction factor y characterizing the degree of imperfection of the crystal; (a) 200 and (b) 400 reflections in LiF, CuK$_\alpha$ radiation [53b]. 1, experiment; 2, calculations including only primary extinction in individual mosaic blocks.

of parameters that describe the polarization properties of each reflection (we have in mind quantities that can be measured with a polarizer and analyzer without measuring the phases of the beams)? It is found that since the σ- and π-polarizations do not mix in each Rayleigh scattering event, only four such parameters are necessary (each is a function of the angle of incidence and angle of reflection). These parameters can be, for example, the coefficients of reflection for σ- and π-polarizations and the complex coefficients of reflection for the off-diagonal components of the polarization tensor, i.e., the quantity relating the off-diagonal components of the polarization tensor of the incident and diffracted beams. Emphasize that the σ- and π-polarizations do not mix if diffraction is of the two-wave (or coplanar multiwave) type, and X-ray polarization anisotropy and rotation of the plane polarization are both absent.

It is important to note that measurements of the polarization properties of diffraction by perfect and, especially, imperfect crystals are also of metrological importance. Since crystals are used as monochromators, it is often important to know the quantitative polarization parameters of radiation after the monochromator (e.g., in the case of structure analysis). The International Union of Crystallography [57] has therefore called for the investigation of the polarization properties of monochromators and methods for their determination [58–60].

8.2   Anisotrophy of X-Ray Susceptibility

The traditional neglect of the anisotropy of X-ray susceptibility in discussions of diffraction by crystals (Sect. 8.1) is often fully justified [61]. However, this anisotropy is the reason for a number of qualitative effects, such as nondiffractive dichroism and birefringence (mostly near absorption edges), which are relatively obvious. A nontrivial qualitative consequence of this anisotropy is the appearance of reflections that are forbidden for symmetry reasons in the case of isotropic susceptibility, i.e., the appearance of so-called "forbidden reflections" [20–23].

Two complementary approaches can be naturally employed to investigate anisotropy in X-ray susceptibility. The phenomenological approach is largely based on symmetry considerations, whereas the microscopic approach takes into account the specific atomic structure of the crystal. It is important to emphasize that symmetry restrictions on the X-ray susceptibility tensor can, in no way, be reduced to properties known in ordinary optics. Thus, when the symmetry of the susceptibility tensor in the optical range is investigated, the crystal is looked on as a homogeneous medium [62,63]. Only the homogeneous part of the susceptibility $\hat{\chi}_0$ is significant, and its symmetry is determined by the crystal point group and is well known in optics [62,63]. In the case of X-ray diffraction, the inhomogeneous (periodic) part of $\hat{\chi}(\mathbf{r})$ becomes significant. Its symmetry is different at different points in the unit cell of the crystal and is restricted by the space group of the crystal. The general properties of X-ray susceptibility are discussed in detail in [61], and the symmetry restrictions on $\hat{\chi}(\mathbf{r})$ are examined in [22,23] (see also the discussion below).

One of the manifestations of the anisotropy of susceptibility is the magnetic scattering of X-rays and magnetic dichroism, which occur in magnetically ordered crystals and are discussed in Chapt. 9.

8.2.1   Birefringence and Dichroism near Absorption Edges

A systematic description of the anisotropy of X-ray susceptibility can be constructed on the basis of quantum-mechanical theory and requires a knowledge of atomic and crystal electron wave functions [26,61,64]. For our purposes, we need only be able to understand the physical reasons for anisotropy and to estimate its magnitude. We shall therefore slightly simplify the true picture of the interaction between X-rays and crystals. It is clear that the anisotropy of susceptibility appears as a consequence of crystal structure and is due to the distortion of the wave functions of free atoms by the crystal field. The anisotropy derives from the dispersion (resonance) corrections to susceptibility, whereas the principal (potential) part of susceptibility is isotropic. The wave functions of the outermost electrons are the most highly distorted, but they provide a very small contribution to the dispersion correction because the binding energy of the outer electrons is small in comparison with the energy of X-ray photons. Appreciable dispersion corrections (on the order of the contribution of several electrons per atom) to the permittivity are provided by the innermost K and L shell electrons when the phonon energy is close to the K or L absorption edge, although these shells remain relatively undistorted by the crystal field. In the simplest dipole approximation, the dispersion correction to the susceptibility of a crystal is given by [61,64].

$$\Delta \chi_{ik} = -\frac{r_e c^2}{m\omega^2} \sum_{j,m,s} \left[\frac{\langle O|p_i^s|m\rangle \langle m|p_k^s|O\rangle}{E_0 - E_m + \hbar\omega - i(\Gamma_m/2)} \right.$$
$$\left. + \frac{\langle O|p_k^s|m\rangle \langle m|p_i^s|O\rangle}{E_0 - E_m - \hbar\omega} \right]_j \delta(\mathbf{r} - \mathbf{r}_j) \qquad (8.29)$$

where $\mathbf{p}^s = -i\hbar\nabla^s$ is the momentum operator acting on the coordinate of the sth charge, $|0\rangle$ is the wave function of the initial state and the final state that coincides with it (our attention is confined to the elastic processes only), and $|m\rangle$ represents the wave function of intermediate states that can lie either in the discrete or continuous spectrum. The origin of the anisotropy of susceptibility can be seen from (8.29): although the wave functions of the initial (and final) states are relatively undistorted, the wave function of the intermediate states $|m\rangle$ can be greatly distorted, so

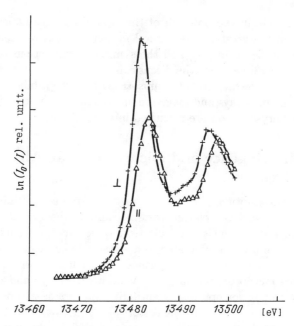

FIGURE 8.9. X-ray absorption coefficient of KBrO$_3$ for radiation polarized parallel and perpendicular to the four-fold axis near K absorption edge of bromine [20].

that the product of the matrix elements $\langle O|p_i|m \rangle \langle m|p_k|O \rangle$ does not reduce to the isotropic component. Note that in the dipole approximation that we are considering here, the eigenpolarizations are found to be linear for the tensor (8.29). The real part of (8.29) is responsible for birefringence and the imaginary part for dichroism (there is no circular birefringence and dichroism). Moreover, the dispersion corrections are not very dependent on the transferred momentum (i.e., on the reflection τ). The last point is related to the fact that spatial dispersion effects are weak for the radii α_K and α_L of the K and L shells with energies in the X-ray range because $\alpha_K, \alpha_L \ll \lambda$ for the corresponding X-ray energies.

Numerous experimental [6,20,21,65,66] (see Fig. 8.9) and theoretical [26, 61,64] investigations have shown that the dispersion corrections and the associated anisotropy of susceptibility reach their maxima in the immediate neighborhood (~ 10 eV) of the absorption edge (the so-called XANES, i.e., X-ray absorption near edge structure). Below the absorption edge, the main contribution to anisotropy is provided by bound excited states, whereas Bloch electron states in the crystal provide this contribution immediately above the edge.

Anisotropy decreases rapidly with increasing photon energy well above the absorption edge (≤ 100 eV) in the EXAFS (extended X-ray absorption fine structure) region, but is still experimentally noticeable [20,21]. In this region we can carry out a semiphenomenological evaluation of $\Delta\hat{\chi}$, for which in intermediate states we can take into account the diffraction of photoelectrons by atoms surrounding a given atom (it is well known that photoelectrons are preferentially emitted in the direction of the polarization vector of the incident X-ray photon). This diffraction is very dependent on the distance to the nearest-neighbor atoms and on the form factors of these atoms. The result of all this is that the anisotropy of the environment of a given atom has an appreciable effect on the anisotropy of $\Delta\hat{\chi}$ in the EXAFS region. Thus, although the physical reason for the onset of the anisotropy of susceptibility near X-ray absorption edges is relatively obvious, a quantitative evaluation is relatively complicated [61,64]. It is therefore natural to perform a symmetry analysis of the susceptibility tensor, which does not involve the detailed consideration of a model.

8.2.2 Symmetry Restriction on the X-Ray Susceptibility Tensor of a Crystal

The X-ray susceptibility of a nonmagnetic crystal is described by the symmetric tensor of rank two $\chi_{ik}(\mathbf{r}) = \chi_{ki}(\mathbf{r})$, which is unaffected by transformations belonging to the space group of the crystal. The easiest way of finding the general form of this tensor is as follows. The atoms in a crystal structure occupy a certain definite regular set of points (one or more). It is therefore sufficient (1) to determine the tensor $\chi_b(\mathbf{r})$ for a single basis atom in a given lattice, taking into account the point symmetry of the atomic site; (2) to obtain the tensor $\hat{\chi}^j(\mathbf{r})$, found for the jth atom in the regular set of points, by transforming the tensor $\hat{\chi}^b(\mathbf{r})$ by a symmetry operation relating the position of the basis and jth atom in the crystal; and (3) to obtain the total tensor $\hat{\chi}(\mathbf{r})$ as the sum over all atoms in the given regular structure and over all the regular sets of points occupied by atoms in a given crystal.

To find the general form of $\hat{\chi}(\mathbf{r})$, we must therefore know how the tensor $\hat{\chi}(\mathbf{r})$ transforms under symmetry operations. Suppose that the tensor $\hat{\chi}(\mathbf{r})$ is given in the space. If we apply to it transformation g that includes the point transformation (rotations or reflections) and translation, the tensor $\hat{\chi}_g(\mathbf{r})$, obtained as a result of this, is related to the original tensor $\hat{\chi}(r)$ as follows (see, e.g., [67,68]):

$$\hat{\chi}_g(\mathbf{r}) = \hat{R}_g\hat{\chi}(\mathbf{r}_g)\hat{R}_g^{-1} \qquad (8.30)$$

where $\mathbf{r}_g = \hat{R}_g^{-1}(\mathbf{r} - \mathbf{a}_g)$, $\hat{R}_g$ is the point transformation matrix, and $\mathbf{a}_g$ is the translation vector. If the tensor $\hat{\chi}(\mathbf{r})$ is invariant under g, then $\hat{\chi}_g(\mathbf{r}) = \hat{\chi}(\mathbf{r})$. If the basis atom lies at a point whose symmetry is described by the point group g, the susceptibility tensor $\hat{\chi}^b(\mathbf{r})$ of this atom will obviously be invariant under any transformation g in this group, i.e., in this case, $\hat{\chi}^b(r)$

must satisfy the following relation that ensues from (8.30):

$$\hat{\chi}^b(r) = \hat{R}_g \hat{\chi}^b(\hat{R}_g^{-1}\mathbf{r})\hat{R}_g^{-1}. \tag{8.31}$$

The tensor satisfying this relation can be obtained by averaging an arbitrary symmetry tensor $\hat{\alpha}(r)$ over the group G [68]:

$$\hat{\chi}^b(r) = \langle\hat{\alpha}(\mathbf{r})\rangle_G \equiv \sum_{g\in G} \hat{R}_g \hat{\alpha}(\hat{R}_g^{-1}\mathbf{r})\hat{R}_g^{-1}. \tag{8.32}$$

Having determined in this way the most general form of $\hat{\chi}^b(\mathbf{r})$ for the basis atom, we can now use (8.30) to find the tensor $\hat{\chi}^j(\mathbf{r})$ for the remaining atoms in the given lattice, taking g to be the symmetry operation g_g relating the position of the jth atom to the basis atom. Of course, all these operations can be performed for any regular set of points occupied by atoms in a given crystal.

All this finally yields the most general form of the tensor $\hat{\chi}(\mathbf{r})$ that is consistent with the space symmetry group of the crystal. This automatically takes into account the possibility of a difference between the electron density of the atom and the spherically symmetric density [the electron density is a scalar proportional to the trace of the tensor $\hat{\chi}(\mathbf{r}]$.

In its symmetry properties, the tensor $\hat{\chi}(\mathbf{r})$ differs radically from the susceptibility tensor in the optical range. In particular, $\hat{\chi}(\mathbf{r})$ does not reduce to a scalar even in cubic crystals. The local symmetry of $\hat{\chi}(\mathbf{r})$ is different at different points of the unit cell. Note that the general form of $\hat{\chi}(\mathbf{r})$ can be found even without resorting to the particular atomic structure of a crystal [22], but the approach employed above is clearer and enables us to identify the contribution to $\hat{\chi}(\mathbf{r})$ due to different species of atoms.

If, in addition to the symmetry properties, we take into account the physical origin of the anisotropy, we can find further restrictions on $\hat{\chi}(\mathbf{r})$. Thus, since the radii of the K and L shell are small, we can substitute $r = 0$ in the dispersive part of $\hat{\chi}^b(\mathbf{r})$ in (8.31, 8.32), which substantially simplifies all the calculations and, for certain special dispositions of the atoms, may modify the form of the tensor $\hat{\chi}(\mathbf{r})$ (examples are given below).

To find the intensity and polarization properties of individual reflections, it is convenient to introduce the tensor form of the structure amplitude $\hat{F}^\tau$, which is proportional to the Fourier component of the susceptibility $\hat{\chi}^\tau$:

$$\hat{F}^\tau = -\frac{\pi V}{r_e\lambda^2}\chi^\tau \equiv \frac{\pi}{r_e\lambda^2}\int \hat{\chi}(\mathbf{r})\exp(-i\tau\mathbf{r})\,d\mathbf{r} \tag{8.33}$$

where all symbols have the same meaning as in (8.2). The amplitudes $\hat{F}^\tau$ are usefully divided into the isotropic and intrinsically anisotropic parts:

$$\hat{F}^\tau = F_\tau\hat{I} + \Delta\hat{F}^\tau \tag{8.34}$$

where $\mathbf{F}_\tau$ is the usual structure amplitude, $\hat{I}$ is the unit matrix, and the anisotropic part is defined so that $\mathrm{Sp}(\Delta\hat{F}^\tau) = 0$.

The symmetry restriction on $\hat{\chi}(r)$ influences the tensor form of $\hat{F}^\tau$ [22,23]. For an arbitrary vector τ, there are no restrictions on the form of $\Delta\hat{F}^\tau$ and, just like any other traceless symmetric tensor, $\Delta\hat{F}^\tau$ has five independent complex components. However if the vector τ points along the symmetry axes, the number of independent components of the tensor $\Delta\hat{F}^\tau$ is reduced (the z-axis is taken along τ): if τ is parallel to axis 2, $\Delta\hat{F}^\tau_{xz} = \Delta\hat{F}^\tau_{yz} = 0$; if τ is parallel to axes 3, 4, 6, $\Delta\hat{F}^\tau_{xy} = \Delta\hat{F}^\tau_{xz} = \Delta\hat{F}^\tau_{yz} = 0$ and $\Delta\hat{F}^\tau_{xx} = \Delta\hat{F}^\tau_{yy} = -\frac{1}{2}\Delta\hat{F}^\tau_{zz}$. If τ is parallel to the mirror reflection plane, then by choosing the axis to be perpendicular to this plane, one has $\Delta\hat{F}^\tau_{xy} = \Delta\hat{F}^\tau_{xz} = 0$.

The intensity and polarization properties of reflections with the tensor amplitude (8.34) can be found in both the kinematic and two-wave approximations by analogy with the analysis of diffraction of Mössbauer radiation [69] (see also Chapt. 3 and 10) and diffraction of light by liquid crystals [70] (see also Chapt. 4). In most cases, $\Delta\hat{F}^\tau$ is small in comparison with F_τ, and the entire analysis can be based on perturbation theory. The most interesting cases are those for which $F_\tau = 0$ by symmetry consideration (forbidden reflections) and $\Delta F \neq 0$, i.e., the inclusion of anisotropy reduces the suppression of reflection (see the next section).

8.2.3   Forbidden Reflections

It is well known that the systematic suppression of reflections is observed in X-ray diffraction by crystal, i.e., the structure amplitudes of some of the reflections vanish systematically because the atoms in the unit cell are situated in a number of symmetry-related positions [32]. The set of these forbidden reflections is determined by the space group of the crystal and is given, for example, in [7]. However, the standard conditions given in [71] were obtained under the assumption that the atomic scattering factors were identical for all atoms in equivalent positions, i.e., it was actually assumed that the atoms forming the crystal were spherically symmetric. In reality, the atoms in the crystal are not spherically symmetric because of the interaction between them, which means that they are not equivalent from the point of view of scattering, and this can give rise to the appearance of the "forbidden" reflections. More precisely, it may be said that atoms situated in equivalent crystallographic positions may be nonequivalent from the point of view of their interaction with electromagnetic (X-ray) radiation, and the scattering amplitudes for such atoms may be different.

There are several physical factors responsible for this difference between the scattering amplitudes corresponding to crystallographically equivalent atoms. The best known are the nonspherical electron density distribution of the atoms, and the anisotropy and anharmonicity of thermal motion of atoms [72–74] (the 222 type reflections in crystals with a diamond structure constitute a well-known example of this). Another example is provided by the dependence of the scattering amplitude on the electron spin, which leads to very weak magnetic reflections when X-rays are diffracted by magnetically ordered crystals (see Chapt. 10). The anisotropy of X-ray susceptibility can also lead to a difference between the scattering amplitudes, since crystallographically equivalent atoms can be related by a symmetry operation containing rotation, and under rotation, the tensor describing some anisotropic physical property (in our case, the anisotropy of X-ray susceptibility) can alter the orientation of its principal axes.

It is well known that the condition for possible Bragg reflections by a crystal with a given space group can be different for general and particular positions of atoms in the unit cell [32,71]. If the nonspherical nature of the atomic electron density and the thermal motion of atoms are taken into account, these conditions for special positions are violated, but those imposed on the general positions are not. However, the latter also may be violated if we consider the anisotropy of susceptibility. Clearly, the conditions associated with the centering of a lattice remain valid even in this case, since the transformation properties of tensors and scalars are the same under pure translations. However, the restrictions on reflections that are due to the presence of glide planes or screw axes are no longer valid [20–24], and we shall prove this rigorously, but let us first give a clear interpretation of this phenomenon.

Suppose the crystal has a glide plane (Fig. 8.10a), i.e., for example, the structure is invariant under reflection in the yz-plane and displacement along the z-axis. Atoms in positions A and B are crystallographically equivalent, but the corresponding susceptibility tensors (indicated symbolically by the ellipses) are rotated relative to one another, and these atoms are polarized differently by the incident X-ray waves. In the case of the 001 reflections with $l = 2n + 1$, these atoms scatter in antiphase. If we ignore the anisotropy of susceptibility, waves scattered by these atoms will extinguish one another, and reflection will be forbidden. When the anisotropy is taken into account, scattering by atoms A and B will be different in both intensity and polarization, and this will mean that the suppression of the reflection will not occur. A similar mechanism can lead to the removal of the suppression of reflections that are "forbidden" by screw axes (Figs. 8.10b and 8.10c).

We now turn to detailed symmetry analysis. A crystal transforms into itself under all symmetry transformations belonging to its space group. Consequently, the structure amplitude $\hat{F}^{\tau}$ must remain unaltered. In the glide-plane reflection examined above, a displacement by half the period along the z-axis leads to multiplication of the structure amplitude of the $0kl$ reflection by $\exp(i\pi l) = (-1)^l$. Reflection by the yz-plane (i.e., $x \Rightarrow -x$) leads to a change in the sign of the xy and xz components of the tensor structure amplitude. Since the structure amplitude must be invariant under this transformation, we find that the xy and xz components must be annulled in the structure amplitude for $l = 2n$ (allowed reflections), whereas for $l = 2n + 1$ ("forbidden" reflections), all the components other

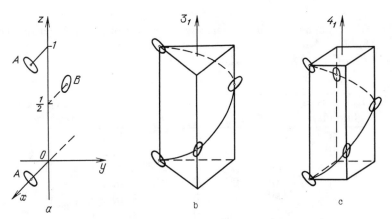

FIGURE 8.10. Effect of symmetry elements on atoms with anisotropic susceptibility: (a) glide plane, (b) 3_1 axis, (c) 4_1 axis.

than xy and xz components are annulled. It is thus clear that symmetry
does not demand the complete annulment of the structure amplitude of
the reflections with $l = 2n + 1$, and the suppression of these reflections is
removed, but only when the susceptibility is anisotropic.

Reflections that are "forbidden" by the presence of screw axes can be
examined similarly. Suppose that the screw axis N_j points along the z-
axis, i.e., the crystal is invariant under the combined operation of rotation
by the angle $\varphi = 2\pi/N$ about the z-axis and displacement along the z-
axis by the fraction j/N of the period ($N = 2, 3, 4, 6$; $j = 1, \ldots, N - 1$).
The tensor structure amplitude $\hat{F}^{00l}$ transforms into $R_\varphi F^{00l} \hat{R}_\varphi^{-1}$ under
rotation and acquires the further factor $\exp(2\pi i l j/N)$ under displacement.
Consequently, the invariance of $\hat{F}^{00l}$ under this operation requires that the
condition

$$\hat{F}^{00l} = \hat{R}_\varphi \hat{F}^{00l} \hat{R}_\varphi^{-1} \exp(2\pi i l j N^{-1}) \tag{8.35}$$

must be satisfied, where the rotation matrix $\hat{R}_\varphi$ has the form

$$\hat{R}_\varphi = \begin{pmatrix} \cos\varphi & \sin\varphi & 0 \\ -\sin\varphi & \cos\varphi & 0 \\ 0 & 0 & 1 \end{pmatrix}. \tag{8.36}$$

If there is no anisotropy, it follows from (8.35) that $\hat{F}^{00l} = 0$ for $l \neq Nn/j$,
where n is an arbitrary integer, i.e., all these reflections are forbidden [71].
When the anisotropy is taken into account, it follows from (8.35) that the
suppression of these reflections is removed, save for some rare exceptions
(Table 8.3). It is clear from Table 8.3 that for each "forbidden" reflection,
all components of the tensor $\hat{F}^\tau$ are expressed in terms of, at most, two
independent parameters, say, F_1 and F_2, which may be complex. To avoid
misunderstanding, we emphasize that these parameters are different for

TABLE 8.3. Components of the tensor structure amplitude $\hat{F}^\tau$ and the index l for "forbidden" reflections (the other components are $F_{yy}^\tau = -F_{xx}^\tau$, $F_{zz}^\tau = 0$, $F_{yx}^\tau = F_{xy}^\tau$, $F_{zy}^\tau = F_{yz}^\tau$, $F_{zx}^\tau = F_{xz}^\tau$), $n = 0, \pm1, \pm2 \ldots$.

Hel. axis or glide plane	F_{xx}^τ	F_{xy}^τ	F_{xz}^τ	F_{yz}^τ	l	Type of reflection
2_1	0	0	F_1	F_2	$2n+1$	I
3_1	F_2	$\pm iF_2$	F_1	$\mp iF_1$	$3n\pm1$	II
3_2	F_2	$\mp iF_2$	F_1	$\pm iF_1$	$3n\pm1$	II
4_1	0	0	F_1	$\mp iF_1$	$4n\pm1$	I
4_1	F_1	F_2	0	0	$4n+2$	II
4_2	F_1	F_2	0	0	$2n+1$	II
4_3	0	0	F_1	$\pm iF_1$	$4n\pm1$	I
4_3	F_1	F_2	0	0	$4n+2$	II
6_1	0	0	F_1	$\mp iF_1$	$6n\pm1$	I
6_1	F_1	$\mp iF_1$	0	0	$6n\pm2$	II
6_1	0	0	0	0	$6n+3$	
6_2	F_1	$\mp iF_1$	0	0	$3n\pm1$	II
6_3	0	0	0	0	$2n+1$	
6_4	F_1	$\pm iF_1$	0	0	$3n\pm1$	II
6_5	0	0	F_1	$\pm iF_1$	$6n\pm1$	I
6_5	F_1	$\pm iF_1$	0	0	$6n\pm2$	II
6_5	0	0	0	0	$6n+3$	
c	0	F_1	F_2	0	$2n+1$	II

different reflections and, in general, they depend on wavelength. Their numerical values can be calculated from microscopic theory. The phenomenological theory we are discussing shows only that these parameters need not be zero if the crystal has only a screw axis or reflecting glide plane. Other symmetry elements can lead to additional relationships between F_1 and F_2 and, in particular, may reduce them (or one of them) to zero. For example, reflections with $|h| = |k| = |l|$ continue to be forbidden in cubic crystals [23]. Moreover, the zero value of these parameters can occur because atoms that contribute to anisotropy are situated in positions of relatively high symmetry.

It follows from Table 8.3 that the tensor $\hat{F}^\tau$ can have a different form for different types of reflection, and it may be expected that the properties of these types of reflection are also different (see below). It is interesting to note that some of the reflections associated with screw axes remain forbidden. For example, in the case of the 6_3 axis, the necessary presence of axis 3 leads to the absence of anisotropy in the xy-plane, so that the suppression of reflections with $l = 2n+1$ is not removed, and $\hat{F}^{00l} = 0$

for them. For axis 6_1 and 6_5, reflections with $l = 6n + 3$ are also found to remain forbidden. When the quadripole interaction is taken into account, these reflections may become allowed, and this can be used to detect the quadrupole mechanism of interaction between X-rays and crystal atoms.

The intensity and, especially, polarization properties of reflections are altered when anisotropy is considered. For allowed reflections, these changes take the form of corrections (the exception is provided by the possibility of $90°$ scattering of π-polarized waves) and can be found in both kinematic and dynamic theories. The polarization properties become radically different for forbidden reflections and are found to be very unusual. For example, a σ-polarized incident wave can produce a π-polarized diffracted wave, and vice versa. Since the structure amplitude of these reflections is relatively small, we can use the kinematic approximation for them. The amplitude of the diffracted wave is then given by

$$\mathbf{E}^d = A\hat{F}^\tau \mathbf{E}^i \qquad (8.37)$$

where the factor A is the same as in (8.3). This expression enables us to evaluate the intensity, and polarization of the incident wave. This can be conveniently carried out in terms of the quantity $\mathbf{I}_{\alpha\beta}$, which describes the intensity of the component with arbitrary (β) polarization in the reflected beam for arbitrary (α) polarization of the incident beam:

$$I_{\alpha\beta} = |A|^2|\beta^*\hat{F}^\tau\alpha|^2 \qquad (8.38)$$

where α and β are corresponding polarization vectors. The polarization of the diffracted wave is determined by the unit vector $\beta^d = \mathbf{E}^d/|\mathbf{E}^d|$. Note that the vector ($\beta^d$) may or may not depend on the polarization vector of the incident beam.

If the incident radiation is σ- or π-polarized or unpolarized, the intensity of the reflection is determined, respectively, by the following expressions:

$$I_\sigma = I_{\sigma\sigma} + I_{\sigma\pi} \qquad (8.39)$$

$$I_\pi = I_{\pi\sigma} + I_{\pi\pi} \qquad (8.40)$$

$$I_{un} = \frac{1}{2}(I_\sigma + I_\pi) \qquad (8.41)$$

where

$$
\begin{aligned}
I_{\sigma\sigma} &= |A|^2|\sigma\hat{F}^\tau\sigma|^2 \\
I_{\sigma\sigma} &= |A|^2|\pi_\tau\hat{F}^\tau\pi_0|^2 \\
I_{\sigma\pi} &= I_{\pi\sigma} = |A|^2|\pi_\tau\hat{F}^\tau\sigma|^2 = |A|^2|\sigma\hat{F}^\tau\pi_0|^2.
\end{aligned}
\qquad (8.42)
$$

Let us examine, for example, the properties of forbidden reflections associated with the presence of a screw axis. It is clear from Table 8.3 that all these reflections can be divided into two basic types—namely, reflections for which F_{xx}^τ, F_{yy}^τ and F_{xy}^τ are all zero (type I) and all the other reflections (type II). We shall show that the polarization properties of these two types of reflections are significantly different.

Type I reflections have the simplest polarization properties. It follows directly from (8.37–8.42) and from Table 8.3 that $I_{\sigma\sigma} = I_{\pi\pi} = 0$ and $I_\tau = I_\sigma = I_\pi = I_{\sigma\pi} = I_{\pi\sigma}$ for all type I reflections, where for axis 2_1 the intensity is given by

$$I_{\sigma\pi} = |A|^2 \cos^2 \theta_B \big[|F_1|^2 \sin^2 \varphi_\tau + |F_2|^2 \cos^2 \varphi_\tau$$
$$- \operatorname{Re}(F_1 F_2^*) \sin 2\varphi_\tau \big] \tag{8.43}$$

in which φ_τ is the azimuthal angle of rotation around the vector $\boldsymbol{\tau}$ (z-axis), measured from the x-axis, whereas for axes 4_1, 4_3, 6_1, 6_3, the intensity has a simpler form

$$I_{\sigma\pi} = |A|^2 \cos^2 \theta_B |F_1|^2. \tag{8.44}$$

A σ-polarized incident wave will therefore produce a π-polarized diffracted wave for type I reflections, and vice versa. An unpolarized beam will produce an unpolarized diffracted beam. Type I reflections disappear in the case of backward diffraction (since $\cos \theta_B = 0$). In the case of the 4_1, 4_3, 6_1, 6_5 axes, the intensity of these reflections does not depend on the azimuthal angle φ_τ.

The intensity of type II reflections is given by the following expressions that ensue from (8.37–8.42) and Table 8.3. For axes 3_1 and 3_3:

$$I_\sigma = |A|^2 \big\{ \cos^2 \theta_B |F_1|^2 + |F_2|^2 (1 + \sin^2 \theta_B)$$
$$+ \sin 2\theta_B \big[\operatorname{Re}(F_1 F_2^*) \cos 3\varphi_t au \mp \operatorname{Im}(F_1 F_2^*) \sin 3\varphi_\tau \big] \big\} \tag{8.45}$$
$$I_\pi = I_\sigma - |A|^2 \cos^2 \theta_B |F_2|^2 (1 + \sin^2 \theta_B).$$

For axes 4_1, 4_2, and 4_3:

$$I_\sigma = |A|^2 \big[|F_1^2| B(\varphi_\tau) + |F_2|^2 C(\varphi_\tau)$$
$$+ \operatorname{Re}(F_1 F_2^*) \cos^2 \theta_B \sin 4\varphi_\tau \big] \tag{8.46}$$

$$I_\pi = |A|^2 \sin^2 \theta_B \big[|F_1|^2 C(\varphi_\tau) + |F_2|^2 B(\varphi_\tau)$$
$$- \operatorname{Re}(F_1 F_2^*) \cos^2 \theta_B \sin 4\varphi_\tau \big]$$

$$B(\varphi_\tau) = 1 - \cos^2 \theta_B \sin^2 2\varphi_\tau$$
$$C(\varphi_\tau) = 1 - \cos^2 \theta_B \cos^2 2\varphi_\tau.$$

For axes 6_1, 6_2, 6_4, and 6_5:

$$I_\sigma = |A|^2 |F_1|^2 (1 + \sin^2 \theta_B)$$
$$I_\pi = I_\sigma \sin^2 \theta_B \tag{8.47}$$

where the upper and lower signs in (8.45) correspond to the two possibilities $F_{xy}^\tau = \pm i F_{xx}^\tau$ (see Table 8.3). In contrast to type I reflections, the type II intensities are different for σ- and π-polarized beams. It is clear from (8.46, 8.47) that for small Bragg angles, the σ-polarized beams have a higher intensity. For three- and six-fold screw axes, the type II reflections have chiral properties, i.e., their intensity is different for right- and left-handed circular polarizations of the incident beams. For example, for backward Bragg reflections ($\theta_B = 90°$), (8.32) shows that only the component with a definite circular polarization (right if $F_{xy}^\tau = -i F_{xx}^\tau$ and left if $F_{xy}^\tau = i F_{xx}^\tau$) will undergo diffractive reflection (the diffracted beam having the same circular polarization). The wave with the opposite circular polarization will not be diffracted. It follows that for $\theta_B = 90°$, the crystal works as a circular polarizer and, for any polarization of the incident beam, the diffracted beam has circular polarization of the definite handiness that depend on l and on whether the screw axis is right- or left-handed (Table 8.3). When $\theta_B < 90°$, the polarization of these type II reflections is elliptic rather than circular. For a six-fold axis, the ratio of the axes of the polarization ellipse is equal to $\sin \theta_B$, where the major axis is parallel to the vector σ and the sign of the polarization (right or left) is determined as for $\theta_B = 90°$. For a three-fold screw axis, the ratio of the axes of the polarization ellipse is a complicated function of θ_B, and we shall not reproduce here the relationship between the parameters F_1 and F_2. For a four-fold screw axis, we have chiral type II reflections if $\mathrm{Im}(F_1 F_2^*) \neq 0$.

Note especially that the intensity of the "forbidden" reflections depends on the azimuthal angle (even for an unpolarized incident beam). By studying this dependence, we can determine the magnitude of the components of (the parameters F_1 and F_2) and their relative phase. It has been shown [23] that the azimuthal dependence can also be used to determine selectively the coordinates of atoms contributing to the anisotropy.

Let us illustrate all this by considering the example of $00l$ ($l = 2n + 1$) reflections in a crystal with space group $P2_13$ and four atoms in special position (a) with point symmetry 3 and coordinates x, x, x; $\frac{1}{2} + x$, $\frac{1}{2} - x$, $\bar{x}$; $\frac{1}{2} + x$; $\bar{x}$, $\frac{1}{2} + x$, $\frac{1}{2} - x$. The anisotropy of susceptibility of atoms in positions of this symmetry is characterized by susceptibility difference $\chi_\parallel - \chi_\perp$ (parallel and perpendicular to axis 3). Consequently, both F_1 and F_2 can be expressed in terms of this difference and the coordinate x. It can be shown that $F_1/F_2 = i\,\mathrm{tg}(2\pi l x)$ and that the intensity of the $00l$ ($l = 2n+1$) reflections is proportional to $1 - \cos 4\pi l x \cos 2\varphi_\tau$. The azimuthal dependence of the intensity of "forbidden" reflections can therefore be used to determine the coordinate x. This was recently demonstrated [21] (Fig. 8.11) for $NaBrO_3$ crystals.

"Forbidden" reflections can also be used in other ways. They are useful in the interpretation of spectra near absorption edges (in contrast to absorption coefficients, these reflections provide information on not only the imaginary, but also the real part of the atomic factors, which is moreover

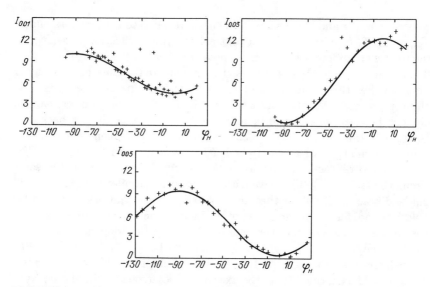

FIGURE 8.11. Observed (points) and calculated (curves) azimuthal dependence of the intensity of "forbidden" reflections $NaBrO_3$ [21]; 1347 eV unpolarized radiation (below the K edge of bromine). The points that deviate substantially from the theoretical curve correspond to Renninger reflections.

not averaged over the unit cell). They can be used to determine the phase of some of the reflection [75] and, together with polarization measurements, they can be used to establish the absolute configuration of enantiomers. In a cubic crystal, these reflections are probably the only way of observing the anisotropy of susceptibility of individual atoms.

The so-called Renninger reflections, i.e., the indirectly excited "forbidden" reflections via allowed reflection [72,76], can impede the observation of "forbidden" reflections. However, this type of multiwave diffraction is possible only for certain azimuthal angles φ_τ, which are readily calculated (the indirect excitation is actually responsible for the individual departures of the points from the smooth curves of Fig. 8.11). Moreover, interference with the Renninger reflections can be used to determine the phase of the tensor structure amplitude of "forbidden" reflections (for the scalar case, this method of solving the "phase problem" has already been used in [77–79]).

8.3   Polarization Effects in X-Ray Optical Components

In this section, we examine methods of producing and transforming the polarization of X-rays. Since polarization-transforming devices are based on diffraction effects, our entire presentation will rest on the theory presented in Sect. 8.1.

8.3.1   Polarization of Radiation Produced by X-Ray Sources

Traditional X-ray sources, i.e., X-ray tubes, produce either bremsstrahlung or characteristic radiation. It is well known that the former is polarized [18,33], but its degree of polarization is relatively low because the direction of motion of electrons emitted by cathode rapidly becomes isotropic as the electrons enter the solid target. The degree of polarization amounts to only a few percent and increases toward shorter wavelengths [80]. The characteristic emission of polycrystalline target is unpolarized. Single crystal produces partially polarized characteristic X-rays (especially in the L and M series), and this serves as a source of information about the anisotropy of environment of atoms, their chemical bonds, and so on (see, e.g., [81]), but these effects are not used to produce polarized radiation. All this means that X-rays from conventional X-ray tubes are usually polarized by diffraction-based devices capable of producing only linear polarization.

The godsend to X-ray optics has been the advent of synchrotron sources, in which radiation is emitted by electrons traveling with ultrarelativistic velocities in a magnetic field [82–85]. Synchrotron radiation from storage rings and wigglers has enormous intensity and a continuous spectrum. It is confined mostly to the orbital plane of the electrons (the divergence in the direction perpendicular to the orbital plane is of the order of $mc^2/E \sim 10^{-4}$ at energies of approximately 5 GeV) and has unique polarization properties. In particular, the synchrotron radiation emitted by an electron traveling on a circular trajectory is 100% linearly polarized in the plane of orbit, and its electric vector lies in this plane. Below and above the plane of orbit (within the divergence angle $\Delta\psi$ at right angles to the plane of orbit), the radiation is right or left elliptically polarized, depending on the sign of the projection of the direction of emission onto the angular velocity of the electron. The intensity distribution is shown in Fig. 8.12 for the linear and circular components of synchrotron radiation. Radiation with a high degree of circular polarization can be produced in specially designed wigglers [86].

Exceptional possibilities for the production of polarized radiation with predetermined type and degree of polarization are provided by a special device for generating synchrotron radiation—namely, undulators [85,87–88] (although we note that it is technically difficult to produce sufficiently

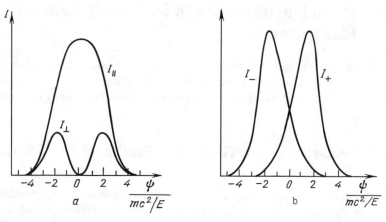

FIGURE 8.12. Intensities of (a) linear and (b) circular polarization components of synchrotron radiation as a function of the angle.

hard X-ray radiation in these devices [89]). In planar undulators, mounted in the straight gaps of electron storage rings, the radiation is always linearly polarized. In helical undulators, the electrons move on helices and the radiation is circular polarized. It has been suggested [90–92] that two planar undulators, rotated and shifted relative to one another, could be used to produce an elliptically polarized beam. The coherent combination of two phase-shifted linearly polarized waves in this system of undulators can produce practically any polarization that can be varied in the course of an experiment. Note also that the radiation emitted by particles channeled in crystals is similar to undulator radiation [93–95].

The synchrotron radiation emitted by existing accelerators is partially depolarized because of the finite size of the beam and spread in the direction of motion of electrons [96–98]. Crystal monocromators are another source of depolarization [99]. Note that because of the difference between the reflection of σ- and π-polarized X-rays, crystal monochromators can be used to alter the polarization of synchrotron radiation, for example, elliptic polarization can be trimmed down to circular polarization. However, as discussed in detail in Sect. 8.1, it must be remembered that the diffraction produces an additional phase difference between σ- and π-polarized X-rays, and partial depolarization may take place. These effects can be estimated and quantitatively taken into account by using the formulas given in Sect. 8.1, but in the case of imperfect crystal monochromators, these estimates are not very reliable because, as a rule, the qualitative parameters describing the imperfection of crystal monochromators are not accurately known.

8.3.2   Production and Analysis of Linear Polarization

Since most X-ray sources do not produce completely polarized radiation, diffraction-based polarization phenomena are widely used to produce, analyze, and transform polarized beams. Diffraction polarizers (and the corresponding analyses) with $2\theta_B = 90°$ are the most widely used (Refs. [2–5,10–14,99–101]), although the most promising polarizers are probably those based on dynamic diffraction effects, such as the Borrmann effect [98,102,103], and on the different widths of the Bragg reflection region for σ- and π-polarized X-rays [104–106] (another possibility is based on the spatial separation of σ- and π-polarized reflected beams [108]). The advantages of the 90° polarizer are its simplicity and the fact that the ability to scatter at 90° exclusively σ-polarized X-rays does not depend on the degree of perfection or the thickness of the crystal. The disadvantage of these polarizers is that a crystal and a reflection with $\theta_B = 90°$ must be chosen for each wavelength. Moreover, for $\lambda \simeq 1\,\text{Å}$, the reflections are found to be of a relatively higher order (e.g., 333 in Ge in the case of CuKα radiation), so that their intensity is relatively low, which reduces the luminosity of the polarizer.

Polarizers based on the Borrmann effect (anomalous transmission effect) exploit the fact that the absorption of Bloch waves in the diffraction region is very different from the absorption of a plane wave. The difference is due to the fact that the Bloch waves, which are superpositions of two or more plane waves, form a standing wave in a crystal. The absorption of this type of standing wave depends on whether its nodes occur at atoms or between them (in the former case, the absorption is greater and, in the second, significantly lower). The important point for us to note is that, for σ-polarization, one can select conditions under which the Bloch wave field can be made to vanish at sites occupied by the atoms, whereas this cannot be done for π-polarized X-rays because the π-polarization vectors have different directions for the direct and diffracted waves, and complete cancellation cannot be achieved. This means that the π-polarized component is absorbed to a greater extent even under the conditions of anomalous transmission, so that by selecting a thick enough crystal, we can achieve complete σ-polarization of both the transmitted and diffracted beams (in the Laue case). The main advantage of the method is that the Borrmann effect is observed for any θ_B and can readily be wavelength-tuned. Moreover, the effect is particularly well defined for strong reflections. The disadvantages are as follows: (a) high-quality crystals have to be employed; (b) for small θ_B the anomalous absorption coefficients for σ- and π-polarization are not very different and one has to use a thick crystal, which leads to the absorption also of "useful" σ-polarization; and (c) the Borrmann effect appears in a small reflection region with angular dimensions $\leq 1''$. All this restricts the range of application of polarizers based on the Borrmann effect.

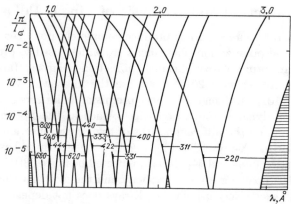

FIGURE 8.13. Polarization ratio $|R_{\pi\pi}|^2/|R_{\sigma\sigma}|^2$ achieved with the Hart–Rodrigues polarizer after four reflections in Si (different reflections with the indices as shown are conveniently used in different wavelength ranges) [105].

The Hart–Rodrigues polarizers [104,105] are probably the most promising. They are based on the difference between the Bragg reflection width for σ- and π-polarized beams (see Fig. 8.1a). These polarizing monochromators consist of two almost parallel crystals offset by an angle $\Delta\theta$ greater than the Bragg reflection range for a π-polarized beam, but less than that for the σ-polarized beam. π-polarization is significantly reduced (Fig. 8.13) after two successive reflections from the crystal. Any desired suppression of π-polarization can be achieved in principle by multiple reflection [104]. In practice, the Hart–Rodrigues polarizers are cut from a single crystal and the offset angle $\Delta\theta$ is conveniently produced by controlled elastic deformation (Fig. 8.14). The polarizers are readily wavelength-tunable, and only a few Bragg reflections are sufficient to cover the entire X-ray range usually employed. The advantages of multiple reflections result from a number of causes, including (a) the suppression of the "tails" of reflection curves; (b) the suppression of the higher harmonics ($\lambda/2$, $\lambda/3$, ...) which have a narrower Bragg reflection range; (c) when the number of reflection is even, there is no change in the direction of propagation of the beam; and (d) the π-polarization is suppressed strongly enough even in the absence of the crystal being offset.

Note that all diffraction-based polarizers have the common and significant disadvantage that the angular divergence of the polarized beam is small in the direction perpendicular to the direction of polarization (it is of the order of the angular width of the Bragg reflection region), whereas the divergence in the direction of polarization is usually greater and is determined by the divergence of the incident beam. This difference must be taken into account in diffraction experiments with polarized beams, in which it can be a source of systematic uncertainty (a detailed analysis of three-crystal diffraction systems is given in [108]. The systems with a

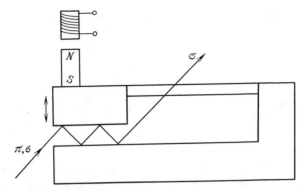

FIGURE 8.14. Hart–Rodrigues polarizer [105].

crossed polarizer and analyzer, examined in detail by Hart [3], are free from many systematic uncertainties and are capable of increasing the precision of polarization measurements.

8.3.3 The X-Ray Quarter-Wave Plate

Despite the fact that practically any polarization can be produced by using synchrotron or undulator radiation, there is a problem of producing and, especially, analyzing arbitrary polarized X-ray beams, including the transformations of linearly polarized radiation into circular polarized radiation and vice versa, i.e., there is the problem of producing the quarter-wave plate. As already noted, ordinary birefringence is very small in the X-ray region and great hopes center on the use of birefringence in the Bragg diffraction region (Sect. 8.1). These hopes have been realized experimentally [3,109,110,111,112]. Note, however, that the direct implementation of this concept in the strong diffraction scattering region involves a number of practical difficulties. The main difficulty is that because of a significant change in the phases and intensities in the diffraction region (Figs. 8.1 and 8.2), beams with a very small $\leq 1''$ angular divergence have to be employed if a high degree of polarization is to be produced. In Sect. 8.1, we already drew attention to the fact that diffractive birefringence decreases relatively slowly as the direction of propagation of the beam deviates from the Bragg angle. We shall show that the use of this birefringence outside the region of strong diffractive reflection offers us a real possibility of transforming the polarization of a beam transmitted by a crystal and, in particular, of producing a quarter-wave plate [7–9].

The magnitude of birefringence, δn, in this angular range is given by (8.17) in both the Laue and Bragg cases, and the intensity of the diffracted wave is very low (of the order of $|\chi_\tau/\Delta\theta|^2$). It follows that, in this case, the situation is, in many ways, analogous to the ordinary optics of anisotropic media, except that the magnitude of birefringence is very dependent on the direction of wave propagation.

Birefringence (8.17) ensures that the phase difference

$$\Delta\varphi = 2\pi \mathrm{Re}(\delta n)L/\lambda$$

is established between the σ- and π-polarized components as the primary wave propagates through the medium (L is the path length in crystal). A wave with linear polarization at $45°$ to the σ and π vectors is transformed into a circular polarized wave for $\Delta\varphi = \pm\pi/2$, i.e., for $L_c = \lambda/|4\mathrm{Re}(\delta n)|$.

In order that this birefringence actually be used to transform polarization, we must ensure that absorption is not too great within the path length L_c, i.e.,

$$\mu L_c \equiv \frac{\mu\lambda\Delta\theta_c}{\sin 2\theta_B \mathrm{Re}(\chi_\tau \chi_{-\tau})} \leq 1 \qquad (8.48)$$

where μ is the absorption coefficient. Since χ_τ is a function of θ_B (via the atomic factor), there must be an optimum angle for (8.48) to be satisfied, and this can be found from the condition for $\mathrm{Re}(\chi_\tau \chi_{-\tau})\sin 2\theta_B$ to be a maximum. Next, since $\mu \sim Z^4$ and $\chi_\tau \sim Z$ (Z is the atomic number), it is convenient to use crystals with low Z atoms to satisfy (8.48).

As an example, consider the propagation of CuKα radiation ($\lambda = 1.54$ Å) in diamond near the 111 reflection. Using (8.17), we find that $L_c = 2.6 \cdot 10^2 \Delta\theta_c$, where L_c is in centimeters and $\Delta\theta_c$ in radians. When $\Delta\theta_c = 0.5' = 1.45\cdot 10^{-4}$, we have $L_c = 0.038$ cm and $\exp(-\mu L_c) = 0.5$. In the wavelength range between about 3 and 0.5 Å, the difference $\Delta\theta_c$ for which $\mu L_c = 1$ is found to change from $\Delta\theta_c \simeq 1'$ to $\Delta\theta_c \simeq 0.13'$, respectively, and the conditions for the validity of this approximation are well satisfied (the intensity of the diffracted wave is less than 1% of the incident intensity). Similar results are obtained for 220 and 200 reflections in LiF (Fig. 8.15); in silicon, $\Delta\theta_c \simeq 1'$ for $\mu L_c \sim 5 - 10$.

We must now consider the effect of a significant angular divergence of the beam (the fact that the beam is not monochromatic can be taken into account similarly). Because of this divergence, $\Delta\theta$ is not strictly fixed so that the wave transmitted by the crystal is a superposition of waves with different polarizations, i.e., it is partially polarized. Suppose that the wave incident on the crystal has a linear polarization at $45°$ to the σ and π vectors, the deviation from the Bragg angle ranged from $\Delta\theta_c - (\Delta\theta'/2)$ to $\Delta\theta_c + (\Delta\theta'/2)$, and all the angles on this interval have equal probability. The mean deviation $\Delta\theta_c$ is chosen so that the transformation of the linear polarization into circular polarization occurs within this difference. It can be shown that the degree of polarization in this case is given by

$$P = \left| \int_{-1/2}^{1/2} \exp\left(i\frac{\pi}{2} \frac{\Delta\theta_c}{\Delta\theta_c + x\Delta\theta'} \right) dx \right|. \qquad (8.49)$$

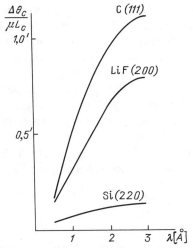

FIGURE 8.15. The difference $\Delta\theta_c$ as a function of wavelength, calculated for the quarter-wave plate for different reflections (the angular difference $\Delta\theta_c$ can be increased by increasing the plate thickness L_c).

When $\Delta\theta' \ll |\Delta\theta_c|$, we can readily show from this that, very approximately,

$$P = 1 - \frac{\pi^2}{96}\left(\frac{\Delta\theta'}{\Delta\theta_c}\right)^2. \tag{8.50}$$

It follows from this expression that $\Delta\theta'$ need not be too small in comparison to $|\Delta\theta_c|$. Even for $\Delta\theta' \simeq 0.5|\Delta\theta_c|$, the polarization calculated from (8.48) is found to be quite high: $P \simeq 0.97$.

We may therefore conclude that the transformation of X-ray polarization (including linear to circular polarization and vice versa) can be carried out with a small loss of intensity. It is important to note that the case we have considered has a number of advantages from a practical point of view, and these are not available when birefringence is used directly in the strong diffraction region. Actually, the diffraction correction to the refractive index varies rapidly within the Bragg reflection range (Figs. 8.1 and 8.2), and the necessary condition for producing a given polarization is that the divergence $\Delta\theta'$ of the incident beam must be much less than the angular width of the Bragg reflection range, i.e., it must be of the order of a fraction of a second of an arc. On the other hand, in the case we are considering, we have to satisfy the less stringent inequality $\Delta\theta' \ll |\Delta\theta_c|$, which is, in fact, more readily satisfied because $|\Delta\theta_c|$ can be of the order of a few minutes of an arc. Moreover, there are several complicating factors in the diffraction region (Sect. 8.1), including the presence of several waves, the Borrmann effect, and so on, whereas in our case, the situation is similar to that of ordinary optics.

A further advantage of this approach is that, outside the Bragg reflection region, birefringence is not very dependent on the degree of perfection of the crystal (Sect. 8.1). The above expressions can therefore be used even for relatively imperfect crystals, provided the deviation from the Bragg angle is much greater than the width of the reflection curve.

To conclude this section, we note that elliptic polarization can be obtained from linear polarization by diffraction in the Bragg geometry if we use the phase difference between diffracted waves with σ- and π-polarizations (Fig. 8.1a) [113,114]. However, the practical implementation of this idea involves a number of difficulties. For example, several successive reflections are necessary to produce circular polarization in this geometry [113].

8.3.4   Pendelosung

This is one of the well-defined dynamic effects arising as a result of interference between Bloch waves. We recall (Sect. 8.1) that for each polarization (σ and π), there are two Bloch eigenwaves with different wave vectors. This difference gives rise to intensity beats (Pendelosung) in both the wave transmitted by the crystal and in the diffracted waves (8.7, 8.8). These beats can be observed, depending on the angle of deviation from the Bragg condition and the path traversed in the crystal. The beats in the dependence on the crystal thickness occur even in integrated parameters (Figs. 8.2 and 8.3). The significant point is that because of the difference between the scattering amplitudes for σ- and π-polarized X-rays, the oscillation periods for these two polarizations are different, and this can give rise to the mutual extinguishing of the oscillations [34,115,116] if the incident radiation is unpolarized, circularly polarized, or polarized at an angle 45° to the σ- and π-polarizations. It is well known that the above beats can be used in the precise measurement of structure amplitudes. It is therefore desirable to use σ- or π-polarized radiation to increase the contrast of the interference pattern (and the precision with which the structure amplitude can be measured) [117]. Pendelosung is also observed in the polarization characteristics of beams [2], such as the angle of rotation and the ratio of the axes of the polarization ellipse (if polarization of the incident beam is different from σ- or π-polarization), but the beats are less well defined in the degree of polarization.

To conclude this section, let us briefly enumerate the applications of polarized X-rays that were not mentioned above. First, there is the topography of almost perfect crystals. In this case, a small distortion of the lattice can be observed if the width of the reflection curve is as small as can be achieved by using π-polarized radiation (it is shown in [118] that a relative distortion of the lattice on the order of 10^{-8} can be observed). The diffraction of π-polarized radiation for angle $2\theta_B$ approaching 90° can be used to eliminate extinction in the case of precise measurement of structure amplitude [119,120]. The absence of scattered π-polarized radiation at 90° is also used to improve the signal-to-noise ratio in X-ray fluorescence analysis [121,122] and in Mössbauer filtration of synchrotron radiation [123]. There is no doubt that X-ray polarization phenomena will find new application and be extended still further.

8.4   Conclusions

X-ray polarization phenomena and their applications (both possible and already implemented) discussed in this chapter show that this method of investigating the properties of solids is informative and promising. Although the practical implementation of the method involves increasingly complicated equipment (including synchrotron radiation sources), all this is justified by the unique character of the data obtained with this approach. A clear picture of possible future developments in this area can be obtained by enumerating some of the unsolved problems. They include, above all, the complete polarization measurements on X-ray beams, i.e., the determination of all three polarization parameters, namely, orientation, and ratio of the axes of the polarization ellipse. At present, the main difficulty is to produce and, especially, analyze circular polarization. This difficulty will be overcome once the quarter-wave plate, which is in practice convenient to use, has become available. "Forbidden" reflections and those connected to the anisotropy of susceptibility (Sect. 8.2.3) can also be used to analyze circular polarizations. Note that the problem of complete polarization measurement can be solved in the case of Mössbauer radiation, which also lies in the X-ray region. Polarizers and analyzers of arbitrary polarization have been developed for Mössbauer radiation [124–128]. X-ray polarization phenomena can also be investigated in ordinary (non-Mössbauer) crystals by using the Mössbauer detection technique in which the source of radiation is a Mössbauer source instead of the X-ray tube (the range of validity of this method is limited by the discrete character of the wavelength employed [129] and, mostly, by the low intensity of Mössbauer sources).

Polarization phenomena under the conditions of multiwave diffraction [130–135] have not been adequately investigated. When nonplanar multiwave geometry is employed, the eigenpolarizations are almost always different from σ and π and, moreover, do not remain constant in the diffraction region. It may be expected, for example, that a particular circular polarization will be preferentially reflected, and there may be some other unusual effects. Note also that in the case of noncoplanar multiwave diffraction by imperfect crystals, the equation for the polarization tensor has to be employed even when the intensities of diffracted waves are calculated. Studies of the polarization of X-ray beams in resonators [136–138] and in diffraction focusing [139–140] have only just begun. The polarization of X-rays that have undergone inelastic scattering or scattering by defects has not been adequately investigated. The polarization of Compton-scattered photons carries information on the momentum distribution of polarized electrons, whereas the polarization of diffusely scattered X-rays can be used to determine the presence of dynamic effects [141]. Relatively little attention has been devoted to polarization effects accompanying diffraction under the conditions of total external reflection [142–145]. Studies of the rotation of the plane of polarization of X-rays in optically active media have only just commanced [3,5,38,146]. Certain technical problems remain unsolved, for example, there is no metrologic basis for polarization measurements in the X-ray range [147,148]. In view of the increasing number of available synchrotron radiation sources and advances in the techniques of traditional X-ray measurements, it is expected that there will be rapid expansion in the study and application of polarization effects in X-ray optics, and rapid progress in the areas enumerated above shall result.

References

1. C.G. Barkla: *Proc. R. Soc. London* **77**, 247 (1606).

2. I.P. Mikhailuk, S.A. Kshevetskii, M.V. Ostapovich, V.P. Shafranyuk: *Ukr. Fiz. Zh.* **22**, 61 (1977).

3. M. Hart: *Phylos. Mag. Ser. B* **38**, 41 (1978).

4. V.A. Belyakov, V.E. Dmitrienko: *Sov. Phys.-Usp.* **32**, 697 (1989).

5. Yu.V. Ponomarev, Yu.A. Turutin: *Sov. Phys. Tech. Phys.* **27**, 129 (1982).

6. L.V. Azaroff, D.M. Pease: *X-Ray Spectroscopy*, ed. by L.V. Azaroff (New York, McGraw-Hill, 1974, p. 284).

7. N.V. Baranova, B.Ya. Zel'dovich: *Sov. Phys. JETP* **52**, 900 (1980).

8. V.E. Dmitrienko, V.A. Belyakov: *Sov. Tech. Phys. Lett.* **6**, 616 (1980).

9. V.A. Belyakov, V.E. Dmitrienko: *Sov. Phys. Crystallogr.* **27**, 6 (1982).

10. S. Chandrasekhar: *Adv. Phys.* **9**, 363 (1960).

11. N.M. Olekhnovich, V.L. Markovich: *Sov. Phys. Crystallogr.* **23**, 369 (1978).

12. L.D. Jennings: *Acta Crystallogr.* **A24**, 584 (1968).

13. N.M. Olekhnovich: *Izv. Akad. Nauk BSSR Ser. Fiz-Mat. Nauk.* **4**, 57 (1980).

14. N.M. Olekhnovich: *Polarization of Diffraction and the Dynamic Scattering of X-Rays in Real Crystals*, Ph.D. Thesis, Ukrainian Institute of Physics, Kiev, 1987 (in Russian).

15. V.A. Belyakov, V.E. Dmitrienko: *Acta Crystallogr.* **A36**, 1044 (1989).

16. S. Chandrasekhar: *Radiative Transfer* (Clarendon Press, Oxford, 1950).

17. G.V. Rozenberg: *Sov. Phys. Usp.* **20**, 55 (1977).

18. V.L. Ginzburg: *Theoretical Physics and Astrophysics* (Oxford, Pergamon Press, 1979).

19. Yu.N. Barabanenkov: *Sov. Phys. Usp.* **18**, 673 (1975).

20. D.H. Templeton, L.K. Templeton: *Acta Crystallogr.* **A36**, 237 (1980); **A41**, 133 (1985).

21. D.H. Templeton, L.K. Templeton: *Acta Crystallogr.* **A42**, 478 (1986).

22. V.E. Dmitrienko: *Acta Crystallogr.* **A39**, 29 (1983).

23. V.E. Dmitrienko: *Acta Crystallogr.* **A40**, 89 (1984).

24. P.M. Platzman, N. Tzoar: *Phys. Rev.* **B2**, 3556 (1970).

25. F. DeBergevin, M. Brunel: *Acta Crystallogr.* **A37**, 314 (1981); M. Brunel, F. DeBergevin: *Acta Crystallogr.* **A37**, 324 (1981).

26. O.L. Zhizhimov, I.B. Khriplovich: *Sov. Phys. JETP* **60**, 313 (1984).

27. M. Blume: *J. Appl. Phys.* **57**, 3615 (1985).

28. D. Gibbs, D.E. Moncton, K.L. D'Amico, J. Bohr, B.H. Grier: *Phys. Rev. Lett.* **55**, 234 (1985).

29. V.I. Iveronova, G.P. Revkevich: *Theory of X-Ray Scattering* (Moscow University, 1978) (in Russian).

30. J.M. Cowles: *Diffraction Physics* (Amsterdam, North-Holland, 1979).

31. Z.G. Pinsker: *X-Ray Crystal Optics* (Moscow, Nauka, 1982) (in Russian).

32. B.K. Vainshtein: *Modern Crystallography* (Berlin, Heidelberg, Springer-Verlag, 1981).

33. A.I. Akhiezer, V.B. Berestetskii: *Quantum Electrodynamics* (New York, Wiley-Interscience, 1965).

34. P. Shalihy, C. Malgrange: *Acta Crystallogr.* **A28**, 501 (1972).

35. S. Annaka: *J. Phys. Soc. Jpn.* **51**, 1927 (1982).

36. S. Annaka, T. Suzuki, K. Onoue: *Acta Crystallogr.* **A36**, 151 (1980).

37. G.G. Cohen, M. Kuriyama: *Phys. Rev. Lett.* **40**, 957 (1978).

38. M. Hart, A.R.D. Rogrigues: *Phys. Mag. Ser. B* **43**, 321 (1981).

39. C.C. Darwin: *Phil. Mag.* **43**, 800 (1922).

40. W.H. Zahariazen: *Acta Crystallogr.* **23**, 558 (1967).

41. P. Becker: *Acta Crystallogr.* **A33**, 243 (1977).

42. V.E. Dmitrienko: *Sov. Phys. Crystallogr.* **27**, 131 (1982).

43. V.A. Belyakov, V.E. Dmitrienko: *Sov. Phys. JETP* **46**, 356 (1977).

44. S. Ramasachann, G.N. Ramachandran: *Acta Crystallogr.* **6**, 364 (1953).

45. S. Chandrasekhar: *Acta Crystallogr.* **9**, 954 (1956).

46. K.S. Chandrasekharan: *Acta Crystallogr.* **12**, 916 (1959).

47. N.M. Olekhnovich: *Sov. Phys. Crystallogr.* **14**, 724 (1970); **I8**, 1240 (1973); *Sov. Phys. Dokl.* **I8**, 728 (1973).

48. N.M. Olekhnovich, M.P. Shmidt: *Izv. Akad. Nauk SSSR Ser. Fiz. Mat. Nauk* **n4**, 110 (1975); **n4**, 115 (1974); **nI**, 118 (1977).

49. N.M. Olekhnovich, V.A. Rubtsov, M.P. Shmidt: *Sov. Phys. Crystallogr.* **20**, 488 (1975).

50. N.M. Olekhnovich, V.L. Markovich, A.I. Olekhnovich: *Acta Crystallogr.* **A36**, 989 (1980).

51. N.M. Olekhnovich, V.L. Markovich, A.I. Olekhnovich: *Sov. Phys. Crystallogr.* **27**, 533 (1982); N.M. Olekhnovich, A.L. Karpei, A.I. Olekhnovich: *Izv. Akad. Nauk. BSSR Ser. Fiz. Mat. Nauk* **n5**, 43 (1983).

52. N.M. Olekhnovich, A.V. Pushkarev: *Dokl. Akad. Nauk BSSR* **296**, 38 (1985).

53. N.M. Olekhnovich, A.L. Karpei, V.L. Markovich: *Krist. Tech.* **I3**, 1463 (1978); N.M. Olekhnovich, A.V. Pushkarev: *Izv. Akad. Nauk BSSR Ser. Fiz. Mat. Nauk* **n6**, 53 (1983).

54. N. Kato: *Acta Crystallogr.* **A32**, 453, 458 (1976); **36**, 171, 763, 770 (1980).

55. S.P. Darbinyan, I.A. Vartan'yants, F.N. Chukhovskii: *Fizika* **8–9**, 67 (1987).

56. V.A. Bushuev: *Sov. Phys. Crystallogr.* **34**, 163 (1989).

57. L.D. Jennings: *Acta Crystallogr.* **A37**, 584 (1981); **40**, 12 (1984).

58. Y. LePage, E.J. Gabe, L.D. Calvert: *J. Appl. Crystallogr.* **12**, 25 (1979); M.G. Vincent, H.D. Flack: *J. Appl. Crystallogr.* **36**, 610, 614 (1980).

59. A.Mcl. Mathieson: *Acta Crystallogr.* **A34**, 404 (1978); H.D. Flack, M.G. Vincent: *Acta Crystallogr.* **A34**, 620; J.L. Lawrence: *Acta Crystallogr.* **38**, 859 (1982).

60. A.V. Vinogradov, I.V. Kozhevikov: Preprint FIAN SSSR n 102 (Moscow 1987) (in Russian).

61. A.V. Kolpakov, V.A. Bushuev, R.N. Kuz'min: *Sov. Phys. Usp.* **21**, 959 (1978).

62. M. Born, E. Wolf: *Principles of Optics* (Oxford, Pergamon Press, 1975).

63. L.D. Landau, E.M. Lifshitz: *Electrodynamics of Continuous Media* (Oxford, Pergamon Press, 1984).

64. S. Ramaseshan, S.C. Abrahams (eds.): *Anomalous Scattering* (Copenhagen, Munksgaard, 1975).

65. J.E. Hahn, K.O. Hodgson: *Inorganic Chemistry: Toward the 21st Century* (Washington, 1983, p. 431).

66. D.H. Templeton, L.K. Templeton: *Acta Crystallogr.* **A41**, 365 (1985).

67. G.A. Korn, T.M. Korn: *Mathematical Handbook for Scientists and Engineers* (New York, McGraw-Hill, 1961).

68. A.V. Shubnikov, V.A. Koptsik: *Symmetry in Science and Art* (New York, Plenum, 1974).

69. V.A. Belyakov: *Sov. Phys. Usp.* **8**, 267 (1975).

70. V.A. Belyakov, V.E. Dmitrienko, V.P. Orlov: *Sov. Phys. Usp.* **22**, 63 (1979).

71. *International Tables for X-Ray Crystallography* (Birmingham, Kynoch Press, 1952).

72. M. Renninger: *Z. Phys.* **106**, 141 (1937).

73. B. Dawson: *Proc. R. Soc. London* **298**, 255, 264, 379 (1967).

74. V.A. Belyakov: *Sov. Phys. Solid State* **13**, 2789 (1972).

75. D.H. Templeton, L.K. Templeton: *Acta Crystallogr.* **A43**, 573 (1987).

76. Yu.S. Terminasov, L.V. Tuzov: *Sov. Phys. Usp.* **7**, 434 (1964).

77. B. Post, J. Ladell: *Acta Crystallogr.* **A43**, 173 (1987).

78. J.Z. Tischler, Q. Shen, R. Colella: *Acta Crystallogr.* **A41**, 451 (1985).

79. E.K. Kov'ev, V.I. Simonov: *JETP Lett.* **43**, 312 (1986).

80. M.A. Blokhin: *Physics of X-Rays* (Moscow, Gostekhizdat, 1957), (in Russian).

81. A. Maizel, G. Leonkhardt, R. Sargan: *X-Ray Spectra and Chemical Bonds* (Kiev, Naukova Dumka, 1981) (in Russian).

82. G.N. Kulipanov, A.N. Skrinski: *Sov. Phys. Usp.* **122**, 369 (1979).

83. I.M. Ternov, V.V. Mikhailin, V.R. Khalilov: *Synchrotron Radiation and Its Applications* (Moscow University, 1980) (in Russian).

84. C. Kunz (ed.): *Synchrotron Radiation* (Berlin, Springer-Verlag, 1979).

85. M.M. Niktin, V.Ya. Epp: *Undulator Radiation* (Moscow, Energoatomizdat, 1988) (in Russian).

86. J. Goulon, P.E. Elleame, D. Raoax: *Nucl. Instrum. Methods* **A254**, 192 (1987).

87. D.F. Alferov, Yu.A. Bashmakov, E.G. Bessonov: *Tr. Fiz. Inst. Akad. Nauk SSSR* **80**, 100 (1975).

88. D.F. Alferov, Yu.A. Bashmakov, P.A. Cherenkov: *Sov. Phys. Usp.* **32**, 200 (1974).

89. E.S. Gluskin, S.V. Gaponov, S.A. Gusev, P. Dhez, P.P. Ilyinsky, Yu.Ya. Platonov, N.N. Salashenko, Yu.M. Shatunov: Preprint INP 83–163 (Novosibirsk, 1983).

90. M.B. Moiseev, M.M. Nikitin, N.I. Fedoseev: *Izv. Vyssh. Uchebn. Zaved. Fiz.* **3**, 76 (1978).

91. K.J. Kim: *Nucl. Instrum. Methods* **A219**, 425 (1984).

92. H. Onuki: *Nucl. Instrum. Methods* **A246**, 94 (1986).

93. V.G. Baryshevskii: *Channeling, Radiation and Reactions in Crystals at High Energies* (Minsk, Belorussian University, 1982) (in Russian).

94. M.A. Kumakhov: *Radiation by Channelled Particles in Crystals* (Moscow, Energoatomizdat, 1986) (in Russian).

95. V.A. Bazylev, N.K. Zhevago: *Radiation by Fast Particles in Matter and External Fields* (Moscow, Nauka, 1987) (in Russian).

96. G. Matterlik, P. Suortti: *J. Appl. Crystallogr.* **17**, 7 (1984).

97. H. Derenbach, R. Malutzki, V. Schmidt: *Nucl. Instrum. Methods* **A208**, 845 (1983).

98. D.H. Templeton, L.K. Templeton: *J. Appl. Crystallogr.* **21**, 151 (1988).

99. F. Riehly: *Nucl. Instrum. Methods* **A246**, 385 (1986).

100. E.S. Gluskin, S.V. Gaponov, P. Dhez, P.P. Ilyinsky, N.N. Salashenko, Yu.M. Shatunov, E.M. Trakhtenberg: *Nucl. Instrum. Methods* **A246**, 394 (1986).

101. P. Dhez: *Nucl. Instrum. Methods* **A246**, 66 (1986).

102. H. Cole, F.W. Chambers, C.G. Wood: *J. Appl. Phys.* **32**, 1942 (1961).

103. J.L. Staudenmann, L.D. Chapman, W.J. Murphy, R.D. Horning, G.L. Liedel: *J. Appl. Phys.* **18**, 519, 724 (1985).

104. M. Hart, A.R. Rodrigues: *Philos. Mag. Ser.* **B40**, 149 (1979).

105. M. Hart: *Lect. Not. Phys.* **112**, 325 (1980).

106. G.G. Avetisyan: *Izv. Akad. Nauk Arm. ssr Tekh. Nauk* **39**, 47 (1986).

107. A.V. Andreev, V.E. Gorshkov, Yu.A. Il'inskii: *Sov. Phys. Tech. Phys.* **32**, 308 (1987).

108. J. Hardy, E. Krousky: *Czech. J. Phys. B* **23**, 966 (1973).

109. D.M. Mills: *Phys. Rev.* **36**, 6178 (1987).

110. G. Schütz, W. Wagner, W. Wilhelm, P. Kienle, R. Zeller, R. Frahm, G. Materlik: *Phys. Rev. Lett.* **58**, 737 (1987).

111. J.A. Golovchenko, B.M. Kincaid, R.A. Levesque, A.E. Meixner, D.R. Kaplan: *Phys. Rev. Lett.* **57**, 202 (1973).

112. O. Brümmer et al.: *Z. Naturforsch.* **37a**, 524 (1982).

113. A.G. Grigoryan, P.A. Bezirganyan, S.A. Aladzhadzhan: Abstracts of All-Union Conference on Methods and Equipment for Research into the Coherent Interaction of Radiation with Matter (Moscow, Institute for Atomic Energy, 1980, p. 66) (in Russian).

114. O. Brümmer, Ch. Eisenschmidt, H.R. Höche: *Acta Crystallogr.* **A40**, 394 (1985).

115. H. Hattory, H. Kuriyama, N. Kato: *J. Phys. Soc. Jpn.* **20**, 1047 (1965).

116. M. Hart, A.R. Lang: *Acta Crystallogr.* **A17**, 73 (1965).

117. K. Utemisov, S.Sh. Shil'shtein, V.A. Somenkov: *Sov. Phys. Crystallogr.* **26**, 711 (1983).

118. U. Bonse et al.: *Nucl. Instrum. Methods* **A208**, 711 (1983).

119. A. Matheson: *Acta Crystallogr.* **A33**, 133 (1984).

120. W.B. Yelon, B. van Laar: *Acta Crystallogr.* **A40**, 261 (1984).

121. T.G. Duzuberg, B.V. Jarrett, J.M. Jaklevic: *Nucl. Instrum. Methods* **115**, 297 (1974).

122. R.W. Ryon, J.D. Zahrt: *Adv. X-Ray Anal.* **22**, 453 (1979).

123. V.A. Belyakov: *Sov. Phys. Usp.* **30**, 331 (1987).

124. V.G. Labushkin, S.N. Ivanov, G.V. Chechin: *JETP Lett.* **20**, 157 (1974).

125. U. Gonser, H. Fisher: in "Mössbauer Spectroscopy II," *The Exotic Side of the Method* ed. by U. Gonser (Berlin, Springer-Verlag, 1981, p. 125).

126. R.Ch. Bokun: Abstracts of All-Union Conference on Methods and Equipment for Research into the Coherent Interaction of Radiation with Matter (Moscow, Institute for Atomic Energy, 1982, p. 77) (in Russian).

127. A.L. Sharma, M. Alimuddin, K.R. Reddy: *Nuovo Cimento D* **5**, 147 (1985).

128. P.P. Kovalenko, V.G. Labushkin, E.R. Sarkisov, A.K. Ovsepyan, I.G. Tolpekin: *Instrum. Exp. Tech.* **29**, 577 (1985).

129. B. Kolk: *Phys. Lett.* **A50**, 457 (1975).

130. S.L. Chang: *Multiple Diffraction of X-Rays by Crystals* (Berlin, Springer-Verlag, 1985).

131. A.M. Afanas'ev, V.G. Kohn: *Phys. Status Solid* **A28**, 61 (1975); *Acta Crystalogr.* **A33**, 178 (1977).

132. I.D. Feranchuk: *Izv. Akad. Nauk SSSR Ser. Fiz. Mat. Nauk* **4**, 109 (1981).

133. S.A. Kshevetskii et al.: *Sov. Phys. Crystallogr.* **30**, 270 (1985).

134. H.J. Juretschke: *Phys. Status Solid B* **135**, 455 (1986).

135. R. Colella: *Z. Naturforsch.* **A37**, 437 (1986).

136. A.M. Egiazaryan, P.A. Bezirganyan: *Izv. Akad. Nauk Arm. SSR Fiz.* **14**, 261 (1979).

137. S.A. Kshevetskii, M.L. Kshevetskaya, et al.: *Ukr. Fiz. Zh.* **25**, 781 (1980).

138. S.A. Kshevetskii, M.L. Kshevetskaya et al.: *Ukr. Fiz. Zh.* **31**, 1059 (1986).

139. V.I. Kushnir et al.: *Acta Crystallogr.* **A41**, 17 (1985).

140. V.A. Baskakov, B.Ya. Zel'dovich: Preprint FIAN SSSR n 191 (Moscow, 1978).

141. V.E. Dmitrienko, V.M. Kagane: *Metallofizika* **9**, 71 (1987).

142. V.G. Baryshevskii: *Sov. Techn. Phys. Lett.* **2**, 43 (1976).

143. M.A. Andreeva, S. Borisova: *Sov. Phys. Crystallogr.* **30**, 849 (1985).

144. A.V. Andreev: *Sov. Phys. Usp.* **28**, 70 (1985).

145. A.M. Afanas'ev, P.A. Aleksandrov, R.M. Imamov: *X-Ray Structure Diagnostics in the Investigation of the Surfaces of Single Crystals* (Moscow, Nauka, 1986) (in Russian).

146. M. Sauvage, C. Malgrange, J.E. Petroff: *J. Appl. Phys.* **16**, 14 (1983).

147. E.A. Volkova: *Polarization Measurements* (Moscow, Izd. Standartov, 1974) (in Russian).

148. J. Fischer et al.: *Nucl. Instrum. Methods* **A246**, 404 (1986).

9

Magnetic X-Ray Scattering

In the previous chapter, we discussed the anisotropy of X-ray susceptibility and the physical effects due to this anisotropy. As was mentioned, one of the sources of that anisotropy is connected to the magnetic properties of solids, i.e., dependence of the interaction of X-rays with solids on the orbital and spin moments of the atoms in solids. Formally the magnetic anisotropy of the X-ray susceptibility was also discussed in the previous chapter, but the magnetic dependence of the X-ray interaction with crystals, as the latest investigations proved, appears to be very promising in the study of crystal magnetic structures. Recently, there has been a burst of interest in the magnetic interaction of X-rays with crystal, especially in connection to the fast growth of synchrotron radiation use in the problems of X-ray optics. There is an opinion that the magnetic dependence of X-ray scattering, due to application for investigations of synchrotron radiation, forms the basis for a new practical method of studying crystal magnetic structures. It is why the questions of magnetic X-ray scattering are addressed in a separate chapter.

9.1   Magnetic Scattering

The principal contribution to the scattering of X-ray is due to the Thomson mechanism, whereby X-rays are scattered by the charge of an electron. Classically speaking, this can be looked on as a dipole emission by a charge accelerated in the electric field of the X-ray wave. Since X-rays are part of the electromagnetic spectrum, one should expect that they will be sensitive to magnetic as well as charge distribution. Indeed, this sensitivity has long been used in analyzing polarization effects in Compton scattering, the cross section of which at a polarized electron is dependent on the sense of circular polarization of a photon [1]. However, because this dependence in the general case is extremely weak for the X-ray energies usually used in X-ray diffraction, it may be ignored in the experiment, and only in 1970 did Platzman and Tzoar [2] first point out the possibility of using this dependence in the study of magnetization densities in solids in the same manner as has been done with neutron scattering.

The interaction of X-ray with magnetic moments can be explained in the following way. In addition to its charge, the electron also has an intrinsic

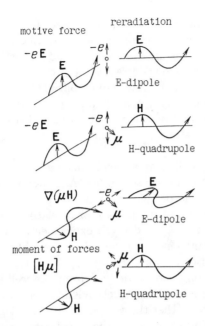

FIGURE 9.1. Magnetic scattering of a photon by an atom in classical electrodynamics [3].

magnetic moment $-2\mu_b s$ and a magnetic moment due to its orbital motion in the atom $-\mu_b l$ (where μ_b is the Bohr magneton; s and l are the
spin and orbital moment, respectively). The interaction of the X-ray wave
with these moments leads to magnetic scattering that is sensitive to the
magnetic structure of solids. In classical language, the principal channels
of this magnetic scattering can be described as follows [3,4] (Fig. 9.1): (1)
magnetic quadrupole emission by the magnetic moment μ moving under
the influence of the force $-e\mathbf{E}$, (2) electric dipole emission of a charge accelerated by the force $-\nabla(\mu\mathbf{H})$, and (3) magnetic dipole emission by the
magnetic moment due to rotation under the influence of the torque $[\mu\mathbf{H}]$.

It is clear from the foregoing that the polarization properties of magnetic
scattering of X-ray are very different from those of Rayleigh scattering. Phe-
nomenologically speaking, the presence of magnetic scattering corresponds
to extra terms of the form $ie_{jkn}s_n$ in the susceptibility (e_{jkn} is a completely
antisymmetric unit tensor) and to extra terms that depend explicitly on
the wave vectors of the incident and scattered waves (equivalent to taking
into account spatial dispersion). These terms lead, in particular, to different scattering of right- and left-circular polarized waves and to the rotation
of the plane of polarization (Faraday effect).

The absolute magnitude of the magnetic scattering amplitude per electron is smaller by the factor $\hbar\omega/(mc^2)$ that the Rayleigh scattering amplitude, and some of the terms are smaller by the factor p/mc where p

is the typical electron momentum in the atom. In the X-ray region, both estimates yield a figure of approximately 0.01. Since, in the atom, only a small number of electrons have uncompensated magnetic moment, and the form factor of these electrons decreases rapidly with an increasing angle of scattering, one finds that the Rayleigh amplitude for an atom is greater by three or four orders of magnitude. Despite the fact that magnetic scattering is weak, the availability of synchrotron radiation sources ensures that it can be a working method for investigation of the magnetic properties of solids.

The quantum-mechanical approach is, of course, essential for the quantitative description of magnetic scattering. Although the corresponding formulas, in which the photon scattering cross section is found as a function of the electron spin, have been known for a relatively long time (for both free and bound electrons [1,4–6]), the experimental observation of X-ray magnetic scattering influenced by the publication of Ref. [2] was performed only in 1972 by De Bergevin and Brunel [7]. The simplest formula describing magnetic scattering can be derived from the well-known formula for Compton scattering [1] in the nonrelativistic approximation. The spin of the free electron ensures that the scattering cross section (Compton effect) is different, depending on whether the circular polarizations are parallel $(+)$ or antiparallel $(-)$ to the spin direction

$$d\sigma^{\pm}/d\Omega = (r_e^2/2)[1 + \cos^2\theta \pm 2\hbar\omega/(mc^2)\cos\theta(\cos\theta - 1)] \qquad (9.1)$$

where ϑ is the scattering angle.

The first experiments on Compton scattering of polarized gamma rays were performed by Sakai and Ono [8], who used 122 keV gamma rays from a radioactive Co^{57} source. This showed that magnetic Compton scattering could be observed and used to investigate the momentum distribution of polarized electrons in a crystal (see also [9,10], in which circular polarized synchrotron radiation was employed). A more detailed theory of magnetic Compton scattering is given in [9,10].

We now turn to a more detailed examination of coherent elastic scattering, in which the atomic system remains in the initial state after the scattering event [1,11–14]. Here we can use the nonrelativistic Hamiltonian for the interaction between the electrons and electromagnetic field:

$$H_1 = \frac{e^2}{(mc^2)}\mathbf{A}^2 - \frac{e}{(mc)}\mathbf{pA} - [\mu\mathbf{H}] = \frac{e^2}{(mc^2)}\mathbf{A}^2 - \frac{e}{(mc)}\mathbf{p'A} \qquad (9.2)$$

where $\mathbf{p'} = \mathbf{p} + (mc/e)[\mu\nabla]$, $\mathbf{p}$ and μ are the momentum and magnetic moment operators of the electron, and $\mathbf{A}$ is the vector potential of the field. The term containing $\mathbf{A}^2$ is responsible for Thomson scattering in first-order perturbation theory and the term $\mathbf{p'A}$ gives the dispersion correction to the scattering amplitude in the second approximation. This automatically includes magnetic scattering [1]:

$$\Delta \mathbf{f} = (e/mc)^2 \sum \left[\frac{\langle O|\mathbf{p}^{s'}\mathbf{A}_1^*|m\rangle\langle m|\mathbf{p}^{s'}\mathbf{A}_0|O\rangle}{E_0 - E_m + \hbar\omega - (i\Gamma_m/2)} \right.$$
$$\left. + \frac{\langle O|\mathbf{p}^{s'}\mathbf{A}_0|m\rangle\langle m|\mathbf{p}^{s'}\mathbf{A}_1^*|O\rangle}{E_0 - E_m - \hbar\omega} \right] \tag{9.3}$$

[compare with (8.1)]. A detailed analysis of magnetic scattering both based on (9.2) and with an allowance for a higher-order relativistic correction is given in [13,14].

As an example, consider the case of high frequencies, for which $\hbar\omega \gg (E_m - E_0)$. Summation over intermediate states $|m\rangle$ can then be carried out in (9.3) in an elementary manner, and Δf contains two terms (spin and orbital) that depend on vectors $\mathbf{S}$ and $\mathbf{P}$, respectively:

$$\Delta f = r_e \left(\frac{\mathbf{PC}}{mc} - i\frac{\hbar\omega}{mc^2}\mathbf{SB} \right) \tag{9.4}$$

where

$$\mathbf{P} = \left\langle O \left| \sum e^{i\mathbf{kr}_s} \mathbf{p}^s \right| O \right\rangle, \qquad \mathbf{S} = \left\langle O \left| \sum e^{i\mathbf{kr}_s} \mathbf{s}^s \right| O \right\rangle$$
$$\mathbf{B} = [\mathbf{e}_1^*\mathbf{e}_0] + [\hat{\mathbf{k}}_1\mathbf{e}_1^*](\hat{\mathbf{k}}_0\mathbf{e}_0) - [\hat{\mathbf{k}}_0\mathbf{e}_0](\hat{\mathbf{k}}_0\mathbf{e}_1^*) - \{[\hat{\mathbf{k}}_1\mathbf{e}_1^*][\hat{\mathbf{k}}_0\mathbf{e}_0]\}$$
$$\mathbf{C} = \mathbf{e}_0(\hat{\mathbf{k}}_0\mathbf{e}_1^*) + \mathbf{e}_1^*(\hat{\mathbf{k}}_1\mathbf{e}_0), \qquad \mathbf{k} = \mathbf{k}_1 - \mathbf{k}_0.$$

The significant point here is that the spin and orbital terms have different dependence on both the polarization $\mathbf{e}_0$ and $\mathbf{e}_1$ and the directions $\hat{\mathbf{k}}_0 = \mathbf{k}_0/|\mathbf{k}_0|$ and $\hat{\mathbf{k}}_1 = \mathbf{k}_1/|\mathbf{k}_1|$ of the incident and scattered beams, so that we can separate the contributions due to these terms and extract magnetic scattering against the background of Rayleigh scattering. For example, in general, magnetic scattering is different for right- and left-polarized beams, so that when this is taken into account, the σ- and π-polarizations are no longer the eigenpolarizations, and so on. A more detailed analysis of the polarization properties of magnetic scattering can be performed in precisely the same way as in Sect. 8.2.3 (see also [15]). Note that from a practical point of view, the kinematic theory almost always suffices for a description of magnetic diffraction, although dynamic analysis has also been carried out [16,17].

To give the reader a general idea of the relative magnitude of the strength of magnetic scattering [14] compared to the other mechanisms of scattering, some estimates are presented below. Pure charge X-ray scattering is larger than pure magnetic scattering by a significant factor:

$$\frac{\sigma_{\text{mag}}}{\sigma_{\text{charge}}} \simeq \left(\frac{\omega}{mc^2} \right)^2 \frac{N_m^2}{N^2} \langle s \rangle \frac{f_m^2}{f} \tag{9.5}$$

where N_m is the number of magnetic electrons per atom, N is the total number of electrons per atom, and f_m and f are the magnetic and charge form factors. For Fe and 10 keV photons,

$$\frac{\sigma_{\text{mag}}}{\sigma_{\text{charge}}} \sim 4 \times 10^{-6} \langle s \rangle^2.$$

Also, the magnetic form factor of atoms falls more rapidly in comparison with the charge form factor because the magnetic density is more diffuse spatially than the charge density. This reduces the ratio even further. Note that the factor $\langle s \rangle$, which goes to zero at the Curie point, is unity only at low temperatures.

The ratio of the magnetic term for X-ray and neutrons is approximately

$$\frac{I_0^x \sigma_{\text{mag}}^x}{I_0^n \sigma_{\text{mag}}^n} \approx \frac{1}{4} \left(\frac{\omega}{mc^2} \right)^2 \frac{I_0^x}{I_0^n} \approx \frac{1}{4} \times 10^{-4} \frac{I_0^x}{I_0^n} \qquad (9.6)$$

where I^x and I^n are the X-ray and neutron densities of the beams, respectively. Neutron sources which give $\sim 10^8$ neutrons/sec on a sample and an X-ray source which gives $\sim 10^{12}$ photons/sec (monochromatic) will yield comparable X-ray and neutron diffraction peaks.

The relative strength of magnetic effects can be increased if the interference between the charge and magnetic scattering is measured. The interference term which may be distinguished at a crystalline reflection (e.g., by means of two measurements with different circular polarization of the incident beam) is linear in the magnetic scattering amplitude and its relative magnitude is $\hbar\omega/mc^2$.

Despite the fact that the cross section for magnetic scattering of X-ray is lower by four to five orders of magnitude than the cross section for magnetic scattering of neutrons, this difficulty can be overcome by the high intensity of synchrotron radiation sources [18]. Thus, while the pioneering work [7,13] on the observation magnetic reflections in NiO and Fe_2O_3, using conventional X-ray tubes, required enormous effort (see [19]), the use of synchrotron radiation sources has resulted in the investigations of even the details of magnetic structure. As an example, we may cite the investigation of the helical magnetic structure of holmium [20,21]. Additional reflections (satellites) appear near the crystal reflections in this case because the pitch of the helix is much greater than the lattice constant. These satellites were investigated in [20,23] and were found to be both magnetic scattering and magnetoelastic effects. The two can be separated on the basis of polarization properties [17,20]. The detailed investigation of positions, widths, and polarization of these satellites, performed in [20–22,24], has revealed a number of nontrivial properties of magnetic ordering in holmium.

9.2   Polarization Properties of Magnetic Scattering

The structure of magnetic scattering amplitude (9.3, 9.4) shows, as was already mentioned, that polarization properties of magnetic scattering are quite different from those of Rayleigh scattering. To describe them, one should use the general approach presented in Sect. 2.1.2 and 3.4.4 or the more specific approach for X-rays given in Chapt. 8. As follows from the sections cited, the polarization properties of scattering are not determined solely by the scattering amplitude. In the general case, they depend on the properties of a sample.

For example, the polarization properties of scattering are different for a thick and thin sample or a perfect and imperfect sample. As we know, different kinds of interference also strongly influence the polarization properties of scattering. This interference can be connected to different mechanisms of scattering (e.g., quite similar to nuclear resonance and Rayleigh scattering examined in Chapt. 3) or interference of the scattering in the bulk of a sample and at its surface (see Sect. 1.3.5). It is why when speaking about the polarization properties of scattering, one should specify the conditions of scattering.

However, before specifying the conditions of scattering, let us overlook the general perspectives connected to polarization measurements and the information about a sample which can be obtained by means of polarization measurements. It has now become apparent that the polarization dependence of magnetic scattering can be exploited in even more detailed studies of solids than Rayleigh scattering. First and most important, the use of polarization dependence provides a natural technique for determining magnetic structures by X-ray scattering. Fortunately, there are novel possibilities which arise from the well-defined polarization characteristics of synchrotron radiation. For example, using the high degree of linear polarization of the incident beam, it has been possible in synchrotron radiation experiments to distinguish between charge reflections, arising from lattice modulation, and magnetic reflections in a spiral magnetic structure [22,24]. This distinction was crucial to the interpretation of pattern in studies of rare-earth metals. Furthermore, currently under discussion is the problem of experimental separation of the spin and orbital magnetic densities by analyzing the polarization of the scattered beam [14]. This separation is not directly possible by the neutron-scattering technique and is important to a fundamental understanding of the electronic properties of magnetic materials.

The first quite reasonable step for the theoretical analysis of the polarization properties of magnetic scattering is to consider the simplest situations. Such simplest situations correspond to the kinematic approximation, i.e., to the scattering at small samples for which a multiple scattering is unessential. Fortunately, for pure magnetic X-ray scattering this situation holds almost always due to the weakness of the corresponding interaction. If, for example, one would like to take into account the magnetic interaction for a crystal reflection, it cannot be more true. In a kinematic approximation polarization properties do not change inside a reflection region and are completely determined by the magnetic scattering amplitude (9.3, 9.4). The degree of polarization of the scattered beam cannot be lower than it is for the incident beam if the sample is perfect and there is no angular or frequency-averaging procedure.

The general expression for the dependence of the scattering cross section on the polarizations of the incident and diffracted beams can be expressed by the amplitude (9.3, 9.4) with the help of the expressions for the cross section and coherent scattering tensor (2.6, 2.7). Very useful are the eigenpolarizations, i.e., those polarizations of the incident and the scattered beams which correspond to the extreme values of the scattering cross section (2.6).

To obtain the eigenpolarizations determined by the amplitude (9.3, 9.4), one needs to construct the coherent scattering tensor (2.7) and with it [or with the help of the polarization tensor (8.10) which in the kinematic approximation is easily expressed by the scattering tensor (2.7)] find the scattering cross section extremes. The eigenpolarizations are, as was mentioned, very useful quantities which help to determine the geometry and polarizations in the experiment, excluding individual terms of the scattering amplitude from the cross section. Note that in some cases eigenpolarizations can be found directly from the symmetry considerations. These cases relate to the propagation and scattering directions coinciding with the symmetry axes in a crystal. For example, if radiation propagates in a crystal along a rotation symmetry axis of third or higher order, the eigenpolarizations are circular ones or any polarization is eigen (this means that X-ray interaction with the crystal in the last case does not depend on polarization at all).

The above-mentioned procedure of X-ray magnetic scattering polarization analysis in the kinematic approximation was carried out by Blume and Gibbs [37]. In their analysis, they preferred to use the Poincaré representation of polarization [36] (which, of course, is equivalent to the method used in this book and was first applied to magnetic X-ray scattering by de Bergevin and Brunel [7]) and examined as well the general case of combined magnetic and charge scattering and some specific cases important to the experiment. A general expression for the polarization-dependent scattering cross section is given by (2.6), with the amplitude being the sum of the usual Rayleigh term and the magnetic term (9.3, 9.4). Because the corresponding expressions are mostly very cumbersome, they are not presented here and the following discussion is restricted by the specific cases only.

For pure magnetic scattering from the expression (9.4), for the magnetic scattering amplitude follows the corresponding explicit form of the coherent scattering tensor (2.7):

$$\langle M_m \rangle = \begin{pmatrix} (\sin 2\theta)S_2 \\ 2(\sin^2\theta)[(\cos\theta)(L_1 + S_1) + (\sin\theta)S_3] \end{pmatrix}$$
$$\begin{pmatrix} -2(\sin^2\theta)[(\cos\theta)(L_1 + S_1) - (\sin\theta)S_3] \\ (\sin 2\theta)[2(\sin^2\theta)L_2 + S_2] \end{pmatrix} \tag{9.7}$$

where an unessential proportionality factor is omitted here. 2ϑ is the scattering angle, and L_i, S_i are the projections of the Fourier transforms of the orbital and spin moments [see (9.4)] at the coordinate axes $\hat{x}_i$ $(i = 1, 2, 3)$; the direction of $\hat{x}_1$ coincides with the direction of $\mathbf{k}_1 + \mathbf{k}_2$, the direction of the second axis is perpendicular to the scattering plane and the direction of the third axis coincides with the $\mathbf{k}$ direction. All of this reminds us also that (9.7) is written on the π- and σ-polarization basis of the incident and scattered beams. On the same polarization basis, the coherent Rayleigh scattering tensor according to (3.13) has the form

$$\langle M_c \rangle = \rho(\mathbf{K}) \begin{bmatrix} 1 & 0 \\ 0 & \cos 2\theta \end{bmatrix} \tag{9.8}$$

where $\rho(\mathbf{K})$ is the Fourier transform of the electronic charge density.

With the help of (9.7), the cross section for pure magnetic X-ray scattering may be presented in the form:

$$\frac{d\sigma}{d\Omega} = \left(\frac{e^2}{mc^2}\right)^2 \left(\frac{\omega}{mc^2}\right)^2 \mathrm{tr}\{\langle M_m \rangle \rho \langle M_m^+ \rangle\}$$
$$\to \left(\frac{e^2}{mc^2}\right)^2 \left(\frac{\omega}{mc^2}\right) \frac{1}{2}\{(1 + \xi_3)(|m_{11}|^2 + |m_{21}|^2)$$
$$+ (1 - \xi_3)(|m_{12}|^2 + |m_{22}|^2)$$
$$+ 2\mathrm{Re}[(\xi_i + i\xi_2)(m_{11}^* m_{12} + m_{21}^* m_{22})]\} \tag{9.9}$$

where m_{ik} are the matrix elements of (9.7) and ξ_i are the Stokes parameters describing polarization of an incident beam. The existence of interference between spin and angular momentum is presented in (9.9) by each of the matrix element products, except $|m_{11}|^2$.

For the interference of the charge and magnetic scattering terms using (9.7, 9.8), one finds

$$\frac{d\sigma}{d\Omega} = \left(\frac{e^2}{mc^2}\right)^2 \frac{\omega}{mc^2}\big[\mathrm{Im}(\rho^* m_{11} + \rho^* m_{22}\cos 2\theta)$$
$$+ \xi_3 \mathrm{Im}(\rho^* m_{11} - \rho^* m_{22}\cos 2\theta) + \xi_1 \mathrm{Im}(\rho^* m_{12} + \rho^* m_{21}\cos 2\theta)$$
$$+ \xi_2 \mathrm{Re}(\rho^* m_{12} - \rho^* m_{21}\cos 2\theta)\big]. \tag{9.10}$$

The polarization-dependent part of (9.10) permits one to obtain information on the magnetic scattering tensor (9.7) by means of polarization measurements. In fact, if the tensor (9.7) is diagonal, the interference term is independent of the circular and linear inclined at 45° to the scattering plane components of the incident beam polarizations. If it is off-diagonal the only dependence on the incident beam polarization is connected to the polarization components just mentioned. The above follows from the way in which Stokes parameters ξ_i enter (9.10) and their connection to polarization components.

In ferromagnets the magnetic and charge scattering give the same Bragg reflections. Since magnetic scattering in the typical case is reduced from the charge scattering by $\sim 10^{-6}$, it is difficult to measure magnetic scattering at the background of the charge scattering. One method to overcome this limitation is to introduce a magnetic field and measure the flipping ratio, i.e., the ratio of the cross sections for the opposite directions of the magnetic field, thereby isolating the magnetic and charge interference term in the cross section [12,38]. It is also possible to isolate the interference term by "flipping" the incident polarization [18]. In addition, because for some geometries magnetic scattering also flips the incident polarization, it is, in principal, possible to measure the pure magnetic scattering from a ferromagnet by analyzing polarization of the scattered beam. Experiments performed to analyze the polarization of the magnetically scattered beam are presented in [22,39].

Using (9.10), it is easy to find the interference term for some specific situations. For example, if $\mathbf{L}$ and $\mathbf{S}$ are perpendicular to the diffraction plane, the scattering tensor (9.7) takes the form

$$\langle M_m \rangle = \sin(2\theta) \begin{bmatrix} S & 0 \\ 0 & 2(\sin^2\theta)L + S \end{bmatrix} \qquad (9.11)$$

and the interference term is independent on ξ_2 and ξ_3. It means that the interference cross sections for pure circular and inclined at 45° to the scattering plane polarizations are identical and equal to the result for an unpolarized beam. Note that for purely linear incident polarization parallel and perpendicular to the scattering plane, it is possible to separate $\mathbf{L}$ and $\mathbf{S}$ by alternately scattering these polarizations.

If $\mathbf{L}$ and $\mathbf{S}$ are in the scattering plane and parallel to $\hat{x}_1$, the magnetic scattering is given by

$$\left.\frac{d\sigma}{d\Omega}\right|_m = \left(\frac{e^2}{mc^2}\right)^2 \left(\frac{\omega}{mc^2}\right)^2 \sin^2 2\theta \sin^2\theta |\mathbf{L} + \mathbf{S}|^2 \qquad (9.12)$$

and is independent of incident polarization. The final degree of polarization is the same as the initial degree of polarization and is described by the Stokes parameters which differ from the initial ones by the sign of ξ_1 and ξ_3. This means that the magnetic scattering flips from one to the other linear components perpendicular and parallel to the scattering plane. The same occurs with the linear polarizations inclined at 45° to the scattering plane.

If **L** and **S** are in the scattering plane and parallel to $\hat{x}_3$, the magnetic scattering is

$$\left.\frac{d\sigma}{d\Omega}\right|_m = \left(\frac{e^2}{mc^2}\right)^2 \left(\frac{\omega}{mc^2}\right)^2 4\,|S|^2 \sin^6\theta \qquad (9.13)$$

and is independent of the incident polarization and orbital angular-momentum density. The final polarization (Stokes parameters) is connected to the initial one in a different way compared to the previous case. The magnetic scattering flips circular and linear (in the scattering and perpendicular to the scattering plane) polarizations (change the signs of ξ_2 and ξ_3). It means again the possibility of measuring the magnetic scattering directly by analyzing the polarization of the scattered beam.

In antiferromagnets and more complicated structures along with crystal reflections where charge and magnetic scattering interfere, pure magnetic reflections exist, for which charge scattering is strictly forbidden for symmetry reasons. One reason for this can be the difference in size of the chemical and magnetic unit cell and the other, when the sizes coincide, the different symmetry of these unit cells (see also Chapt. 8). In the helical structures with a large pitch compared to the crystal unit cell size, these magnetic reflections are called satellites and their positions are quite close to those of crystal reflections.

The description of the intensities and polarization properties of reflections in these more complex magnetic structures is quite similar to the description given above for ferromagnets. However, if in the former case, as a unit in all expressions enters atomic scattering amplitude (9.3, 9.4), then in the last case the structure scattering amplitude $F_s = \sum \Delta f_i \exp(i\mathbf{k}\mathbf{r}_i)$ enters as a unit where the sum runs over all atoms in the unit cell. The example of the corresponding results for Mössbauer diffraction which are quite similar to those of the case examined here is presented in Sect. 3.4, and we will not go into the details of the more complex corresponding expressions [37].

It is worthwhile to emphasize that a general technique for determining unknown magnetic structures by X-ray scattering is to study the angular dependency of the intensities and polarization properties of the magnetic and interference cross sections and, in particular, these dependencies for rotations of the sample around the reciprocal lattice vector, connected to the corresponding reflection.

As an excellent example of using X-ray polarization measurements for the investigation of magnetic and crystal structure discuss in short the results of Ref. [22,24], where holmium was observed and the magnetic structure and modulation of the crystal lattice due to incommensurate effects were interpreted. The X-ray magnetic scattering measurements were performed at a synchrotron radiation source and reveal, along with the magnetic satellites, additional reflections corresponding to some long-wave modulation of the helix in holmium (Fig. 9.2). By measuring the polarization properties

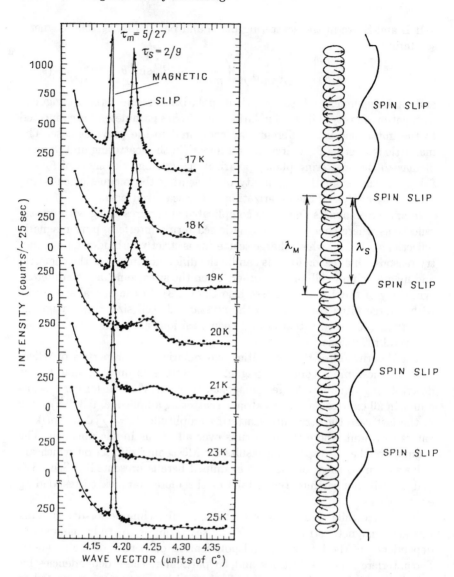

FIGURE 9.2. Synchrotron X-ray diffraction patterns of the satellite above (004) Bragg point of holmium, studied for decreasing temperature from 25–17 K. In addition to the sharper magnetic satellite, a second broad satellite appears at the slip position $\tau_s = 12\tau_m - 2$. When the temperature is lowered and τ_m approaches 5/27, then the interplanar lattice modulation sharpens and approaches $\tau_s = 2/9$ [23].

of these additional reflections, it was established that their origin is not connected to magnetic scattering but is due to the crystal lattice deformation originating with the magneto-elastic interaction in the crystal. The origin of long-range magneto-elastic modulation was explained by the spin slip model of the holmium helix structure, in which there is one spin slip for several consequent atomic layers that yields the corresponding lattice modulation (see the right side of Fig. 9.2).

9.3 X-Ray Resonance Magnetic Scattering

In the discussion above, we have not looked for the frequency dependence of magnetic scattering, but as was shown by the results of Refs. [3,40], magnetic scattering reveals a resonance dependence on X-ray frequency and at the frequencies close to absorption edges can be strongly enhanced. An especially strong enhancement of the magnetic scattering at the resonances was observed in [39] for holmium. Moreover in this work, some unusual properties of magnetic X-ray scattering were observed. Namely, when the energy of the incident X-ray was tuned through the L_{III} absorption edge, along with the first, the second, third, and fourth harmonic satellites were additionally observed. Complex polarization dependence was found, with the resonance peaks for the $\sigma - \sigma$ and $\sigma - \pi$ components of the magnetic scattering being separated by about 6 eV for the first two harmonics, and occurring at the same resonance energy for the third and fourth harmonics.

Hannon and coauthors [41] proposed explaining this behavior on the basis of electric quadrupole (E2) electron transitions to $4f$ levels and electric dipole (E1) transitions to $5d$ levels. It is noteworthy to emphasize that this "magnetic" scattering results from electric multipole transitions. This is due to the exclusion principal allowing only transitions to unoccupied orbitals, resulting in an "exchange interaction" which is sensitive to the magnetization of the f and d bands.

To obtain strong resonance enhancement, the scattering must involve a low-order electric multipole transition between a core level and an unfilled atomic shell (or a narrow band). In this case, the atomiclike nature of the transition is increased because the core hole yields an additional binding of the excited electronic state.

The authors of [41] predict that for the rare earths, enhanced magnetic resonance scattering will occur at the L_{II}, L_{III}, M_{II}, and M_{III} absorption edges, involving an E2 transition to the tightly bound $4f$ shell and E1 transition to the $5d$ band. The L_{II} and L_{III} resonances lie in the 1–2 Å region and are well suited for diffraction studies of magnetism in crystals.

The method of resonant magnetic scattering, in particular, the description of the dependence of the scattering amplitude on the atomic magnetic moment orientation, is quite similar to the description of the Mössbauer resonance scattering presented in Chapt. 3. It is based on scattering amplitude expansion through the spherical harmonics corresponding to the multipolarity of the electron transition and taking into account the terms with the smallest energy denominators [compare with (3.6–3.9)]. The above-mentioned expansion for the electric dipole resonance scattering amplitude may be presented in the following form [41,42]:

$$f_{E_1}^{(x\,res)} = \frac{3}{4}\lambda\{\mathbf{e}^* \cdot \mathbf{e}_0[F_{11}^{(e)} + F_{1-1}^{(e)}]$$

$$- i(\mathbf{e}^* \times \mathbf{e}_0) \cdot \hat{\mathbf{z}}_j[F_{11}^{(e)} + F_{1-1}^{(e)}]$$

$$- (\mathbf{e}^* \times \hat{\mathbf{z}}_j)(\mathbf{e}_0 \cdot \hat{\mathbf{z}}_j)[2F_{11}^{(e)} - F_{11}^{(e)} - F_{1-1}^{(e)}]\} \qquad (9.14)$$

where $\hat{\mathbf{z}}_j$ is the direction of the quantization axis defined by the local moment of the atom, and F_{1M}^e are the factors proportional to the square of the corresponding matrix element of transition between the ground and an excited atomic state and inverse proportional to the energy denominator.

From (9.14), it follows that there are only three distinct polarization responses for E1 scattering. The first term $\mathbf{e}^* \cdot \mathbf{e}_0$ is independent of the direction of the magnetic moment. The second term $-i(\mathbf{e}^* \times \mathbf{e}_0)$ depends linearly on the direction of the magnetic moment and will give first harmonic satellites in an antiferromagnet. In these satellites, incident σ-polarization scatters only to π-polarization, whereas incident π-polarization scatters to both σ- and π-polarization. The third term depends quadratically on the moment direction and gives second harmonic satellites in a spiral antiferromagnet (and also a contribution to the lattice reflection). This term scatters σ- and π-polarizations to both σ- and π-polarizations of the scattered beam. The existence of the first and second satellites is a characteristic signature of E1 and M1 transitions.

In an analogous way, the scattering amplitude for an electric quadrupole transition may be constructed. The corresponding expression [41] contains 13 distinct terms from zero to fourth order in the moment direction. In a spiral antiferromagnet, each order will give rise to a separate magnetic satellite. The appearance of four harmonic satellites is a characteristic signature of quadrupole resonance. The polarization dependencies of the satellites of any order for the quadrupole resonance allow all possible combinations of scattering between σ- and π-polarizations.

The ideas presented explain the complex resonance spectra obtained at the L_{III} edge in holmium [39]. The double-peak structure represents the superposition of resonance arising from different transitions. The lower resonance several eV below the edge is the E2 transition $2p_{3/2} \leftrightarrow 4f$, giving rise to four harmonics. The appearance of only two harmonics with high-energy resonance and the absence of σ to σ scattering in the first harmonic identify this as an E1 resonance, presumably the $2p_{3/2} - 5d$ transition. In Fig. 9.3, theoretical calculations of the energy dependence of the different satellite intensities, conducted with the simplest approximation [41], are presented; they can serve as a good qualitative illustration of the expla-

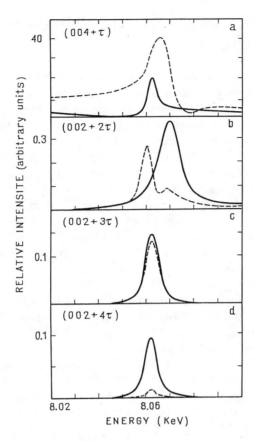

FIGURE 9.3. Relative scattering intensities (theoretical) vs. X-ray energy for the L_{III} edge in holmium: (a) $(0\ 0\ 4 + \tau)$, (b) $(0\ 0\ 2 + \tau)$, (c) $(0\ 0\ 2 + 3\tau)$, and (d) $(0\ 0\ 2 + 4\tau)$. The solid lines give $\sigma \Leftrightarrow \sigma$ scattering, and the dashed lines $\sigma \Leftrightarrow \pi$ [41].

nation given above for the experimental results. The analogous effect of magnetic scattering enhancement in the actinide series predicted by Hannon et al. [41] was observed recently in [44] for UAs at the M_{IV} and M_V absorption edges with the maximal enhancement factor 10^7.

The discussion above of the sizable resonance enhancement of X-ray magnetic scattering presents a very promising future for the study of crystal magnetic structures in general and especially for the investigation of the magnetic structure of thin film, surfaces, and separation boundaries in solids [14,38,43].

9.4   X-Ray Magnetic Birefringence and Dichroism

Apart from magnetic scattering, the X-ray region reveals the presence of magnetic dichroism and birefringence and, in particular, magnetic rotation of the plane of polarization (Faraday effect). These phenomena are described by magnetic corrections to X-ray susceptibility that are proportional to the magnetic correction to the forward-scattering amplitude of magnetic atoms. It is readily verified that the approximation used to derive (9.3) is then insufficient because the amplitude (9.3) vanishes for forward scattering [6,25]. Nonzero effects are obtained either in the case of high-order corrections [on the order of $\alpha^2 = (e^2/hc)^2$] or in the resonance case (near the absorption edge) when the dispersion corrections become significant. The former case has been examined theoretically by Baryshevskii et al. [26] (see also [2]) and the results provide a quantitative description of the Faraday rotation of gamma rays with energies of about 200 keV [27]. For the X-ray range (CuKα), theory predicts a rotation of the plane of polarization in iron by approximately 0.008°/cm, whereas recent experiments show that only the upper limit in iron is less than 50°/cm [28]. The observed rotation in nickel is of the order of 2–5°/cm [29].

The dispersion correction [30–34] discussed in Chapt. 8 may be more significant in the X-ray range. Partially circularly polarized synchrotron radiation has been used [33] to show that circular dichroism in iron undergoes a considerable change (including a change of sign) near the absorption edge and may reach $5 \cdot 10^{-4}$ in the immediate neighborhood of the edge (~ 10 eV) for a thickness of 2 mg/cm^2, i.e., 12.5 mμ. A dichroism of the order of 10^{-4} has been seen in the EXAFS region (~ 100 eV). Since dichroism is observed in the form of individual bands, it follows from the dispersion relations that the rotation of the plane of polarizations in these bands should be of the same order as dichroism, i.e., ~ 100°/cm.

In the EXAFS region, magnetic scattering and dichroism are a source of information about the magnetic structure of the nearest-neighbor environment of an atom (the atom itself may, in fact, be nonmagnetic). Recall that oscillation in EXAFS arises because of the back scattering of photoelectrons, which depends on the spin of the photoelectron and magnetic moment of the scattering atom. Additional information can be obtained from the interference of magnetic scattering with Rayleigh scattering and the discussion above (Chapt. 8) scattering by the anisotropic part of the X-ray susceptibility.

To conclude this chapter, we note that the successful investigation of X-ray magnetic scattering will depend on advanced polarization measurement techniques, especially circular polarization measurements. The real hope is to obtain in the near future a new method for crystal magnetic structure investigation connected to the following. Magnetic scattering of X-rays has a number of advantages in the study of magnetic properties compared with neutron scattering (discussed in Chapt. 7, see also the most recent works [45]). For example, SR sources can be used to achieve better resolution in k space (10^{-4}Å^{-1} is the prospect), which is particularly important in the study of long-period structures [14,20,21–24]. Resonance effects (see above) have a particular advantage, e.g., in the selective investigation of the magnetic properties of atoms of a particular species, including the magnetic properties of unfilled electronic states. Polarization measurements can be used to separate spin from orbital contribution to magnetization distributions and, in the case of resonance effects, to examine the structure of atomic and crystal states [13,20–22,24].

References

1. A.I. Akhiezer, V.B. Berestetskii: *Quantum Electrodynamics* (New York, Wiley-Interscience, 1965).

2. P.M. Platzman, N. Tzoar: *Phys. Rev. B* **2**, 3556 (1970).

3. F. DeBergevin, M. Brunel: *Acta Crystallogr.* **A37**, 314 (1981); M. Brunel, F. DeBergevin: *Acta Crystallogr.* **A37**, 324 (1981).

4. M. Gell-Mann, M.L. Goldberger: *Phys. Rev.* **96**, 1433 (1954).

5. V.A. Belyakov, V.E. Dmitrienko: *Sov. Phys.-Usp.* **32**, 697 (1989).

6. F.E. Low: *Phys. Rev.* **96**, 1428 (1954).

7. F. DeBergevin, M. Brunel: *Phys. Lett.* **A39**, 141 (1972).

8. N. Sakai, K. Ono: *Phys. Rev. Lett.* **37**, 351 (1976).

9. M.J. Cooper, D. Laundy, D.A. Cardwell, D.N. Timms, R.S. Holt, G. Clark: *Phys. Rev.* **B34**, 5984 (1986).

10. D.M. Mills: *Phys. Rev.* **36**, 6178 (1987).

11. S.W. Lovesey: *J. Phys. C: Solid State Phys.* **20**, 5625 (1987).

12. D. Gibbs: *Physica B* **159**, 145 (1989).

13. O.L. Zhizhimov, I.B. Khriplovich: *Sov. Phys. JETP* **60**, 313 (1984).

14. M. Blume: *J. Appl. Phys.* **57**, 3615 (1985).

15. S.W. Lovesey: *J. Phys. C* **20**, 5625 (1987).

16. F.A. Babushkin: *Dynamic Theory of Magnetic Scattering of X-Rays by Antiferromagnetic Materials* (Leningrad University, 1979), (in Russian).

17. S.M. Durbin: *Phys. Rev. A* **36**, 639 (1987).

18. M. Brunel, G. Patrat, et al.: *Acta Crystallogr.* **A39**, 84 (1983).

19. N.N. Faleev, A.A. Lomov, V.G. Labushkin: *Acta Crystallogr.* **A37**, C-374 (1981).

20. D. Gibbs, D.E. Moncton, K.L. D'Amico, J. Bohr, B.H. Grier: *Phys. Rev. Lett.* **55**, 234 (1985).

21. D. Gibbs, D.E. Moncton, K.L. D'Amico: *J. Appl. Phys.* **57**, 3619 (1985).

22. D.E. Moncton, D. Gibbs, J. Bohr: *Nucl. Instrum. Methods* **A246**, 839 (1986).

23. J. Bohr, D. Gibbs, D.E. Moncton, K.L. D'Amico: *Physica* (Utrecht) **A140**, 349 (1986).

24. J. Bohr, D. Gibbs, J.D. Axe, et al.: *Physica B* **159**, 93 (1989).

25. V. Kamersky, J. Kaszer: *Acta Phys. Slov.* **30**, 329 (1980).

26. V.G. Baryshevskii, O.V. Dumbrais, V.L. Lyuboshits: *JETP Lett.* **15**, 78 (1972).

27. V.M. Lobashov, L.A. Popeko, L.M. Smotritskii, A.P. Serebrov, E.A. Kolomenskii: *JETP Lett.* **14**, 373 (1971).

28. M. Hart, A.R.D. Rodrigues: *Phys. Mag. Ser. B* **43**, 321 (1981).

29. J. Hardy, E. Krousky, O. Renner: *Phys. Stat. Sol.* **A53**, 143 (1979).

30. B.T. Those, G. van der Laan, G.A. Sawatzky: *Phys. Rev. Lett.* **55**, 2086 (1985).

31. V.G. Baryshevkii, S.A. Maksimenko: *Opt. Spectroscop.* (USSR) **61**, 606 (1986).

32. G. van der Laan, B.T. Thole, G.A. Sawatzky, J.B. Goedkoop, J.C. Fuggle, J.M. Esteva, R. Karnatak, J.P. Remeika, N.A. Dabkowska: *Phys. Rev.* **B34**, 6529 (1986).

33. G. Schütz, W. Wagner, W. Wilhelm, P. Kienle, R. Zeller, R. Frahm, G. Materlik: *Phys. Rev. Lett.* **58**, 737 (1987).

34. J.B. Goedkoop, B.T. Thole, G. van der Laan, G.A. Sawatzky, F.M.F. de Groot, J.C. Fuggle: *Phys. Rev.* **B37**, 2086 (1987).

35. Yu.A. Izyumov, V.E. Naish, R.P. Ozerov: *Neutron Diffraction by Magnetic Materials* (Moscow, Atomizdat, 1981) (in Russian).

36. U. Fano: *Rev. Mod. Phys.* **29**, 74 (1957).

37. M. Blume, D. Gibbs: *Phys. Rev. B* **37**, 1779 (1988).

38. C. Vittier, D.B. McWhan, E.M. Gyorgy, J. Kwo, B.M. Buntschuh, B.W. Batteman: *Phys. Rev. Lett.* **56**, 757 (1986).

39. D. Gibbs, D.R. Harsshman, E.D. Isaacs, D.B. McWhan, D. Mills, C. Vittier: *Phys. Rev. Lett.* **61**, 1241 (1988).

40. K. Namikawa, M. Ando, T. Nakijima, H. Kawata: *J. Phys. Soc. Jpn.* **54**, 4099 (1985).

41. J.P. Hannon, G.T. Trammel, M. Blume, D. Gibbs: *Phys. Rev. Lett.* **61**, 1245 (1988).

42. P. Carra, M. Altarelly, F. de Bergevin: *Phys. Rev. B* **40**, 7324 (1989).

43. M. Weinert, A.J. Freeman, S. Ohnishi, J. Davenport: *J. Appl. Phys.* **57**, 3641 (1985).

44. E.D. Isaac, D.M. McWhan, C. Peters, G.E. Ice, D.P. Siddons, J.B. Hastings, C. Vettier, O. Vogt: *Phys. Rev. Lett.* **62**, 1671 (1989).

45. *Magnetic X-Ray Scattering*, Proceedings of the 1990 Daresburg Workshop, Science and Engineering Council, Daresburg Laboratory, 1991.

10

Mössbauer Filtration of Synchrotron Radiation

Although the first proposal to use synchrotron radiation (SR) for the excitation of a Mössbauer transition was made by S.L. Ruby in 1974 [1], only in 1983 was a paper published [2] on successful Mössbauer investigation with the use of SR, carried out in the Soviet Union at the VEPP-3 storage ring in Novosibirsk (Institute of Nuclear Physics, Siberian branch of the Academy of Sciences of the USSR). In 1985 a similar investigation [3,4] was carried out in the Federal Republic of Germany at the SR source in Hamburg (HASYLAB). These investigations demonstrated the effectiveness of using SR in yet another promising area of research and they were milestones in a number of Mössbauer investigations with SR [4–6]. The essence of this work is that they have succeeded in extracting (filtering) out of the continuous SR spectrum an extremely highly monochromatized and tightly collimated beam of gamma radiation of rather high intensity. The physical principal underlying the filtration process is the properties of Mössbauer diffraction (i.e., discussed in Chapt. 3 as the diffraction of radiation when it is scattered resonantly by the nuclei of the crystal through the Mössbauer transition). This diffraction makes it possible to separate spatially and energetically the radiation that has undergone coherent resonance scattering. The degree of monochromatization $\Delta E/E$ and the energy E of the beam are determined by the corresponding parameters of Fe^{57} Mössbauer transition used in both experiments, i.e., $E = 14.4$ keV and $\Delta E/E \rightarrow \sim 3 \cdot 10^{-13}$. Its collimation is determined by the sharp directionality of the SR and the low divergence of the beams of radiation diffracted by perfect crystals. All these factors have made it possible, especially the first time in the work reported [3], to generate the highest beam brightness attained at that time (by comparison with diffracted beams using ordinary Mössbauer sources). In combination with the pulsed time structure that the beam derives from the SR source and its controllable polarization characteristics, these beams open up new possibilities in experimental gamma-ray optics.

Since 1985 there has been a sharp rise in the interest in SR Mössbauer experiments, and several groups have reported on the experiments of such a nature. These investigations are based on both interesting physical effects and sophisticated modern experimental techniques, they promise interesting results as well as an explanation of the fundamental physics so in its applications. The main part of the theory underlying the investigation was already presented in Chapt. 3. However, the specifics of SR make it necessary to present more details than offered in Chapt. 3. It is why the following sections, along with a description of SR Mössbauer experiments, contain the results of further development of the theory presented in Chapt. 3.

10.1   On Synchrotron Radiation Sources and Mössbauer Filtration

Most of the experimental works on Mössbauer scattering cited in Chapt. 3 represent a unique physical investigation, and broad extensions of these studies are hindered by technical problems, mainly connected to the low activities and brightness of existing ordinary Mössbauer radiation sources based on radioactive decay. Therefore, along with attempts to increase the activity of ordinary Mössbauer sources, suggestions have been made to create sources for the directed emission of gamma rays which, although of rather low total activity, could provide high brightness in a selected direction. In particular, investigators have discussed the possibility of extracting a narrowly directional, highly monochromatized beam of gamma rays from the continuous spectrum of a powerful X-ray source by diffracting the radiation into nuclear diffraction peaks [7,8] or by coherent Coulomb excitation of the Mössbauer transition by charged particles in crystals [9–11]. Time has shown, however, that the practical separation of Mössbauer gamma rays has come about only with the advent of synchrotron radiation in the X-ray region, whose brightness is much greater than that of conventional X-ray sources. These SR sources are of a circular electron (positron) accelerators design for high-energy physics, or dedicated SR storage rings with high-enough energy for the electrons [12]. The most practical method of Mössbauer filtration (i.e., separation of a highly monochromatized and sharply directed beam) was proven to be [2,3] that based on pure nuclear Bragg reflections (see Chapt. 3) of synchrotron radiation.

For all the conceptual simplicity of separating from SR a line with an energy width on the order of the width of the Mössbauer level, the realization of this idea presents an extremely complex experimental problem. The complexity stems from the fact that we are dealing with the separation of a line on the order of 10^{-8} eV wide out of the continuous synchrotron radiation spectrum. If one realizes only a line on order of 1–10 eV width, it can be separated out of SR continuum comparatively easily with the use of ordinary diffraction monocromators, then it becomes clear that further monochromatization utilizing the Mössbauer effect (some time with additional premonochromatization) must reduce the width of the line by another eight to nine orders of magnitude, and here the background radiation problem becomes very acute.

TABLE 10.1. Synchrotron radiation sources [13]. Part 1. Third generation sources.

Location	Ring (inst.)	Electron energy (GeV)	Notes
BRAZIL			
Campinas	LNLS-1	1.15	Dedicated*
CHINA (ROC-TAIWAN)			
Hsinchu	SRRC(Synch.Rad.Res.Ctr.)	1.3	Dedicated*
ENGLAND			
Daresbury	DAPS(Daresbury Lab.)	.5–1.2	Proposed/Ded.
FRANCE			
Grenoble	ESRF	6	Dedicated*
	Super ACO(LURE)	0.8	Operating
GERMANY			
Dortmund	DELTA(Dortmund Univ.)	1.5	Ded./FEL Use*
Berlin	BESSY II	1.5–2	Proposed/Ded.
INDIA			
Indore	INDUS-II(Ctr.Adv.Tech.)	2	Proposed/Ded.
ITALY			
Trieste	ELETTRA(Synch.Trieste)	1.5–2	Dedicated*
JAPAN			
Hiroshima	HISOR(Hiroshima Univ.)	1.5	Proposed/Ded.
Kyushu	SOR(Kyushu Univ.)	1.5	Proposed/Ded.
Nishi Harima	SPring-8(Sci.Tech.Agcy.)	8	Dedicated*
KOREA			
Pohang	Pohang Light Source	2	Dedicated*
SWEDEN			
Lund	MAX II(Univ.of Lund)	1.5	Dedicated*
USA			
Argonne, IL	APS(ANL)	7	Dedicated*
Berkeley, CA	ALS(LBL)	1.5	Dedicated*
USSR			
Kharkov	HP-2000	2.0	Dedicated*

Present rings cannot achieve the full performance of insertion devices, particularly the high brightness that can be produced by undulators. The third generation of synchrotron storage rings addresses this problem with rings designed to have even lower emittance ($< = 40$ nanometer radians) and long straight sections for insertion device installation. Note that PEP, PETRA and TRISTAN operated at lower energies become 3rd. generation sources and that although there is no commitment to their use, there are beamlines already at PEP and design studies of low emittance optics and beamline arrangements have been done at PETRA and TRISTAN.

*In construction as of 5/91.

TABLE 10.1. Synchrotron radiation sources [13]. Part 2. Second generation sources.

Location	Ring (inst.)	Electron energy (GeV)	Notes
CHINA (PRC)			
Hefei	HESYRL(USTC)	0.8	Dedicated
ENGLAND			
Daresbury	SRS(Daresbury)	2	Dedicated
GERMANY			
Berlin	BESSY	0.8	Dedicated
INDIA			
Indore	INDUS-1(Ctr.Adv.Tech.)	0.45	Dedicated*
JAPAN			
Okasaki	UVSOR(Inst.Mol.Sci.)	0.75	Dedicated
Osaka	KANSAI SR	2.0	Proposed/Ded.
Tokyo	SOR-Ring(U of Tokyo)	0.38	Dedicated
Tsukuba	TERAS(Elec.Tech.Lab.)	0.6	Dedicated
Tsukuba	Photon Factory(KEK)	2.5	Dedicated
SWEDEN			
Lund	MAX(Univ. of Lund)	0.55	Dedicated
USA			
Baton Rouge, LA	CAMD(Lous.State U.)	1.2	Dedicated*
Stoughton, WI	Aladdin(SRC)	0.8–1	Dedicated
Upton, NY	NSLS I(Brookhaven Lab.)	0.75	Dedicated
	NSLS II(Brookhaven Lab.)	2.5	Dedicated
USSR			
Kharkov	N-100(KPI)	0.1	Dedicated
Moscow	Siberia I(Kurchatov Inst.)	0.45	Dedicated
	Siberia II(Kurchatov Inst.)	2.5	Dedicated*
Zelenograd	TNK(F.V. Lukin Inst.)	1.2–1.6	Dedicated*

There are defines as *storage rings* which were designed for use as dedicated sources. They have emittances in the range 40–150 nanometer radians. Many of these have immense capacity, some supporting over 50 beamlines and over 2000 users. Recognizing the need for even brighter sources, most of these second generation machines have one or several insertion devices incorporated.

*In construction as of 5/91.

TABLE 10.1. Synchrotron radiation sources [13]. Part 3. First generation storage ring sources.

Location	Ring (inst.)	Electron energy (GeV)	Notes
CHINA	BEPC		
Beijing	(Inst. High En.Phys.)	1.5–2.8	Partly Ded.
DENMARK			
Aarhus	ASTRID(ISA)	0.6	Partly Ded.
FRANCE			
Orsay	DCI(LURE)	1.8	Dedicated
	ACO(LURE)	0.54	**
GERMANY			
Bonn	ELSA(Bonn Univ.)	3.5	Partly Ded.
Hamburg	DORIS II(HASYLAB)	3.5–5.5	Partly Ded.
	PETRA(HASYLAB)	6–13	Planned Use
ITALY			
Frascati	ADONE(LNF)	1.5	Partly Ded.
JAPAN			
Sendai	TSSR(Tohoku Univ.)	1.5	Proposed/ Partly Ded.
Tsukuba	Accumulator Ring(KEK)	6	Partly Ded.
	Tristan Main Ring(KEK)	6–30	Planned Use
NETHERLANDS			
Amsterdam	AmPS	0.9	Planned Use*
Eindhoven	EUTERPE (Tech.Univ.Eind.)	0.4	Planned Use*
USA			
Gaithersburg, MD	SURF II(NIST)	0.28	Dedicated
Ithaca, NY	CESR(CHESS)	5.5	Partly Ded.
Stanford, CA	SPEAR(SSRL)	3–3.5	Dedicated
	PEP(SLAC/SSRL)	5–15	**
Stoughton, WI	Tantalus(SRC)	0.24	**
USSR			
Novosibirsk	VEPP-2M(Inst.Nucl.Phys.)	0.7	Partly Ded.
	VEPP-3(Inst.Nucl.Phys.)	2.2	Partly Ded.
	VEPP-4(Inst.Nucl.Phys.)	5–7	Partly Ded.

The first generation of storage ring sources were built as part of high-energy physics programs and initially used parasitically and then in most cases as dedicated sources. The emittances of these rings are generally in the hundred to several hundred nanometer radian range. However, the large colliders (PEP, PETRA and Tristan) can achieve very low emittance when operated in a dedicated mode at low energy.

*In construction as of May 1991.
**Withdrawn from operation.

The solution of this problem is supplied by another important property of synchrotron radiation, its pulsed nature. The typical duration of a burst of SR is on the order 1–0.1 ns, with an interval (depending on the length of the electron orbit in a storage ring) on the order of 1 μs between pulses. This time structure of SR make it possible to use time-delay techniques to detect the monochromatized radiation that is essentially free of background radiation that has not undergone resonance scattering from Mössbauer nuclei. The principle of this separation lies in the slowness of the Mössbauer scattering, the time of which is on the order of the lifetime of the Mössbauer level (for Fe^{57}, this is $\sim 10^{-7}$s), which is by far longer than the time of interaction of synchrotron radiation with the sample by any other channel.

At the present time, there are nearly 40 SR sources in the world which are operating, being constructed, or are in a design stage. However, not all of them are capable of producing hard enough radiation to be used in Mössbauer experiments. The number of SR sources suitable for Mössbauer experiments will increase in the near future though, and along with the currently operating SR sources in Tsukuba (Japan), Novosibirsk (USSR), Hamburg (Germany), Frascatti (Italy), Orsay (France), Daresbery (UK) and Cornell, Berkeley, Upton, and Stanford (U.S.) will be available the so-called third generation dedicated synchrotron radiation sources enumerated in Table 10.1 [13].

Concluding this section, we note another possibility of Mössbauer filtration which had not been previously realized but which has recently attracted the attention of experts. It is associated with the phenomenon of total external reflection of an X-ray beam from a material containing Mössbauer isotope [14,15] and was noted directly in connection to the use of synchrotron radiation for this purpose. At grazing angles of incidence onto a sample of these materials, there exists a small range of angles φ near a zero grazing angle, $(1 - n_e)^{1/2} < \varphi < (1 - n_r)^{1/2}$ (where n_r and n_e are, respectively, the indices of refraction of the medium for resonance Mössbauer radiation and X-rays of nearly the same energy as Mössbauer radiation), in which the only photons that undergo total reflection are those that have undergone nuclear resonance interaction with the sample, and as a result, the reflected radiation is highly monochromatized. Outside of this range, at smaller grazing angles, photons which interact only with electrons also undergo total external reflection, and in this case, monochromatization by total external reflection does not occur. In order to get rid of this parasitic reflection, the authors of [14,15] have suggested deposing on top of the layer containing Mössbauer nuclei a film of material with dielectric properties which allow for the quenching of the nonresonant part of radiation. The details of their calculations show that using a few (up to four) reflections from such layers under the condition of total external reflection also makes possible the separation of beams of gamma rays monochromatized

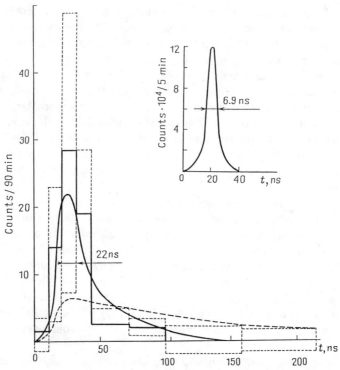

FIGURE 10.1. Time distribution of gamma quanta obtained by Mössbauer filtration of synchrotron radiation in a hermatite crystal [2]. The solid curve is the calculation for a nonexponential decay [18,19]. The dashed line corresponds to the calculation of decay for an isolated Fe^{57} nucleus. The points on the histogram show the statistical error of the measurements, the insert shows the exciting synchrotron radiation pulse, measured using the nonresonance scattering of the radiation.

to the level of Mössbauer radiation from continuous synchrotron radiation spectrum.

10.2   The First Observations of Mössbauer Filtration

The first experimental observations of Mössbauer synchrotron radiation filtration were reported in [2] (Fig. 10.1). The authors of the work [2] performed in Novosibirsk detected it resonantly scattered in a pure nuclear reflection component of SR (with an enriched by Fe^{57} hematite sample). Using a time-delay technique, they also observed a decline in the decay of the Mössbauer level of Fe^{57} excited by SR from the exponential decay law.

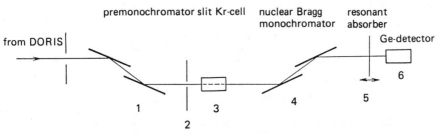

FIGURE 10.2. Experimental arrangement for observing filtration of synchrotron radiation [3]: 1) monochromator, 2) slit, 3) krypton cell, 4) nuclear Bragg monochromator, 5) resonance absorber, and 6) germanium detector.

The next observation of Mössbauer filtration of synchrotron radiation with much better statistics was reported in [3], and we will discuss it in more detail. The work was performed on the DORIS storage ring at the HASYLAB in Hamburg. A diagram of the experiment is shown in Fig. 10.2. The SR beam from the storage ring was first monochromatized by two germanium single crystals, and then (after passing through a collimating slit on a krypton cell, the latter being used for calibrating the energy of the radiation) it was directed onto the nuclear monochromator. The intensity of the beam incident on the nuclear monochromator was $(3\text{--}6)\cdot10^{10}$ photons per second. For additional monochromatization, the investigators used the principal described in the previous section that involves the separation of the Mössbauer line by scattering the beam into a purely nuclear (magnetic) diffraction peak. For the nuclear Bragg monochromator, they used two single crystal films of yttrium iron garnet (YIG) enriched to 88% in Fe^{57}. The films were arranged in the proper positions for the consecutive two-fold purely nuclear scattering into the peak corresponding to strong (200) nuclear reflection.

The radiation monochromatized in this way was detected with a germanium semiconductor detector (with an energy resolution around 1 keV), connected to the storage ring pulse circuit by a delay coincidence circuit. The spectrum of delayed coincidences of the germanium detector is shown in Fig. 10.3. For the lower curve, the energy distribution of the beam incident on the nuclear monochromator did not include the energy of the Mössbauer line at 14.4 keV, whereas for the upper curve, the energy of the beam was centered on that line. A conspicuous difference between the two curves immediately strikes the eye. This difference is accounted for by the presence, in the first case (the lower curve), of only the fast component of the scattering, due to scattering of the radiation by the electron of the crystal. In the second case (the upper curve), on the other hand, besides the fast scattering peak that is also present in the first case, there is a delayed component that is due to nuclear resonance scattering via the Mössbauer level. In their work, these same authors [3] were also able for

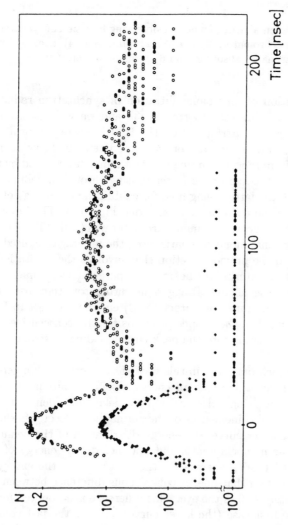

FIGURE 10.3. Time dependence of scattering intensity of synchrotron radiation in resonance (upper curve) and out of resonance (lower curve). The peak around 100 ns does not correspond to the time characteristics of noncoherent decay of the Mössbauer level [3]. Time is measured in ns.

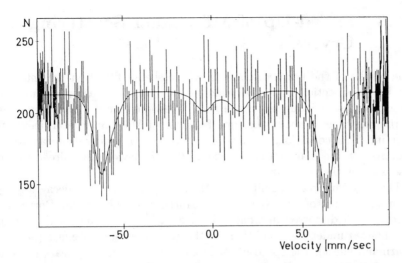

FIGURE 10.4. Mössbauer spectrum of radiation after scattering of synchrotron radiation by two YIG crystals. Solid curve: theoretical calculation of the spectrum [3].

the first time to record a Mössbauer spectrum of the synchrotron radiation beam monochromatized in this way, using for this purpose a Mössbauer absorber placed in front of the germanium detector (Fig. 10.4). This spectrum reproduces the characteristic features of the well-known Mössbauer transmission spectrum of YIG and is in accord with the theoretical calculations of Mössbauer diffraction in YIG.

The result of this investigation show that the final result of the Mössbauer filtration of synchrotron radiation is a beam of Mössbauer photons with an angular divergence less than 70 mrd and an intensity of one photon per second. The authors also noted that under the conditions of their experiment, they believed it realistic to anticipate increasing the intensity of the monochromatized beam about 20 times, to say nothing of the enormous possibilities of increasing the intensity of a beam monochromatized by Mössbauer filtration of synchrotron radiation sources, specifically by the use of undulators and wigglers.

The results of the lastest investigations bear out the optimism of the work under discussion, and we shall discuss this later after first presenting the corresponding theoretical background.

10.3   Speed-Up of Mössbauer Scattering in Crystals

The results of the first experiments on Mössbauer filtration presented in the previous section not only suggested new perspectives on gamma-ray optics, but in general also raised some questions about the theory. One of the questions is about the time behavior of Mössbauer level deexcitation under the conditions of Mössbauer filtration of synchrotron radiation. Fortunately, the main ideas relating to time behavior have already been discussed by the theorists.

The very important result about the speed-up of nuclear decay follows from the papers of Yu. Kagan and coauthors on the nuclear Borrmann effect [16,17] discussed in Chapt. 3. In fact, because coherent scattering is enhanced under the conditions of the Kagan–Afanas'ev effect (nuclear Borrmann effect), the speed-up of Mössbauer scattering appears to be a natural phenomenon. However, at the time of the works mentioned, hopes to observe this phenomenon seemed completely unrealistic and only after the proposal [1] to use synchrotron radiation for Mössbauer experiments was it realized that such experiments are quite plausible with the use of synchrotron radiation.

The physics of the observed speed-up of nuclear decay can be understood with the following simple explanation. When one observes coherent Mössbauer scattering of synchrotron radiation (i.e., radiation with constant spectral density on the scale of the Mössbauer linewidth) at a perfect crystal, the spectral distribution of scattered quanta does not coincide with the spectral distribution in the Mössbauer line. The reason for this is connected to the fact that at dynamical diffraction (diffraction with perfect crystals), the reflection coefficient does not depend (in the ideal case) on the strength of interaction and can be equal to unity far away from the resonance energy. At resonance (strong interaction), SR quantum is scattered by the thin layer of the crystal; far from the resonance (weak interaction), the reflection coefficient may be the same but the reflection is connected to the much thicker layer of the crystal. In real crystals, of course, the dependence of the reflection coefficient on the deviation from the resonance frequency exists, but it is not as sharp as the resonance Mössbauer line. It means that the frequency linewidth of the scattered radiation may be significantly wider than the Mössbauer linewidth. From the well-known relationship between the spectral width of the pulse $\Delta\omega$ and its duration Δt, $\Delta\omega \cdot \Delta t \sim 1$, it follows that the duration of the scattered pulse can be much shorter than the lifetime of the Mössbauer level.

To find the explicit time behavior of the decay, one needs to find the exact spectral distribution of the scattered radiation, for example, with the help of the dynamical theory of Mössbauer diffraction presented in Chapt. 3. Below the corresponding results will be presented in the same manner as they were derived in the original papers [18,19].

Consider the scattering of a SR pulse of duration T at a crystal containing nuclei of a Mössbauer isotope with the lifetime of the Mössbauer level τ_0. Assume also that the angle of incidence is inside of the strong Bragg reflection region or is close to this region. The amplitude of the SR pulse electric field has the form

$$E_\omega(t) = \epsilon_\omega e^{-i\omega t}\psi(t) \tag{10.1}$$

where $\psi(t)$ equals unity in the interval $0 < t < T$ and zero out of the interval. The amplitude of the scattered wave field may be presented in the form

$$E'_\omega(t) = \epsilon_\omega \int_{-\infty}^{+\infty} dt' G(t-t')\psi(t')e^{-i\omega t'}$$

$$G(t) = \int_{-\infty}^{+\infty} \frac{d\omega}{2\pi} R(\omega)e^{-i\omega t} \tag{10.2}$$

where $R(\omega)$ is the amplitude of crystal reflection of the monochromatic wave of the frequency ω.

Suppose the quanta are within the frequency interval $\Delta\omega \gg 1/T$ near the resonance frequency ω_0. For the intensity of the scattered wave, one easily finds with the help of (10.1)

$$I_\alpha(t) = 2\pi \frac{N_\alpha}{\Delta\omega} \frac{1}{T} \int_0^T dt' |G_\alpha(t-t')|^2. \tag{10.3}$$

Here we use the parameter α describing the deviation of the incidence angle from the Bragg angle and determined by (2.30). We suppose also that the divergence of the incident beam with respect to α is small because it is determined by a small divergence of the SR incident beam. The quantity N_α in (10.3) is defined by $\int N_\alpha \, d\alpha = N_0$ where N_0 is the total number of quanta per one pulse of SR.

Assume that the hyerfine structure of the Mössbauer spectrum of the isotope under consideration is well resolved or consists of a single isolated line, and polarization separation in the dynamical equations (2.26) takes place, i.e., the equations have the form of (3.23). Then the Bragg reflection amplitude for a thick crystal corresponding to the individual component can be written as

$$R_\alpha(\omega) = -\frac{b_{10}}{|b_{10}|} \frac{\eta}{\tau_0} \frac{1}{\omega - z_0 + \sqrt{(\omega-z_0)^2 - (\eta/\tau_0)^2}} \tag{10.4}$$

$$z_0 = \omega_0 - (\eta/\pi + i/2)/\tau_0$$
$$\eta = 1/(y - iy_0), \qquad y_0 = 2\chi_0''/|b_{10}|$$
$$p = 2|b_{10}|/(b_{00} + b_{11}).$$

Here we use the abbreviations $b_{\alpha\beta}$ ($\alpha, \beta = 0, 1$) and $\chi_0 = \chi_0' + i\chi_0''$ which can be defined by representing the coefficients $(\hat{\epsilon})_{\alpha\beta}$ in the dynamical theory equations (3.23) in the form

$$(\hat{\epsilon})_{\alpha\beta} = b_{\alpha\beta} \frac{\Gamma/2}{(\omega - \omega_0) + i\Gamma/2} + \chi_0 \delta_{ab} \tag{10.5}$$

where the second term describes the interaction with electrons. Then

$$y = \frac{\alpha - 2\chi_0'}{|b_{10}|} = \frac{2\sin 2\theta_B}{|b_{10}|} \left(\psi - \frac{\chi_0'}{\sin 2\theta_B} \right) \tag{10.6}$$

and the condition for a maximum of Bragg reflection intensity is $y = 0$ instead of $\alpha = \psi = 0$.

Substituting (10.4) into (10.2), one finds after some calculations

$$G_\alpha(t) = \frac{i}{\tau_0} \frac{b_{10}}{|b_{10}|} \exp\left\{ -i\omega_0 t - \frac{t}{2\tau_0} \right\}$$
$$\times \frac{J_1[\eta(t/\tau_0)]}{(t/\tau_0)} \exp\left\{ i\frac{\eta t}{p\tau_0} \right\} \theta(t) \tag{10.7}$$

where $J_1(x)$ is the Bessel function.

Consider the times greater than T. In this case, one finds approximately for the intensity (10.3)

$$I_a(t) \approx \frac{2\pi}{\tau_0} \frac{N_\alpha}{(\Delta\omega\tau_0)} \exp\left\{ \frac{-t}{\tau_0} \right\}$$
$$\times \left| \frac{J_1[\eta(t/\tau_0)]}{(t/\tau_0)} \right|^2 \exp\left\{ -\frac{2\eta'' t}{p\tau_0} \right\}. \tag{10.8}$$

Usually, $y_0 \ll 1$ for the crystal enriched by the Mössbauer isotope. Let the deviation from the Bragg condition be small, i.e., $y_0 < |y| \ll 1$. Then at $t > |y|\tau_0$, the time dependence of the intensity of delayed radiation (relative to the electron fast scattering process) takes the form

$$I_\alpha(t) \sim |y| \frac{e^{-t/\tau_0}}{(t/\tau_0)^3} \cos^2\left(\frac{t}{|y|\tau_0} - \frac{3\pi}{4} \right). \tag{10.9}$$

The principal change in the decay law is connected to the collective character of the scattering of each photon by the system of resonance nuclei in a crystal. The collective state (created during the course of the scattering process and spread over the crystal) decays through gamma quantum emission essentially faster than in the case of an isolated nucleus. The time-integrated reflected intensity Q_α is concentrated then within the interval $t \le |y|\tau_0$ and is equal to

$$Q_\alpha \approx \frac{N_\alpha}{(\Delta\omega\tau_0)} \frac{1}{|y|}. \tag{10.10}$$

Let us consider the case of a strictly satisfied Bragg condition, $y = 0$, and suppose at first that electron absorption is absent ($\chi_0'' = 0$). Then it is easy to see from (10.7) that $G(t) \sim \delta(t)$, i.e., the delay effect is completely absent. This is connected to the fact that a decrease of the nuclear scattering amplitude when deviating from the resonance and an increase in the effective thickness of the reflecting layer compensate each other. The existence of insignificant electron absorption, limiting the contribution of the far wings of the resonance, restores the small delay $\sim y_0 \tau_0$.

When $|y| > 1$ (i.e., out from the strong Bragg reflection region) for the entire time region, the usual law of delay restores

$$I_\alpha(t) \sim \frac{1}{y^2} e^{-t/\tau_0}. \tag{10.11}$$

However, in this case the reflected intensity is strongly decreased. One sees now that it is necessary to use the angle interval which satisfies the condition $|y| \sim 1$ (i.e., the region of strong Bragg reflection) to realize the time delay associated with conserving the intensity.

In the case of a large beam, angle divergence or mosaic crystal integration in (10.8) should be carried out over α. After this integration, one finds

$$\overline{I}(t) \approx \frac{16}{3\tau_0} \frac{N_{res}}{\Delta y} \frac{e^{-t/\tau_0}}{(t/\tau_0)}, \qquad N_{res} = \frac{\int N_\alpha d\alpha}{(\Delta\omega\tau_0)}. \tag{10.12}$$

Once again, the acceleration of the decay takes place owing to the collective character of Mössbauer scattering by the crystal.

The acceleration of the decay and enhanced role of resonance wings lead to a broadening of the frequency distribution of the delayed reflected radiation compared with the usual Mössbauer line. Because of this, one is faced with the problem of loss of radiation in using this radiation. The effective broadening may be on the order of a 100 natural Mössbauer linewidths. However, even in this case, the monochromatization of the beam is high enough to be effectively used in X-ray spectroscopy of solids.

The presented theoretical results explain the acceleration in the decay of the Fe[57] Mössbauer level first observed in [2]. The following experimental observations revealed some new interesting features of the Mössbauer filtration of SR which are connected to the hyperfine splitting of Mössbauer isotope levels. These are temporal quantum beats in the delayed decay which along with the acceleration of decay, were predicted theoretically [15,20] before an experimental observation took place. The corresponding theory will be presented in the next section.

10.4   Temporal Quantum Beats in Mössbauer Filtration of SR

Examine SR Mössbauer filtration in the case where the levels of the Mössbauer nucleus (ground and Mössbauer) are split by a hyperfine interaction. In this case, the SR pulse excites several nuclear sublevels of the Mössbauer level suddenly and coherently, which then "oscillate" at their various natural frequencies, thereby yielding beats at the different frequencies in the decay probabilities. The excitation amplitude of this initial state is, for a particular nucleus, independent of the state of atomic binding and other low-energy perturbations since the SR spectrum is white over many electron volts. However, if the probability of coherent scattering is appreciable, then the quantum beats resulting from the interference of waves emitted by different nuclei can also be appreciable in coherent scattering (the great probability of incoherent processes leads to smearing of the interference in waves emitted by different nuclei).

The important element in SR pulse experiments is that the very well collimated pulse produces a spatially coherent excitation of the Mössbauer nuclei in a crystal, and if the beam direction is set for a Bragg reflection, then the diffracted wave will exhibit beats, the measurements of which will give the different frequencies of the Mössbauer hyperfine transitions. At just the same conditions, the Kagan–Afanas'ev effect (nuclear Borrmann effect) can take place, so even for a small crystal, coherently scattered gamma rays may exceed in number those normally expected (in $\sim 4\pi$ sr) for incoherent scattering and become comparable to or larger then the number of internal conversion electrons.

To describe the temporal beats, consider following [20] an excitation of the Mössbauer level with the same SR pulse as in the previous section, but assume now that the ground and Mössbauer levels of the nucleus in the crystal are split by hyperfine interaction. The emission of Mössbauer quantum initiated by the absorption of SR quantum proceeds through the SR process: $+|g, m_i\rangle|\chi_0\rangle \rightarrow |e, n_i\rangle|\chi_0\rangle \rightarrow |g, m_i\rangle\chi_0\rangle + \gamma$ where the quantum numbers m and n describing the hyperfine sublevels of the ground and Mössbauer levels are marked by the subscript i, specifying their value for the different cites of the crystal lattice. We assume here that the process is recoilless and the state of the phonon system of the crystal does not change. If one neglects the retardation time connected to the size of the sample, the expression for the amplitude of the electric field at time t and point R due to the elastic scattering by individual nuclei of the sample may be written:

$$\mathbf{F}_{el}(\mathbf{r}, t) \approx \frac{i}{\hbar} \sum_{n_i m_i} \frac{e^{-i(\omega_{n_i m_i} - i\Gamma/2)t^*}}{r}$$
$$\cdot k_0^2 \hat{M}^e_{m_i n_i} \hat{M}^a_{n_i m_i} \mathbf{F}_{\omega_0} e^{-i(\mathbf{k}_j - \mathbf{k}_0)R_1} \psi(t^*) \qquad (10.13)$$

where ω_0 is the mean transition frequency, $\omega_{nm} = \epsilon_{en} - \epsilon_{gm} = \omega_0 + \Delta_{nm}$, and ϵ_{en} and ϵ_{gm} are the sublevel energies of the excited and ground states. The time t^* entering the right side of (10.13) is the retarded time at the sample $t^* = t - r/c$, and $\hat{M}^a_{pq}$, $\hat{M}^e_{pq}$ are the matrix elements of absorption and emission of quantum (their explicit form is given in Sect. 3.2).

If the crystal unit cell contains several sites occupied by the Mössbauer nucleus, one should sum up (10.13) over all Mössbauer nucleus in the cell. If one now takes into account the fact that the coherent component of the field is determined by the quantities averaged over the hole sample (see Sect. 3.2), then the coherent field related to one unit cell may be presented in the form:

$$\mathbf{F}_{coh}(r, t) = (k_0^2/r)\mathbf{M}^{\perp}_{coh}(t^*, \mathbf{k})S(\mathbf{k}) \qquad (10.14)$$

where

$$\mathbf{M}^{\perp}_{coh}(t, \mathbf{k}) = -\frac{C}{2j_0 + 1} \sum_{\alpha, n, m} e^{-i[\omega_{nm}(\alpha) - i\Gamma/2]t}$$
$$\cdot \hat{M}^e_{mn}\hat{M}^a_{nm}\mathbf{F}_{\omega_0}e^{-i\mathbf{kr}_\alpha} \qquad (10.15)$$

where α denotes summation over all sites occupied by the Mössbauer nucleus and other notations are the same as in Sect. 3.2.

If a crystal sets at a Bragg angle, then the reflected intensity in the kinematic approximation will be proportional to the square of (10.15). This will result in temporal beats corresponding to the difference frequencies of all allowed transitions ω_{mn}. It is why in the coherent process the time evolution of the nuclear decay is determined by both the excited and ground-state hyperfine splitting, and from experiments both splittings can be, in principal, determined. This point distinguishes coherent and incoherent processes. The beats in incoherent scattering and internal conversion electron intensity yield only excited-state splitting and are described by the usual time differential perturbed angular correlations [21].

The kinematic approximation, mentioned above, works for small or imperfect crystals, so in the general case for the description of beats, one should use the dynamical theory, for example, in the same manner as it was used in the previous section.

The first experimental observation of quantum beats was reported in [22] with a two-nuclear-reflection SR Mössbauer filtration experiment with a yttrium-iron-garnet (YIG) crystal. The next observations of the beats [23,24] occured in a single-nuclear-reflection experiment conducted by the same group with YIG and $FeBO_3$ crystals. Because the structure of $FeBO_3$ is simpler than that of YIG, the results of the $FeBO_3$ experiment will be briefly discussed below.

In this experiment, Mössbauer filtration measurements were made for the pure nuclear reflection (111) of a perfect $FeBO_3$ antiferromagnetic single crystal. As is known (see Chapt. 3), the hyperfine spectrum of radiation

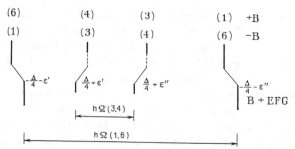

FIGURE 10.5. Hyperfine splitting between four $\Delta J_z = \pm 1$ transitions in iron borate. Here the transitions are $(1) = [+1/2 \leftrightarrow +3/2]$, $(3) = [+1/2 \leftrightarrow -1/2]$, $(4) = [-1/2 \leftrightarrow +1/2]$, $(6) = [-1/2 \leftrightarrow -3/2]$. In the drawing the ϵ shifts of the full Hamiltonian are added. The room temperature values are $\hbar\Omega(1.6) \approx \hbar\Omega_M(1.6) = 110.5\Gamma$, $\hbar\Omega(3,4) \approx \hbar\Omega_M(3,4) = 17.5\Gamma$, $\Delta = -3.9\Gamma$, $\epsilon' = +0.06\Gamma$, $\epsilon'' = -0.06\Gamma$ with $\Gamma = 4.655 \cdot 10^{-9}$ eV [24].

coherently scattered in a pure nuclear reflection with an antiferromagnet is formed only by the lines corresponding to transitions with $\Delta m = \pm 1$. The scheme of the corresponding lines is presented in Fig. 10.5 and the results of the experiment in Fig. 10.6. The observed spectra show pronounced beats which are very well described by theory (dynamical calculations). The same spectra also reveal the speed-up of decay, discussed in detail in the previous section, and its dependence on the angle of SR incidence with the sample.

Almost all features of the quantum beat spectrum can be understood in the kinematic approximation, although for a quantitative description of the experiment, one should use the dynamical theory of Mössbauer diffraction presented in Sect. 3.4 in the same way as was done for a single line spectrum in [18,19]. In the kinematic approximation, the intensity of the filtrated radiation is proportional to the squared modulus of (10.15). If one takes into account the relative strengths of the transitions and the splitting of the spectrum shown in Fig. 10.5, the quantum beats may be presented in the form

$$I(t) \sim e^{-\Gamma t/\hbar} \left| \sin\left[\frac{1}{2}\Omega_M(1,6)t\right] \right.$$
$$\left. - \frac{1}{3}e^{-i\Delta\Omega t}\sin\left[\frac{1}{2}\Omega_M(3,4)t\right] \right|^2 \qquad (10.16)$$

where the notations are given in Fig. 10.5.

The expression (10.16) gives the characteristic quantum beat pattern for any simple antiferromagnet. Moreover, the quantum beat modulation of the coherent decay rate presented by this expression yields the initial complete suppression of coherent decay. This suppression is characteristic not only for any antiferromagnet, but also for the arbitrary pure nuclear reflection of any magnetic structure or more generally for any hyperfine field

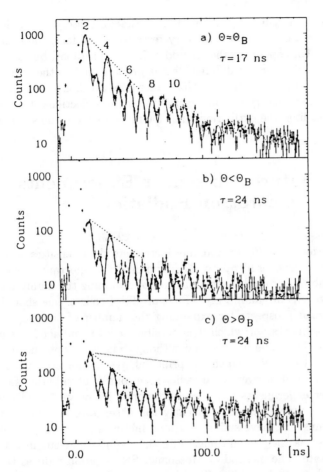

FIGURE 10.6. Time spectra of the coherently scattered radiation measured at different angular positions: (a) exact Bragg angle, (b) ≈ 20 μrad below Bragg angle, and (c) ≈ 30 μrad above Bragg angle. The solid lines are the calculated theoretical spectra averaged over a Gaussian spread of 30 μrad, plus a constant background contribution. The dotted lines show a natural exponential decay $\exp(-\Gamma/ht)$, and the dashed lines indicate the initial enhanced decay with lifetime τ. The ratio of delayed to prompt counts is about 10^{-3}. All data were accumulated within a day [24].

structure with symmetry different from that of the crystal lattice. These phenomena are based on symmetry restrictions. In fact, the pure nuclear reflections correspond to the forbidden X-ray reflections, but at $t = 0$, the sum in (10.16) over the different line is proportional to the corresponding sum for the case of the unsplit Mössbauer line, which is identically equal to zero by the symmetry reasons for the forbidden reflections (for the unsplit Mössbauer line, the extinction rules for reflections are the same as for X-rays).

10.5   Future Mössbauer Experiments with Synchrotron Radiation

The results reported in [3] and later papers [22–33,47–56] confirm the earlier optimism for [3,34–36] for a further increase in the intensities of beams obtained by Mössbauer filtration of SR and make it important to examine the new possibilities raised by this method of obtaining the highly monochromatized beams of gamma rays. First of all, of course, one should remark on the broad prospects for expanding the number of isotopes of various chemical elements, for which it is possible now to conduct Mössbauer experiments. The issue here is that although many isotopes have low-lying excited nuclear levels which, in principal, permit an observation of the Mössbauer effect, conventional radioactive sources of radiation suitable for Mössbauer experiments have been developed for only a small number of them. The impediment is frequently the absence of suitable radioactive decay chains that have sufficiently long lifetimes for the experiments and which have as their penultimate link the required nucleus in a low-lying excited state. The method of Mössbauer SR filtration reduces the source problem, because to achieve filtration, it is sufficient to have the required isotope with the level for Mössbauer studies only in the ground state.

For instance, Ruby [1] (see also [34]), in proposing the idea of Mössbauer filtration of SR, immediately suggested creating by this method a source for carrying out a Mössbauer experiment on the isotope K^{40}. Of course, this is only one such example, and the number of nuclei that are candidates for entering the list of Mössbauer isotopes by these means is quite comparable to the number of isotopes in which the Mössbauer effect has been observed to the present time.

Another promising direction for study is easily discerned in the analysis of a series of Mössbauer diffraction experiments which are unique and are frequently performed at the very limits of what is possible. This was already mentioned above when studying with the help of SR the acceleration of decay of a nuclear level upon its coherent excitation under diffraction conditions and studying the Kagan–Afanas'ev effect (the work in [17] is an example of research performed with the use of conventional Mössbauer source). The corresponding experiments are, of course, being carried out more effectively with SR [25–34].

From an applied point of view, broad and important prospects are opening up in connection to the study of the structure of magnetic fields and an electric field gradient in crystals. In this direction, in addition to the exploitation of these possibilities for determining magnetic and crystal structure afforded by the observation of purely nuclear reflections, one should also bear in mind the results possible with magnetic X-ray scattering discussed in Chapt. 9, which, of course, are essentially connected to SR.

The potential exists for using SR (through the technique of Mössbauer filtration) in the study of surface properties [35,36] and, in particular, to observe inelastic channels of decay in Mössbauer nuclear levels. These possibilities have been demonstrated in [37], where electron internal conversion was observed under conditions of Mössbauer diffraction, and it was shown that the method is highly sensitive to the perfection of the surface layers.

Let us note also the possibility of using SR for carrying out new Mössbauer diffraction experiments which would be very problematic if standard sources of Mössbauer radiation were used. For instance, the coherent SR-induced excitation of a cascade of gamma transitions in a single crystal has been studied theoretically [38]. It was found that, in comparison to the case of coherent excitation of the first nuclear level, which has been the main topic of discussion in this chapter, the coherent excitation of a cascade of low-energy nuclear gamma transitions exhibits interesting features, especially in the angular distribution of the coherent emission of the gamma rays of the cascade. In [38,39] the case of the arbitrary number of gamma transitions was treated and in [39] it was additionally assumed that a hyperfine splitting of the Mössbauer line is present. However, since the case of practical interest involves only two or three transitions in the cascade, we shall discuss the results used by way of example for a cascade of two Mössbauer gamma transitions and assume also that the Mössbauer line is unsplit. Note that observation of the coherent excitation of a cascade of two Mössbauer gamma transitions would evidently be simplest to carry out experimentally using the same iron isotope Fe^{57}, in which the coherent excitation of the first Mössbauer level by SR was observed. In this isotope, there is another low-lying level at 136.4 keV in which the Mössbauer effect was also observed. It is also important that this level decays with overwhelming probability through the first Mössbauer level with the emission of a 122 keV gamma quantum.

Thus, as applied to the isotope Fe^{57}, the coherent excitation of the cascade means the following. A 136.4 keV beam of SR excites in the crystal the second Mössbauer level, which decays via a cascade of gamma quanta of energy 122 and 14.4 keV and returns the nuclei and crystal as a whole to its initial state. This sequence satisfies the conditions for a coherent process (the amplitudes are summed, and not the cross sections of the process in the individual nuclei of the crystal). In connection to the coherence of the process under discussion, the spatial periodicity of the crystal lattice has a significant effect on the process. In particular, the angular distribution of the gamma rays turns out to be different from the distribution described by the well known gamma-ray angular correlations in a cascade, which are observed in noncoherent nuclear decay [21].

Analysis of the process [38,39] shows that the crystal lattice introduces an additional correlation only upon the last quantum of the cascade (in the case of Fe^{57}, this is the 14.4 keV quantum). In the kinematical approximation—that is neglecting scattering of the quanta of the cascade in the crystal—the angular distribution of the radiation emitted coherently in the second transition of the cascade is given by the function $W(\mathbf{k}_s, \mathbf{k}_2)$, which describes the strong correlation between the direction of the incident SR beam (wave vector $\mathbf{k}_s$) and the direction of emission of the cascade's last quantum (wave vector $\mathbf{k}_2$):

$$W(\mathbf{k}_s, \mathbf{k}_2) = f \int \int \sigma(E_s) I(E_s) \tilde{W}(\mathbf{k}_s, \mathbf{k}_1, \mathbf{k}_2) \, d\Omega_{k_1} \, dE_s \delta(\mathbf{k}_p - \mathbf{k}_1 - \boldsymbol{\tau}) \quad (10.17)$$

where $\sigma(E_s)$ is the cross section for excitation of the second level and depends on the SR energy E_s [see (3.11)], f is the Lamb–Mössbauer factor for the second transition in the cascade, $I(E_s)$ is the SR intensity, $\mathbf{k}_1$ is the wave vector of the quantum in the first transition of the cascade, $\mathbf{k}_p = (\mathbf{k}_s/|\mathbf{k}_s|)(\omega_2/, c)$, ω_2 is the quantum frequency of the second transition of the cascade, c is the velocity of light, $\boldsymbol{\tau}$ is a reciprocal vector of the crystal, $\tilde{W}(\mathbf{k}_s, \mathbf{k}_1, \mathbf{k}_2)$ is the well-known angular correlation function in the noncoherent channel of the same cascade, and the integration in (10.17) is carried out over the directions of the emission of the first quantum of the cascade.

Note that the connection between the direction of the SR beam and the direction of the last quantum emission is given by the relation which is completely analogous to the Bragg condition in typical diffraction (i.e., when the energies of the initial and scattered quanta are the same) and which requires that the argument of the delta function in (10.17) be equal to zero. There is an important difference, however, and that is that in this case, the analog to the Bragg condition does not contain the wave vector of the initial quantum $\mathbf{k}_s$, whose energy coincides with the energy of the second Mössbauer level, and the modulus of the wave vector $\mathbf{k}_s$ differs substantially from that of the wave vector $\mathbf{k}_2$ of the final quantum and $\mathbf{k}_p$. Therefore, what is of concern in the orientation of the crystal and the angles of incidence and scattering is that they are the same as for diffraction through the first excited level, i.e., geometrically, the corresponding Bragg conditions are the same as for the usual Mössbauer diffraction.

Another similarity between the excitation of a coherent cascade and typical Mössbauer diffraction reveals itself in the case of magnetically ordered crystals or crystals with the complex structure of an electric field gradient [39]. For these cases in the coherent cascade excitation, as well as Mössbauer diffraction, pure nuclear reflections exist which are effectively used for the Mössbauer filtration of SR [2,3,22–33]. And it is why the crystals for which pure nuclear reflections exist look most suitable also for coherent cascade excitation by SR.

The theoretical proposal of another kind of experiment (coherent inelastic Mössbauer scattering [40]), which demands not only a source of high brightness but also the high-energy resolution of a detector, is presented in the next section.

10.6 Coherent Inelastic Mössbauer Scattering

The development of Mössbauer technique and, in particular, the successes in Mössbauer filtration of SR made quite real the performance of the experiments which previously seemed unrealistic. The detection and energy resolution of coherent inelastic Mössbauer scattering belong to such types of experiments and perhaps, it was for this reason that until now only a few theoretical papers had slightly touched on this problem [39,41,42]. Recently some new theoretical predictions relating to the angular distribution of quanta in Mössbauer coherent inelastic scattering were made [40]; they can be verified in experiments on the Mössbauer filtration of SR. Below a short discussion of the corresponding theoretical results is presented.

At first glance, it seems that coherent inelastic phonon Mössbauer scattering (i.e., nuclear resonance scattering accompanied by the creation or annihilation of phonons in the crystal lattice) does not differ much from the scattering of X-rays or thermal neutrons. However, the large resonant nuclear scattering time of gamma rays, as compared to the inverse characteristic phonon frequencies, and the division of the scattering process into two stages, absorption and emission of gamma rays, leads to a substantial difference in the angular distribution of coherent inelastic phonon Mössbauer scattering from that observed for X-rays and neutrons. The main difference is that the angular distribution of coherent Mössbauer inelastic scattering does not depend on the phonon wave vector, with the exception of very low phonon frequencies. Therefore, the angular distribution of gamma rays in coherent inelastic scattering practically coincides with that of coherent elastic Mössbauer scattering.

Let us consider the diffraction of gamma rays by a crystal containing nuclei of a Mössbauer isotope in its lattice sites. The beam incident on the crystal is assumed to have a continuous frequency distribution. Consider the coherent inelastic phonon nuclear scattering in which the final state of the nuclei coincides with their initial state, whereas the final state of the crystal lattice differs from its initial state in the number of phonons. The process of the scattering proceeds in two stages, due to the long lifetime

τ_n of an excited Mössbauer level. This is why the amplitude of coherent inelastic scattering by an individual nucleus takes the form [41]

$$f_{n_f n_0(\mathbf{k},\mathbf{k'})} = \frac{\Gamma_i}{2k} \sum_n \varphi^N \frac{\langle \chi_{n_f} | e^{-ikr} | \chi_n \rangle \langle \chi_n | e^{ikr} | \chi_{n_0} \rangle}{E_k - E_R - (\epsilon_n - \epsilon_{n_0}) + i\Gamma/2} \qquad (10.18)$$

where the notations of the crystal wave functions are the same as in Sect. 10.5, n_p specifies the state of the phonon system, ϵ_p are the corresponding energies of the phonon system, E_k is the energy of the absorbed quantum, φ^N is the factor determined by the nuclear matrix elements which explicit form, as well as the other notations may be found in Chapt. 3.

In a kinematical approximation, the scattering cross section at the crystal may, with the help of (10.18), be presented in the following form:

$$\frac{d\sigma(\mathbf{k},\mathbf{k'})_{n_f n_0}}{d\Omega_{k'}} = \left| \sum_p f_p \exp[i\mathbf{k}\mathbf{r}_p(t_a)] \exp[-i\mathbf{k'}\mathbf{r}_p(t_e)] \right|^2 \qquad (10.19)$$

where the summation is taken over all the crystal sites containing Mössbauer nuclei. Here t_a and t_e are the moments in time of the quantum absorption and emission. Equation (10.19) leads to different expressions for the angular distribution of scattering, depending on the magnitudes of the characteristic times in the scattering system. The simplest formulas correspond to the nuclear time τ_n being larger than all other characteristic times and to the case of scattering without a change in the phonon system.

In the case of scattering without a change in the phonon system, in Eq. (10.18) the phonon energy difference $\epsilon_{n_0} - \epsilon_n$ turns to zero and (10.19) leads to the well-known formulas for the scattering cross section given in Chapt. 3. In particular, all the above-mentioned formulas are proportional to delta-function $\delta(\mathbf{k} - \mathbf{k'} - \boldsymbol{\tau})$ where $\boldsymbol{\tau}$ is the reciprocal lattice vector of the crystal.

In the case of fast photon relaxation ($\tau_n \gg \tau_p$), temperature averaging is carried out in the matrix elements of (10.18), and as a result, the points of quantum absorption $\mathbf{r}_p(t_a)$ and emission $\mathbf{r}_p(t_e)$ become noncorrelated and the crystal-site equilibrium positions should be taken as emission and absorption points. Thus, the scattering cross section is again proportional to $\delta(\mathbf{k} - \mathbf{k'} - \boldsymbol{\tau})$.

In the general case, the energies of absorbed and emitted photons as well as the corresponding wave vectors $\mathbf{k}$, $\mathbf{k'}$ do not coincide with the values determined by Mössbauer transition energy. The difference is determined by the energy transferred to the phonon system during the absorption and emission of a quantum by the nucleus. As the corresponding changes of the wave vectors are small compared to their resonant values ($\Delta k/k \approx 10^{-6}$), the values of $\mathbf{k}$ and $\mathbf{k'}$ in the delta-function can practically always be considered equal to their resonant values.

If the process involves a high-frequency phonon ($\tau_n \gg \omega^{-1}$), the matrix elements in (10.18) are to be averaged over the time interval T with

$T \approx \tau_n \gg \omega^{-1}$. Again, as the photon absorption and emission points the equilibrium positions of the crystal lattice sites should be taken. As a result, the cross section (10.19) is again proportional to $\delta(\mathbf{k} - \mathbf{k}' - \boldsymbol{\tau})$.

If the process involves a very-low-frequency phonon $(\tau_n \ll \omega^{-1})$, the points at which the quantum is absorbed and emitted become dependent on the quantum wave vector. They are not just determined by the equilibrium positions of crystal sites, but are also influenced by the wave of displacement $\exp(i\mathbf{qr})$ of the crystal sites originated by the corresponding phonon with wave vector $\mathbf{q}$. The condition $\tau_n \ll \omega^{-1}$ then allows us to consider the case $t_a = t_e$ in (10.19). Taking into account the already mentioned spatial modulation of the crystal sites, one sees that one phonon coherent inelastic scattering cross section is proportional to $\delta(\mathbf{k} - \mathbf{k}' - \mathbf{q} - \boldsymbol{\tau})$. However, from the condition $\tau_n \ll \omega^{-1}$ and the dispersion relation for phonons, $\omega = c_p q$, where c_p is the velocity of sound, it follows that in the delta-function, $|\mathbf{q}|$ is small $(|\mathbf{q}| \ll |\mathbf{k}|, |\boldsymbol{\tau}|; |\mathbf{q}|/|\mathbf{k}| \leq 10^{-6})$, provided that the characteristic value $\tau_n \gg 10^{-8}s$ is considered. This means that under the condition $\tau_n \ll \omega^{-1}$, although the phonon wave vector appears in the corresponding angular distribution of inelastic coherent photon scattering, the contribution of the relevant process is small (as q^3/τ^3), due to the small phase volume of such phonons. Moreover, the small value of q/k signifies that the angular distribution of scattering is practically identical to the distribution determined by the delta-function $\delta(\mathbf{k} - \mathbf{k}' - \boldsymbol{\tau})$.

We have discussed above the scattering process with fixed energy of incident quanta. This situation first of all corresponds to experiments with a conventional Mössbauer source. In the general case (e.g., the SR source), the cross section (10.19) is to be averaged over the initial energy distribution of quanta and integrated over their final energies if not only the elastic part of the scattering is sought. Now we examine the cross section (10.19) for the different experimental situations.

For elastic resonance, scattering (10.19) results in a well-known expression described by (2.5) and (3.11).

If the energy of the incident quantum is fixed, whereas the energy of the scattered quantum is not, the situation corresponds to scattering experiments with a conventional Mössbauer source where the energy of the scattered quanta is not measured. For this case, one obtains from (10.18, 10.19) the following expression for the cross section:

$$\frac{d\sigma_i(\mathbf{k}, \mathbf{k}')}{d\Omega_{k'}} = \sum_{n_f n_0} \left| \frac{\Gamma_i \langle \chi_{n_f} | e^{-i k' r} | \chi_n \rangle \langle \chi_n | e^{i k r} | \chi_{n_0} \rangle}{E_k - E_R - (\epsilon_n - \epsilon_{n_0}) - i\Gamma/2} \right|^2$$

$$\cdot \sigma^N(\mathbf{k}, \mathbf{k}') \delta(\mathbf{k} - \mathbf{k}' - \boldsymbol{\tau}). \tag{10.20}$$

By carrying out an approximate summation over the final states in (10.20) and taking into account that $\mathbf{k}'$ practically does not depend on the final state n_f, one obtains (see also [41,42]):

$$\frac{d\sigma_i(\mathbf{k},\mathbf{k}')}{d\Omega_{k'}} = \sigma^N(\mathbf{k},\mathbf{k}')\left|\sum_n \frac{\langle\chi_n|e^{ikr}|\chi_{n_0}\rangle}{E_k - E_R - (\epsilon_n - \epsilon_{n_0}) - i\Gamma/2}\right|^2$$
$$\cdot\delta(\mathbf{k}-\mathbf{k}'-\boldsymbol{\tau}). \qquad (10.21)$$

This expression shows that the cross section integrated over the energy of the scattered quanta does not depend on the characteristic of the phonon transition accompanying quantum emission. This is quite natural because an excited nucleus goes to the ground state with the same rate no matter what phonon processes accompany the nuclear transition (our discussion here is limited by the kinematical approximation and does not take into account the dynamical effects considered in Sect. 10.4,5). If now in sum of (10.21), one takes only the term containing $n = n_0$ (which corresponds to the conventional Mössbauer source), the scattering cross section takes the form which differs from the corresponding expression for elastic Mössbauer scattering only by the absence of the factor $f^2(k')$, i.e., it is $[f(k')]^{-2}$ times larger than the elastic scattering cross section.

Now consider a situation corresponding to the nuclear coherent scattering of SR. In this case, the expressions (10.19, 10.20) for the cross sections should be integrated over the energies of incident quanta, both in the case of measuring the scattered-quanta energy and in the case of the absence of such measurements. For the cross section integrated over the energies, both the incident and scattered quanta one obtains using (10.21) are

$$\left[\frac{d\sigma_i(\mathbf{k},\mathbf{k}')}{d\Omega_{k'}}\right]_{syn} = \int \frac{d\sigma_i(\mathbf{k},\mathbf{k}')}{d\Omega_{k'}}\rho(E_k)\,dE_k$$
$$= \sigma^N(\mathbf{k},\mathbf{k}')\int\rho(E_k)\,dE_k$$
$$\cdot\left|\sum_n \frac{\langle\chi_n|e^{ikr}|\chi_{n_0}\rangle}{E_k - E_R - (\epsilon_n - \epsilon_{n_0}) - i\Gamma/2}\right|^2 \delta(\mathbf{k}-\mathbf{k}'-\boldsymbol{\tau})$$
$$= 2\pi\sigma^N(\mathbf{k},\mathbf{k}')\frac{\Gamma_i^2}{\Gamma}\rho(E_R)\delta(\mathbf{k}-\mathbf{k}'-\boldsymbol{\tau}), \qquad (10.22)$$

where $\rho(E)$ is the SR spectral density. To obtain the coherent scattering cross section corresponding to the resonant energy of scattered quanta, one should assume that the final state coincides with the intermediate one in (10.20). As a result, one gets

$$\left[\frac{d\sigma_i(\mathbf{k},\mathbf{k}')}{d\Omega_{k'}}\right]_{syn} = f^2(\mathbf{k}')\sigma^N(\mathbf{k},\mathbf{k}')\delta(\mathbf{k}-\mathbf{k}'-\boldsymbol{\tau}). \qquad (10.23)$$

A comparison of (10.22, 10.23) and (3.11) shows that the intensities of scattered quanta in experiments with a SR source are higher than in the case of a conventional Mössbauer source, because in the former case, Lamb–Mössbauer factors $f^2(k)$ or both $f^2(k')$ and $f^2(k)$ are absent. The energy distribution of quanta in inelastic coherent scattering has a simple form in the case of fast phonon relaxation $(\tau_n \gg \tau_p)$. In this case, the energy spectrum of coherently scattered quanta coincides with the emission spectrum of an excited Mössbauer nucleus in the crystal lattice [43]. The relative probability of a scattered quantum having energy different from the resonant one by ΔE is determined by the probability of a quantum emission with transfer of energy ΔE from the phonon system. The corresponding probability for the most probable one-phonon process is given by the formula [43]

$$P\big[E_R - (\epsilon_{n_f} - \epsilon_{n_0})\big] = |\langle \chi_{n_f} | e^{-ikr} | \chi_{n_0} \rangle|^2. \tag{10.24}$$

Let us now discuss possible experiments concerning the above-mentioned effects. The most straightforward experiment is an energy analysis of the gamma quanta resonantly scattered in the direction of the elastic-diffraction reflection [44]. Another possibility is a temperature analysis of the scattering cross section temperature dependence. In accordance with the above presented formulas, temperature dependence may be proportional to the square of the Lamb–Mössbauer factor or its first power, or it may be totally independent of temperature, depending on the energy resolution of the measurements applied to the incident and scattered beams.

The independence of inelastic coherent scattering of the phonon momentum can be revealed also in absorption Mössbauer spectra [45] because out of the Bragg condition, inelastically coherently scattered quanta are directed almost always in the incident beam direction (see also [42]). For the corresponding analysis, Mössbauer spectra with hyperfine splitting are the most suitable, because for some hyperfine lines, incoherent scattering is forbidden (see, e.g., [46]) and incoherent additions to the others lines alter the usually assumed ratios of individual line intensities. Coherent inelastic scattering should also reveal itself in a comparison of electron conversion spectra with Mössbauer spectra measured at the exit side of a sample.

Up to now, there had been no direct experiments for detecting inelastic coherent nuclear scattering, with the exception of [44] which represents the energy investigation of coherent inelastic scattering. The energy analysis of scattered quanta in Mössbauer diffraction in [44] proved the existence of spin incoherence in Mössbauer scattering. However, this analysis seems to be insufficiently thorough for a conclusive affirmation about the absence of the inelastic scattering component in the direction of the diffraction maximum. Therefore, it seems that for the observation of effects discussed, it is desirable to once again conduct an experiment similar to that of [44]. Analysis of the already completed experiments along the lines of the above discussion also seems to be useful.

10.7 Recent Experiments on Mössbauer Filtration

The years following the first experiments on Mössbauer filtration of SR [2,3] were actively used by different groups in Europe, U.S., and Japan to perform a new experiment on Mössbauer filtration and to improve the corresponding technique—in particular, to reach a higher intensity for the filtered beam. In this section, the recent experimental results relating to Mössbauer filtration of SR will be briefly discussed.

First of all, the results of the U.S. collaboration should be mentioned [46–49] (to be more exact, it should be noted that in the field of Mössbauer SR experiments, very strong cooperation exists inside one country, as well as on an international scale). In the cited works, achievements were reached both in an improvement of experimental technique and investigation into the nature of Mössbauer filtration. Nuclear Bragg scattering of SR from a hematite crystal with a signal-to-noise ratio of 100/1 without the time-delay technique was observed in [47]. The success of this work was determined by the use of a six-reflection crystal premonochromator assembly providing extremely high-energy resolution, close to 5 meV, and small angular width, close to 0.4″, which unfortunately cannot be used to the full extent because of the imperfection of the hematite crystal (its rocking curve width is 3″). In the subsequent work of the same collaborators [48], the full time evolution of the delayed component in Mössbauer filtration at hematite was measured and the quantum beats discussed in Sect. 10.6—in particular, the zero intensity of the filtrated beam in the initial moment of filtration were observed. The next important step in the Mössbauer filtration technique was reported by the same group in [49]. This step consists of performing Mössbauer filtration of SR at a reflection in hematite allowed in X-ray diffraction with the help of a [46,47,49,50] high angular and energy resolution achieved early on. The significance of this result lies in the removal of restrictions on sample crystallography, making the stimulation of low-lying nuclear excitation by SR a more versatile and generally applicable technique. The last publication of this group [51] cited here reports on the observation of quantum beats for forward scattering in a Mössbauer filtration experiment with hematite. This result also suggests the broad applicability of Mössbauer filtration in the studies of gamma-ray optics phenomena.

The group from Hamburg continued their Mössbauer filtration experiments at YIG [52] under the conditions of nuclear level crossings for a nucleus occupying two crystallograpically nonequivalent sites in the unit cell. The cited work confirmed that for the existence of a pure nuclear reflection, it is essential to have nonequivalence of the hyperfine fields at different sites (e.g., as in this work, a different orientation of the principal axis of the hyperfine field), and even if the positions of the hyperfine line coincide in Mössbauer spectra, pure nuclear reflections arise for symmetry reasons.

The recent publications stemming from Soviet–German collaborations [53–56] are devoted to the Mössbauer filtration experiments with a $FeBO_3$ crystal at room temperatures to the Neel temperature of this antiferro-magnet. The results reported in these papers are promising from the point of view of applications, as well as from that of using Mössbauer filtration in physical investigations. A narrow single-line filtrated radiation was observed in [53] slightly above the Neel temperature of $FeBO_3$. The corresponding temporal quantum beats [55] reveal strong temperature dependence close to the Neel point, with the beat frequencies approaching zero near the Neel point. To produce a simple wave packet in filtration experiments, the authors of [54] introduced a Mössbauer absorber to SR scattered into the reflection (333) of $FeBO_3$ at room temperature. They managed to obtain practically a single line of radiation behind the absorber and observed changes in the temporal quantum beat patterns behind the absorber in comparison with those in the absence of the absorber. These results show that quantum beats can be successfully used for the investigation of magnetic properties of crystals near the transition points and that there are different ways to produce a single-line source for the subsequent analysis of Fe^{57}-containing materials.

In concluding this chapter, it should be noted that Japanese researchers also have reached quite high levels of experimentation in the field of Mössbauer filtration of SR, and their results in this field will be awaited in the near future. The basis for our expectations is a preliminary report about a Mössbauer filtration experiment with hematite [57] performed with a very high premonochromatization of SR compared with that in [3] and the latest publication [58].

References

1. S.L. Ruby: *J. Phys. Colloq.* **35**, C-6, 209–212 (1974).

2. A.I. Chechin, N.V. Andronova, M.V. Zelepukhin, A.N. Artem'ev, E.P. Stepanov: *JETP Lett.* **37**, 531 (1983).

3. E. Gerdau, R. Rüffer, H. Winkler, W. Tolksdorf, C.P. Klages, J.P. Hannon: *Phys. Rev. Lett.* **54**, 835 (1985).

4. E. Gerdau, R. Rüffer, H. Winkler, et al.: Abstracts of International Conference on Mössbauer Eff Effects, (Belgium, Leuven, 1985, p. 6).

5. R.L. Cohen, G.L. Miller, K.W. West: *Phys. Rev. Lett.* **41**, 391 (1978).

6. A.N. Artem'ev, V.A. Kabanik, Yu.N. Kazakov, V.A. Kulipanov, G.N. Meleshko, V.V. Sklyarevsky, A.N. Skrinsky, E.P. Stepanov, V.B. Khlestov, A.I. Chechin: *Nucl. Instrum. Methods* **152**, 325 (1978).

7. V.A. Belyakov: *Sov. Phys. Usp.* **18**, 267 (1975).

8. R.N. Kuz'min: *Apparatus and Methods of X-Ray Analysis*, Vol. 10 (Moscow, Mashinostroenie, 1972, pp. 85–89) (in Russian).

9. V.A. Belyakov, V.P. Orlov: *Phys. Lett.* **A44**, 463 (1973).

10. V.G. Baryshevsky, I.D. Feranchuk: *Phys. Lett.* **A57**, 183 (1976).

11. V.A. Belyakov: *Diffraction Optics of Periodic Media with a Complex Structure* (Moscow, Nauka, 1988, Chapt. 3) (in Russian).

12. K. Witte: *Synchrotron Radiation News* **3**, 6 (1990).

13. H. Winick, G.P. Williams: *Synchrotron Radiation News* **?**, N5, 23 (1991).

14. J.P. Hannon, N.V. Hung, G.T. Trammell, E. Gerdau, M. Muller, R. Rüffer, H. Winkler: *Phys. Rev. B* **32**, 5068 (1985); *Phys. Rev. B* **32**, 5081 (1985); *Phys. Rev. B* **32**, 6363 (1985); *Phys. Rev. B* **32**, 6374 (1985).

15. G.T. Trammell, J.P. Hannon, S.L. Ruby, P. Flinn, R.L. Mössbauer, F. Parak: *AIP Conf. Proc.* **38**, 46 (1977).

16. A.M. Afanas'ev, Yu. Kagan: *Sov. Phys. JETP* **21**, 215 (1965).

17. U. Van Bürck, H.J. Maurus, G.V. Smirnov, R.L. Mössbauer: *J. Phys. Ser C* **17**, 2003 (1984).

18. Yu. Kagan, A.M. Afanas'ev, V.G. Kohn: *Phys. Lett. A* **68**, 339 (1978).

19. Yu. Kagan, A.M. Afanas'ev, V.G. Kohn: *J. Phys. Ser. C* **12**, 615 (1979).

20. G.T. Trammell, J.P. Hannon: *Phys. Rev. B* **18**, 165 (1978).

21. E. Karlsson, E. Mathias, K. Sigbahn: *Perturbed Angular Correlations* (New York, Humanities Press, 1964).

22. E. Gerdau, R. Rüffer, R. Hollatz, J.P. Hannon: *Phys. Rev. Lett.* **57**, 1141 (1986).

23. R. Rüffer, E. Gerdau, R. Hollatz, J.P. Hannon: *Phys. Rev. Lett.* **58**, 2359 (1987).

24. U. Van Bürck, R.L. Mössbauer, E. Gerdau, R. Rüffer, R. Hollatz, G.V. Smirnov, J.P. Hannon: *Phys. Rev. Lett.* **59**, 358 (1987).

25. J. Arthur, G.S. Brown, D.E. Brown, S.L. Ruby: *Phys. Rev. Lett.* **63**, 1629 (1989).

26. E. Gerdau, R. Rüffer: *Nucl. Instrum. Methods A* **246**, 362 (1986).

27. G. Faigel, D.P. Siddons, J.B. Hastings, P.E. Haustein, J.R. Grover, J.P. Remeika, A.S. Cooper: *Phys. Rev. Lett.* **58**, 2699 (1987).

28. M. Grote, R. Röhlsbereger, E. Gerdau, et al.: *Hyp. Int.* **58**, 2439 (1990).

29. U. van Bürck, G.V. Smirnov, R.L. Mössbauer, Th. Hertrich: *J. Phys. Condensed Matter* **2**, 3989 (1990).

30. H.D. Rüter, R. Hollatz, R. Rüffer, et al.: *Hyp. Int.* **58**, 2477 (1990).

31. V.A. Belyakov, Yu.M. Aivazian: *Khim. Fiz.* **9**, 3 (1990).

32. V.A. Belyakov: *Sov. Phys. Usp.* **30**, 331 (1987).

33. E. Gerdau, R. Rüffer, U.D. Rüter: *Hyp. Int.* **40**, 49 (1988).

34. S. Ruby: *Hyp. Int.* **40**, 63 (1988).

35. P.A. Aleksandrov, A.M. Afanas'ev, M.K. Melkonyan: *Sov. Phys. Solid States* **25**, 578 (1983).

36. M.A. Andreeva, R.N. Kuzmin: *Solid State Commun.* **49**, 743 (1984).

37. G.V. Smirnov, A.I. Chumakov: *Sov. Phys. JETP* **62**, 673 (1985); *Sov. Phys. JETP* **62**, 1044 (1985).

38. V.A. Belyakov, Yu.M. Aivazian: Proceedings of 6th National Conference on Synchrotron Radiation (Novosibirsk, Institute for Nuclear Physics, 1984, pp. 318–320) (in Russian).

39. V.A. Belyakov, Yu.M. Aivazian: *Nucl. Instrum. Methods A* **261**, 322 (1987).

40. V.A. Belyakov, Yu.M. Aivazian: *Nucl. Instrum. Methods A* **282**, 628 (1989).

41. G.T. Trammell: *Phys. Rev.* **126**, 1045 (1962).

42. Yu. Kagan, A.M. Afanas'ev: *Pis'ma Zh. Eksp. Teor. Fiz.* **5**, 51 (1962).

43. H.J. Lipkin: *Quantum Mechanics* (Amsterdam, North-Holland, 1973).

44. A.N. Artem'ev, V.V. Sklyarevskii, G.V. Smirnov, E.P. Sepanov: Proceedings of 5th International Conference on Mössbauer Spectrum, Part 3 (Prague, Czechoslovakian Atomic Energy Commission, 1975, pp. 707–709).

45. V.A. Belyakov: Abstracts of Latin American Conference on Applied Mössbauer Effect, (Havana, Cuba, 1990, p. 8.3); *Hyp. Int.* **66/67** (in print).

46. G.V. Smirnov, Y.V. Shvyd'ko: *Pis'ma Zh. Eksp. Teor. Fiz.* **34**, 4095 (1982).

47. G. Faigel, D.P. Siddons, J.B. Hastings, P.E. Haustein, J.R. Grover, L.E. Berman: *Phys. Rev. Lett.* **61**, 2794 (1988).

48. J.B. Hastings, D.P. Siddons, P.E. Haustein, L.E. Berman, J.R. Grover: *Phys. Rev. Lett.* **63**, 2252 (1988).

49. D.P. Siddons, J.B. Hastings, G. Faigel, L.E. Berman, P.E. Haustein, J.R. Grover: *Phys. Rev. Lett.* **62**, 1384 (1989).

50. D.P. Siddons, J.B. Hastings, G. Faigel: *Nucl. Instrum. Methods* **A266**, 329 (1988).

51. J.B. Hastings, D.P. Siddons, U. van Bürck, R. Hollatz, U. Bergan: *Phys. Rev. Lett.* (in print).

52. R. Rüffer, E. Gerdau, H.D. Rüter, W. Sturhahn, R. Hollatz, A. Schneider: *Phys. Rev. Lett.* **63**, 2677 (1989).

53. A.I. Chumakov, M.V. Zelepukhin, G.V. Smirnov, U. van Bürck, R. Rüffer, R. Hollatz, H.D. Rüter, E. Gerdau: *Phys. Rev. B* **41**, 9545 (1990).

54. U. Van Bürck, R.L. Mössbauer, E. Gerdau, W. Sturhahn, H.D. Rüter, R. Rüffer, A.I. Chumakov, M.V. Zelepukhin, G.V. Smirnov: *Europhys. Lett.* **13**, 371 (1990).

55. H.D. Rüter, R. Rüffer, E. Gerdau, R. Hollatz, A.I. Chumakov, M.V. Zelepukhin, G.V. Smirnov, U. Van Bürck: *Hyp. Int.* **58**, 2473 (1990).

56. A.I. Chumakov, G.V. Smirnov, M.V. Zelepukhin, V. Van Bürck, E. Gerdau, R. Rüffer, H.D. Rüter: *Europhys. Lett.* **17**, 269 (1992).

57. S. Kikuta, et al.: Abstracts of 1st Soviet–Japanese Seminar on Crystal Characterization by X-Rays and Synchrotron Radiation, (Kaluga, USSR, Sept. 21–26, 1990, p. 6).

58. S. Kikuta et al.: *Jap. J. Appl. Phys.* **30**, L1686 (1991).

Index